Practical and Professional Ethics

VOLUME TWO
Environmental Ethics

Practical and Professional Ethics Series

Vol. 1: The Primer of Applied Ethics
Vol. 2: Environmental Ethics
Vol. 3: Bio-Medical Ethics
Vol. 4: Educational Ethics
Vol. 5: Economic and Business Ethics
Vol. 6: Socio-Political Ethics

About the Author

Mr. Debashis Guha, teaches Ethics and Phenomenology and Existentialism in the Post Graduate Department of Philosophy, Ranchi University, Ranchi. After doing masters course in Philosophy from Delhi University in 1980, he was conferred the Ph.D. degree of Jadavpur University in 1988. Dr. Guha takes interest in Philosophy of Religion, Social and Political Philosophy, Philosophy of Science and Analytic Philosophy. He taught at the undergraduate level in a constituent college of Ranchi University from 1981 to 1989 and was then asked to serve the Post Graduate Department. Dr. Guha is teaching Ethics for twenty-four years. He was awarded the Visiting Associateship of the U.G.C. (1994-96) in Jadavpur University, Calcutta, where he worked out *An Inquiry into the cases of Pain and Suffering,* which awaits publication. He was further awarded U.G.C.'s Major Research Project (1997-2000) on *Applied Ethics in Indian Context,* which has been now shaped in Six volumes, of which the present work is the second one under the series: **Practical and Professional Ethics**. Dr. Guha has fourteen papers to his credit. He has presented philosophical papers in different Seminars, Symposia and Congress, two of them were presented in International Congress; one in the Platinum Jubilee Celebrations of the Indian Philosophical Congress, 2000-01, and the other in the XVth International Congress of Aesthetics in Tokyo, Japan in 2001. During his Major Research Project work, Dr. Guha has talked to many people from different walks of life, interviewed them and also arranged ethical dialogues leading to resolution of value-laden practical problems of urgency. Dr. Guha envisions peoples' participation in Practical and Professional Ethics.

Practical and Professional Ethics

Volume-Two

Environmental Ethics

DEBASHIS GUHA

CONCEPT PUBLISHING COMPANY, NEW DELHI-110059

Cataloging in Publication Data—DK
Courtesy: D.K. Agencies (P) Ltd. <docinfo@dkagencies.com>

Guha, Debashis.
Practical and professional ethics / Debashis Guha.
v. cm.
Includes bibliographical references.
Includes indexes.
Contents: v. 1. The primer of applied ethics — v. 2. Environmental ethics
ISBN 8180693147 (set)
ISBN 8180693155 (v. 1)
ISBN 8180693163 (v. 2)

1. Applied ethics. 2. Professional ethics. I. Title.

DDC 170 22

ISBN 81-8069-314-7 (set)
ISBN 81-8069-316-3 (Vol. II)

First Published 2007

Published and Printed by
Ashok Kumar Mittal
Concept Publishing Company
A/15-16, Commercial Block, Mohan Garden
New Delhi-110059 (INDIA)

Phones : 25351460, 25351794
Fax : 091-11-25357103
Email : publishing@conceptpub.com

Bratati

my caring and sharing wife

Foreword

It gives me immense pleasure in introducing the book written by Dr. Debashis Guha of the Department of Philosophy of Ranchi University who possesses an enviable command over the subject in general and ethics in particular.

The present book forms the part of the outcome of his major research project which impressed us much because of its relevance to the need of the hour. Philosophy as such is often misunderstood as a subject that has hardly any relevance to the ordinary life and worthy an ideal pastime of the idle brain. The major research project undertaken by Dr. Guha has wiped out this misconception to a considerable extent. I had the privilege to inaugurate the first among the six series of intersubjective corroboration dialogues on environmental ethical issues organized by Dr. Guha and honesty requires it to be admitted that I was not only impressed but also enlightened by the meaningful and thought-provoking outcome of the discussions. It is not for nothing that Dr. Guha was invited to present a paper on the ethics of Bonsai, an important aspect of his research project, in the XII international congress of aesthetic in Japan which was highly appreciated and acclaimed.

The methodology of the work is attractive and meaningful as it stresses on people's participation in ethical resolution of social problems thus bringing a so-called utopian subject at our doorstep. It is needless to say that the work on environmental ethics will pave the way for interdisciplinary research while arising awareness for ethical consideration in all matters related to man and nature.

Last but not the least, the growing ecological imbalance threatening our very existence on this beautiful mother earth, the ethical regard for environmental safeguard is a must and I congratulate Dr. Guha for being the pioneer of this great cause from philosophical point of view.

I am confident that the students of philosophy and environmental science will be highly benefited by this book and the scholars as well as the reading public will enjoy it and thank the author.

Ranchi
August 25, 2005

Dr. C.R. LAHA
Rtd. Professor and Head,
Dept. of Bengali, Ranchi University
Ex Director, Academic Staff College,
Ranchi University
Ex Vice-Chancellor, Ranchi University

Preface

On a pleasant cold morning in December 1981, I decided to join the job of my life-teaching philosophy to the undergraduates in a constituent college under the Ranchi University of the state of Bihar (now Jharkhand) situated in a place called Chakradharpur which is nearly 120 km from Ranchi, my hometown (now the state capital of Jharkhand). I was completely unaware of the place, its location and the people or the economy there. I was at sea, where to stay, what kind of treatment I will get there and so on. Most frightening was the idea of thrown into a hot, dingy, filthy place with filthy people.

Only then, one of the friends of my father assured me that "go, it is a beautiful place". Is it more beautiful than Ranchi, the summer capital of Bihar? He said, "Leave Ranchi and then you see what the real beauty of the Chotanagpur plateau is. As you snake down the 1200 metres altitude to reach the bottom of the bowl to Chakradharpur, you will know all about beauty". Moreover, he assured, "it is a township important for railway sub-division, people are nice too".

On the ninth day of December, I took a government bus, which was for a starter, a complete mess. I managed a window-side back seat to be soon amused at the jerks and swerves to inch through a land I never knew to be so beautiful. Hardly after 10 to 15 km, there were riot of colours, as there were countless trees, big and small, numberless bushes and sudden spurts of white, crystal clear ribbon-like streams bouncing way through huge and small boulders. Once we reached a place called Bandgaon, I knew what was ahead. I could see the head of hills to my front and all around, and here and there, everywhere the green soldiers made a thick veil of mystery to human existence. Beside the one-way narrow, ill-maintained road, one can suddenly find a village bazaar, barren except two or three days when most of the trade is done. People emerge from the mysterious green, I do not know from where, who are the most beautiful I have seen thus far. The tribes of different kinds come here with their goods in small and big bamboo baskets to sell and to buy for an honest and simple living. Though these dark-skinned, uncombed people with most simple dresses to cover them up in utter carelessness, bear an electric smile telling you the unmatched ethic of living in pure and serene beauty. Among them, during the weekly bazaars could be seen the people of towns, the middle-men buying the jungle products as cheap as pebbles to take them to towns.

Leaving Bandgaon you have the first *ghat* (a slowly winding valley), called Hirni. Hirni is an unforgettable beauty. On your right, you have the head of hills and on your left, deep valley. Wherever you see, you find wave of hills all dotted thickly in green. As you wind down to cross the Hirni stream, which flows out of a beautiful waterfall, you move ahead to climb up and then come down quickly through another *ghat*, called the Chaki. Here you have dark and deep woods with a feast of colours. After crossing another stream, you finally climb up and then go down through the steepest *ghat*, called the Tebo. At the final lap of Tebo, elevations and slopes are dangerous and you feel that at times your bus could suddenly roll into a limbo.

The climbing up and going down through three *ghat* full of woods takes the major time of your journey. Sometimes it is tiresome but all the while, you are experiencing beauty of your life-time. I was a testimony to this journey and beauty for long eight and half years till I was asked to come back to teach philosophy to the post-graduate students in the Ranchi University department in early 1990. During these eight years and half I was also a dumb spectator of loot and plunder—the devastation of the land, the woods and the people by the greedy town-folks, the trouble-shooters who are best expressed as '*dikus*'. These people for comfort, luxury and so-called modernity devastated the green and its people. In 1990, while I travelled, I could see the core of the jungle or even beyond it and the veil stood, no more, denuding the concealed beauty of tribal life. Gradually looting the travellers, and arms upheavals which were insulted as "extremism", started. Now the case is frightening, people are scared of travelling late in the afternoon and at night. Most pitiable to me is that most of the colour and the cover are gone. Believe me, it is no more beautiful. Who is responsible? Who took away the elegance and the peace?

There are number of questions in my mind. I feel an urgency to find the unethicality of plundering the earth, the sky, the water, the air and the people by looters of modern times so that wherever a little bit of Bandgaon or Hirni or Chaki or Tebo is left, be in India or elsewhere, may be saved. The discovery of the unethicality would reveal the ethic of environmental and ecological protection but that is not enough; ethic ought to be covered up by action. In writing this book, I am trying my bit for unravelling unethicality and ways to become a moral being by doing something worthwhile.

This work is not merely a theoretical churning, therefore, it is a work of action. Here I am not merely interested in raising ethical questions about ecological crises in modern times, I am trying to force a course of action as has been suggested by people most concerned and most hurt. Though the action part is heavily based on Indian experience, it may be useful elsewhere as the methodology of practical ethics as has been founded in the Primer of Applied Ethics (Volume one of the series of six works in Practical and Professional Ethics), which I call, "Intersubjective Corroboration".

DEBASHIS GUHA

Contents

Introduction

I have a strong conviction, (which is not an insane *blik*), that Ethics as we now know cannot be treated as a handmaid of trained ethicists' (for that reason philosophers') theoretical jugglery. Readers of the Volume One of the Series of Practical and Professional Ethics come to know about the different reasons for having this *blik*. To be precise, Ethics as I see, it is a branch of philosophy, having at least four important tasks to perform—the metaethical, the normative and the practical being three most important ones, and another task of some value is to unravel the psychological foundation of ethics. Yet, foundational framework, that is, in so far as its basic tasks are concerned ethics appears to me as a spreading beam of star, reaching out to new horizons. Ethics never appears to me as static any more—it is not closed, so to say. I have been harping in the entire Primer (Volume One) and in the Preface of this work that theoretical juggling does not satisfy the whole of ethics. Because ethics as such need not and should not be isolated from the life humans live in a realm where earth, sky, water, air and small and big non-human animals and plants and trees mingle in an organic whole or the realm of existence is crucial in our present realization of ethics as moving like beams emanating from a star. We did realize in the Primer that being dynamic, ethics has already surpassed its limit of the four tasks, particularly the practical task to embrace another dynamic way to do things in the name of Professional Ethics. The reasons are simple because our professional life within the short or relatively longer biological life we live in the society and in this vast ecological community is not one of a moral islander rather closely knit with the entire social and ecological fabric. The decisions we make as professionals and act accordingly do have direct and indirect impact on people, society and may be too far reaching to make an impact on the life and things around us. These influences of our intentional actions on different things are further subjected to evaluations of different kinds. We ask whether or not they are economical, pragmatic, and judicious and so on. One extremely important evaluation of our actions is whether or not they are ethical, that is, whether or not they are ethically good and desirable. The point is whether or not our decision-making and actions are good merely as being desired or they are good as being intrinsically *desirable*. We also ask whether or not such actions are right and just, and also done in the sense of duty and obligation. Here also, rightness, justice, duty and obligation bear a moral or ethical sense, not purely legal or to use another notion, "official". We are concerned about the *moral basis* of calling certain actions of our professionals' right, just, dutiful and obligatory and for that matter we are in search for calling some actions as morally right, just as so on. Our professional life, therefore, is full of professional decision-making and actions needing ethical basis for their proper evaluation in a civil society, which is not eager only about acts being judicious or pragmatic or economical. The far-reaching question of ethicality is thus one of basic desirability and goodness or "of ought" and "should". Why such and such things "should be done" by the professionals are basic, it is beyond the "official is" which is more concerned about why such and such things "are done". Not only intrinsic goodness, we also, speak about the extrinsic or external goodness of professional acts while we are concerned about their moral evaluation because such acts apart from answering why they ought to be done, also answer or should answer, how the desirable end needs to be approached. Hence, ethical questions in professional life cover both the good of actions as means and as end or those of desirability as means and end.

Interesting enough is to notice (and we did notice in the Primer) that a search for the basis of moral evaluation of our intentional actions, professional as well as non-professional in nature, often ends up in pressing one or some ethical standard or standards, or norm or norms. Equally important is to ask whether or not one can resolve several questions or trifles of professional life and non-professional life by calling these norms into service. An inquiry into possibility of application of norms for ethical resolution of practical problems leads us to what we know as Practical Ethics (elaborated in the Primer). Here it is sufficient to notice that ethics, if it includes questions about evaluation of professional and non-professional decision-making and actions, should include both normative and practical inquiries. We thus find that the whole gamut of ethics stands out as astral but within it is a coherence of tasks normative, practical and professional, each supporting the other as legs of tripod.

Notwithstanding such coherence in the system called ethics, there is a strong urge for moving beyond norms or at least urge to underplay need of norms than embracing inculcation of some basic *virtues* in man qua man. The revival and growth of the ethics of virtue is thus natural within ethics. It is however an old story in the eastern and the western ethical traditions, its contemporary revival stresses a virtuous ethical agent over and above all things we know as important in ethics. Do we now have to conceive of imparting virtues to build up virtuous decision makers? This is yet another dynamics to be considered for its viability in ethical life. Whatever may be the case, ethics does not end here.

Beyond profession, we have people having occupation and several occupations do have concerns for both decision-making and actions and the desirability concern of decision-making and actions. Hence, ethics of occupation (discussed in the Primer) stands out as another dynamic venture within ethics. It is with attending to the moral problems in occupations as well to find out ways out of those problems. Ethics is thus growing and it has been a good lesson to some old hats in philosophy that sarcastically but myopically hold that contemporary philosophers are worried about saving their face by finding new ways to be practical. Ethics has been thoroughly practical or dynamic in its approach. It has never been away from life, society and value-laden questions and their resolution in the east and the west. Let us accept that tiny bubble in the history of ethics which called ethics as purely metaethical has been overthrown in its peak period by the synthetic forces in ethics because people felt and still feel that it had a bad habit of throwing the baby with the wash water.

It is in this work that I now move on with this holistic nature of ethics to unfold yet another story. It is about Environmental Ethics, which is again a testimony to free ethics from the bondage of theory biasness because we strongly feel that among many practical concerns, which are value (read ethical-value) laden, concerns about environment and ecology are extremely important. It is for the simple reason that occupational or professional or whatever life we live and thus face value-problems for resolution, we are in an environment and an ecological system which might not be disregarded altogether. We cannot disregard that our intentional actions have far-reaching effects, positive or negative, desirable or undesirable on non-animals, trees, lakes and rivers, the air we breathe and the sky under which we are. Can we for instance, disregard the fact that questions of desirability may be raised if in the capacity of decision-makers we install a mega thermal power station and start causing irreparable damage to life and things around us and in that way harm ourselves to great extent? Don't we face a big "ought" or "should" question? I believe we face them than not for it is extremely non-sense to be both foolhardy and neutral towards value of ecology and value of value-laden problems in ecology.

Once we start giving at least some importance to these things we come to realize that ethical concern about environmental and ecological problems is closely related to practical ethics, professional ethics and occupational ethics. Arguably, it may have close affinity with virtue ethics as well. Interesting is that ethical concern in environmental issues is closely related to performing a metaethical task as well. Keeping these interrelations in mind the structure of the present work has been erected. In Chapter 1, we perform metaethical task of considering the definition and nature of "environment", "ecology" and "environmental and ecological ethics". We will also analyze the tasks of environmental ethics and its value. In Chapter 2, we further consider some metaethical problems like the justification of an ethic of environmental matters and the problems of value, liberty, rights, duties and obligations in an ethic of environment. Another important problem to be considered is the one of proper model of resolving value-laden environmental problems. In the Third Chapter, we move on to the application of the model established by us to find out real dynamics of ethical resolution of four types of value-laden environmental problems. We would also pronounce the points of resolution and analyse the interplay of ethical norms in the real dynamics. The Fourth Chapter considers the nature, need and validity of metaphysical basis of environmental ethics with special reference to Hindu environmental ethics. Finally, in Chapter 5, the possibility of a peoples' environmental ethics to stop ecological terrorism along with other questions will be discussed.

The work has been completed with the generous help of a number of people. Mr. G. Kumarappa of the National Library, Calcutta, Sister Mary Paul of Nirmala College and the number of people who came forward to participate in ethical dialogues. The students and teachers who participated in these dialogues need to be thanked in particular. One cannot forget the help of three young people Mr. Purushottam Kumar, Mr. Rajendra Prasad Singh and Ms Sushma Priyadarshana in completing the fieldwork and dialogues. I thank Mr. Sanjay Singh for carefully typing the Manuscript and the entire family of Concept Publications, New Delhi.

1

The Nature of Environmental Ethics

If philosophers have learnt their basic lesson properly, it is expected of them to take care of the analysis of meaning of central concepts and fundamental problems before embarking in the realm of widespread technicalities, which become a part of any discipline, philosophy being no exception. Ethics, particularly environmental ethics, needs some care in so far as clarification of meaning of concepts and central questions are concerned because people not so conversant in this field might find it too cumbersome to digest the so-called rich food for thought. There are, however, queer differences among the readers of a work on environmental ethics. In so far as philosophers or ethicists are concerned, they need to be extremely clear in their minds about the nature of the concepts like "environment", "ecology" and "concepts" like "environmental ethics" and "ecological ethics". Non-philosophers, such as biologists or environmentalists or ecologists, aspire to know what "environmental ethics" and "ecological ethics" are. Interestingly, the scientists of unique training would like to know the reasons behind more fashionable intrusion of ethics in their field in the name of 'interdisciplinary study'. They would like to know apart from the justificatory part of the study, the way non-scientists analyse the notions of 'environment' and 'ecology', and the study related to environment and ecology. It may not be necessarily skeptical on part of the scientists to inquire about such things, but the way some issues are analysed for their meaning by the non-scientists does at times transcend the limits of stipulative definitions of the sciences because analysing the meaning of a scientific notion need not be necessarily within a given fixity of definition. Yet, it should not be subjectively imputed meaning either. Therefore, it would be always interesting to discuss whether or not some notions like "environmental ethics" should be strictly logically defined for their meaning or should we peel off the performative part or the functional part of the expression to know what its meaning consists in. Interesting question about the validity of using the terms "environment" and "ecology" almost synonymously in "environmental ethics" and "ecological ethics" may be raised. Apparently a lay problem (because to scientists the notions are obviously different), it does not remain so when we reconsider the profuse "overlapping" in nature, content, methodology, findings and most importantly, the ethical implications in the study of environment and ecology. It appears that an inquiry about using "environmental ethics" in place of "ecological ethics" (though deemed to be different but overlapping), stands a useful one. Finally, when one reaches a point after the analysis of important concepts, one finds that rather than a fixed definition or a strict logical definition of these concepts, there are sufficient grounds to construct "normative definitions" of "environment", "ecology", "environmental and ecological sciences or studies" and "environmental and ecological ethics".

There are obvious reasons for considering the term "ecology" for its meaning clarification. We shall gradually reveal the reasons for considering "ecology" first. It is too simplistic to say that "ecology" is "eco" + "logos" that is, "the logic of the house" or "the study or science of the house" or "the scientific study of the nature, structure and functions of the household", where "house or household" is "where we live". Certainly, "where we live" is not simply "this world or the earth" rather, "what covers us, a huge expanse as if around the earth, giving us an idea of 'an inside' of a house or that we are '*in* a household'. But this simplistic analysis of meaning of the term has its own vagueness unless the "eco" in "ecology", *i.e.*, the so-called "household" has not been satisfactorily deconstructed for its meaning. The "eco" in "ecology" comes from the root "oikos" and, therefore, "ecology" is basically one particular "oikos" notion, the other being "economics". This root, "oikos" basically reminds us of a 'whole system' of 'sustainability', a system in which various discrete parts are 'interrelated'. This gives us a feel of "integrity" among "what exists", and to unfold this "exists" would be childish because in the words of Larry L. Rasmussen, "The numbers are stunning, both the big ones and one very small one. There may be fifty billion galaxies. Each has billions and billions of stars… earth is the only life-form in the universe"[1]. 'Rasmussen goes on to clarify that from this root, we do have the idea of "oikamene" or "the whole inhabited earth" or "the inhabited globe", which is again referring to an all-inclusive system, a 'whole' those envelopes life forms. Hence, all life forms depend on each discrete entity within the 'whole'. Keeping this in mind, we may say that the root "oikos" refers to "habitat" or to be more clear, "the habitable system". This 'habitable' system is one of a 'habitable household' that is, one single but vast place to live in or to sustain life. All other words, derived from this root, having 'eco' as a prefix refers to this place where life could be sustained along with other things'. Rasmussen depicts it nicely as: "Habitat is the core meaning of all *eco* words: economy, ecology, ecumenicity itself. Oikos—earth as a vast but single household of life—means the capacity for survival, that is, sustainable habitat. It means space and the means for the living of all living things… all life forms need carefully filled habitats"[2].

This amount to saying that "ecology" refers to dynamic systems' knowledge, that is, wholesome logic of how the entire life of the household and the living together in that household is made possible. The English term "ecology" was used initially by Ernst Haeckel in 1866 in a book regarding animal morphology much influenced by Darwin's new theory of evolution by natural selection. Haeckel has acknowledged that the theory of evolution explained and formed the basis of ecology as an independent science or study. Clearly, ecological study had a primary affinity with theory of biology. S. A. Forbes (1875) as well as J. C. Arthur (1895) did accept the truth that the whole Darwinian doctrine belonged to ecology. Arthur opined that Darwin was the first among all to write about what he called, "vegetable ecology". H. C. Cowles (1904) was also emphatic that ecology among other tasks as a modern biological science has to clarify the intricacies of "adaptation". Even the more recent ecologists John Hopkins (1967) thinks that theory of "natural selection" is in fact an ecological theory, though ecologists in large numbers neglect Darwinian contribution and find it more comforting to treat it as a mere offshoot of physiology and has little or nothing to do with an "independent theory" and "philosophy". We will come to these misgivings in a short time.

For the present, what we have is a nearly clear meaning of term "ecology" as an independent science. If we are not too negligent towards the legacy of Haeckel we would realize the meaning of the term in the line in which we have been deconstructing the root of the term. He says, (as quoted in Rasmussen's book), "by Ecology, we mean the body of knowledge concerning the economy of nature—the investigation of the total relation of the animal both to its inorganic and to its organic environment; including, above all, its friendly and inimical relations with those animals and plants with which it comes directly or indirectly into contact"[3]. In this sense, ecology is one unique way to embark in the realm of "Oikos"; the uniqueness is of course in dealing with the knowledge of the varied relations among life systems necessary for a habitable earth. However, it would be a little too much to restrict our analysts on the basis of living system because what envelopes us all, the whole that creates the super-structure of our grand household, need not be forgotten. We are thus embedded in a household the building blocks of which are inanimate as well.

Haeckel understands and our deconstruction of "ecology" reveals that "ecology" is essentially a "polymorphic" discipline. No doubt at its nascent stages it cared much for plant and animal interrelations with those of others like human beings, it gradually spread its periphery covering limnology, marine ecology, physiological population and community ecology and even forestry, fisheries, agronomy, pest control and wildlife management. Allied discipline like parasitology has also been induced after some hitches. Keeping these things in mind, one can say that 'ecology' is a new natural history based on biological acumen.

Ecology as a scientific natural history has been developed on three major lines. One by Glacken (1967) is most interesting wherein we find the reference to "environmental studies" dealing with a designed earth, environmental influences on man and man's influence on the earth. We shall come to this given compass of earth and man as we proceed on next about "environmental studies".

One interesting feature of the Haeckelian legacy of looking at ecology has been much helped by an older study by Linnaeus

(1749) and later by Edgerton (1973), which may be termed as the "oldest ecological theory" and fathered philosophical ecology as the "balance-of-nature" theory. It has some "providential" flavour to it, which lingers till this date, though Edgerton's line of the development of ecological science has been one moving away from providentialism to an evolutionary, *naturalistic* conception of nature. So we have a transition from the "formalistic" to the "naturalistic" conception of "ecology" as a science though Edgerton has to be bold enough to move away from the so-called father of modern theoretical ecology, S. A. Forbes (1887) who did transform the providential balance-in-nature to "the beneficent power of natural selection". Yet, Forbes was extremely influential in building a philosophical ecology in his seminal essays like 'Lake as a microcosm' (1887) and 'On some interactions of organism' (1880), and was perhaps set an ethical seed to germinate as the balance-in-nature was looked upon by him as the "highest good of all the species". The evolutionary ethics was thus in the stage of germination. Forbes did not stop at his ethical venture, he did more with his romantic ideas like "a just equilibrium" in nature and also "harmony in brute nature" mixed with a cautionary man-centred ethic of "sympathy" and "self-sacrifice".

Though with Forbes (*et al.*) the germs of eco-philosophy and eco-ethics have started germinating. It was in 1953 after the publication of the excellent work titled "The Fundamentals of Ecology" by Eugene Odum the interdisciplinary collaboration of views, particularly the one between ecological and environmental studies and ethics got a spurt with the initiation of the concept to "ecosystem" or "ecosystems ecology". Odum propagated the idea that our habitable earth is in fact organized in the shape of an interlocking 'ecosystems' which exists in a movement towards orderliness through a turmoil of early competition so that the habitable earth shows more and more interrelatedness and co-operation among its basic components. Odum thought that in this process of shift from turmoil to interrelatedness through a number of definitely significant stages, the ecosystems does achieve what we call equilibrium or what Odum called "homeostasis". This dynamics of homeostasis persists until some crucial erosion or damage breaks up the fundamental patterns. In this act of breaking up the ecosystems, thus tending to entropy, the Homo sapiens are the most responsible one. The modern *avatar* of Homo sapiens in the view of Odum has turned the ecosystems to a 'junkyard' rather than the *oikos*, our habitable home which are the best place to live in. It is here that the ecological formalism which took a naturalistic turn in Edgerton further ethically reshaped or at least trying to shape like that by Forbes got a serious rethinking about the ethical implications of human being's 'junkyard' form of life. Ecosystems ecology is thus one place where we find ethics to thrive on a solid footing though we have to keep in mind that contemporary ecology has challenged the conception of ecosystems being inextricably moving towards a state of homeostasis until an entropic situation does not prevail. It is rather believed that there is a cyclic pattern of disturbances with some crystallization of back up damage control as well.

Understandably, if the whole of ecosystems is dynamic and not static, and that evolutionary and other changes apart, human intervention is a must not merely for its own sustenance but for the whole dynamics of the economy of production and distribution at a sturdy rate, the best conceived "equilibrium" seems to be a far cry. And most unrepentantly said, man cannot help out this dynamics to stop at any rate. So the lull is there, provided unharnessed economical development itself does not face a lull. Yet, "sustainable" economical development may be treated as one sweet-pill if ecosystems' orderliness is happily wed to the deliberate junkyard form of life of human beings. So the ecosystem's economy and ecosystem's ecology with eco-ethical forbearing are gradually brought together for a further mind twister in the history of knowledge. We cannot stop this twister as we further consider another one—"environmental" concerns in ethics. But for that let us delve into the nature of "environment".

"Environment" refers to the natural world infested by human beings and landscapes, which act as per the evolutionary natural progresses. However, the human beings have significantly manipulated and altered many natural systems and, therefore, "environment" also includes the several manipulated and altered entities and systems including landscapes consisting of towns, cities, factories, agriculture, and agro-forestry and so on.

Keeping these things in mind one can say that "environment" means that which surrounds us, and in that sense, it is something that engulfs us, is beyond us or permeates us. We, the existential entities all around are in it, we are all extremely intimately related to it. There is no respite in being going away from its cover. However, "environment" may be more clearly and more technically analysed as either and "existential entity" for instance a person, a living being, a work place and so on, the sense in which we often speak of family, social, natural and business or work environment; or generally as "natural environment" meaning thereby air, water, soil, mountains, seas, trees and so on. Taken in any of these sense, "environment" implies an intricate system of interacting entities, that is, each entity constituting a system is influenced by and influences other entity or entities positively or negatively or in both ways. In this sense, we speak of not merely a human environment and merely an environment for the human beings rather, an environment in which all entities are, or are given to them. Yet, there is no denying the fact that in a more restricted sense, we speak of our environment or that which is given to us or that which exists with us but certainly not like us but beyond ourselves. From this *constricted view* of "environment" often the intimacy among existential or the element of "wholeness" is never revealed, and to many of us who take care of its ethic have to rely heavily on this "primary" meaning.

We may now explain in scientific terms that elements of the environment are at least two: The components supplying energy and building blocks and the components for necessary regulations. We may further elaborate to find that environment is constituted of some *material factors* like oxygen, carbon, hydrogen etc. whose presence is crucial in the survival and development of living organisms. These factors are the energy supplying and building factors therefore. The other one, the *regulatory factor* of the living processes are present as compounds making part of the organism in small quantities which includes catalysts and inhibitors of metabolic processes. If we now speak of "human environment" in particular, it also comprises of the same building and energy supplying and regulatory elements. However, in case of regulatory components, we speak of climatic and environmental pollution as well as social, economic and political environment. The whole of the living conditions of human beings may be thus termed as regulatory, and this is what is crucial in determining the nature of environmental study in the context of human environment from an overall ecological study, which is emphatically biological. Some elaboration in this regard may be helpful in realizing a talk more specific about environment and human environment vis-à-vis ecological concerns. Though man is a product of a biosphere, which determines his properties significantly, the radical changes in biosphere itself ought to be within the adaptation limits of man. We call these factors "ecological" ones, which is nothing but external factors in the environment in which the man is and induce some changes in the rate of development or degradation of an individual. However, the number of ecological factors in an environment differs according to both the environmental changes and the changes in sensitivity of the organism and in its needs, which induced changes in individuals over a long time becomes inherent properties of the individuals. In any case, all these factors remaining the same, in case of human beings, we have to take into account the cultural adaptation, which are relatively unchanging. Yet, in many instances, social and cultural requirements exceed biological needs. Hence, ecological studies of human environment may not be sufficient enough, it needs some unique back up from other factors in environmental studies.

Now if "environmental crises" is a dreaded expression and "environmental ethics" is an aid to the crises, we ought to remember that the way of looking at "environment" from *our* way of looking at things, its "beyondness" amounts to some kind of a "disconnectedness" whereas the truth may be an "involvedness" or an "organic relation"—some sort of "oneness" if we like to be more metaphorical but not metaphysical. From this angle, "environment" appears not as an Indian or Pakistani or American would explain, or in the way you and me or he or she or they might explain, for *us* "environment" is an "integrity" notion. We shall further see that this "integrity" notion does help *us* in empathizing with agonies of the environment or of earth at least, lest, "environmental crises" is a dreaded notion.

Keeping these technical (not so metaphorical) and implicit (most hopefully metaphorical but not metaphysical) meaning of "environment" we may now explain "environmental science" as a systematic and scientific study of human environment and their role in it. On the other hand, "Environmental studies", a little different from the science, deals with several environmental topics as broad as including social aspects of the environment. Such concern is not unscientific rather, *trans-scientific* as it covers the concerns of non-scientists as well. Now if we bring in "ecology" as a science here for a contrast, it studies the intricately manifold relationships between all living entities and their environment. We have seen that such a study is predominantly biological. Environmental science, in its extremely "broad" notion does reveal the close affinity with ecological science and at times one has a feeling that 'trans-human' notion of environmental science is hardly any different from ecological science. Yet, a thin line a difference may be that the broadest sense of environmental science need not be taken as predominantly biological for it covers social and other non-biological aspects as well. In this sense, environmental science may be looked upon as more broad in its task or subject matter than ecological science. Truly, environmental studies as we have seen just now, appears to be broader than a purely scientifically oriented ecological and environmental science. We do realize that environmental issues and environmental crises may be so broad as to what purely scientific issues might not touch upon. Hence, we maintain henceforth that within the broader context of environmental studies in "environmental ethics" makes a special mark because it is often thought as one good aid to resolve crises in ecological and environmental sciences. It seriously takes up the crises caused by the great manipulator and marauder—the Homo sapiens. What we get is that the deliberate or intentionally acting human beings with their 'junkyard' philosophy of life are placed crucial in serious environmental crises in modern times. It is here that we doubly realize why ethical matters creep in environmental science and environmental studies. We will now analyze the notion, "environmental ethics" and also "ecological ethics".

In order to perform our task of clarifying the meaning of "environmental ethics" we can either propose a definition of the same, which we believe is more or less satisfactory enough or else we do not propose logical or other types of definition. Instead, we try to analyse first the meaning of the components of the expression: "environmental ethics". A statement of "environment" has been unambiguously obtained without many technicalities of scientific definitions. We need not repeat the same here for it does not serve any purpose. However, a statement of the same might be equally important when we are set to explain finally what amounts to "environmental ethics". There is no doubt that we should first start analysing the meaning of "ethics" and then "ethics" in "environmental ethics".

To clarify what is "ethics", a strict or logical definition is not needed at all (see Volume I) because formal or non-formal definitions are either incomplete assertions of what are the tasks of ethics or are unsatisfactory statements of what one thinks ethics should be like. The latter efforts are more injurious to ethics as they resulted into several misconceptions like, ethics is nothing but moral science and study of social and personal morality. Whereas the former effort, a little less subjective but myopic, ends up in leaving out same task or the other. A careful reader of the subject finds quite often than not that the practical; the professional and the virtue ethical tasks are left out. It is desirable, therefore, that we become more careful qua analysts of the various tasks of ethics and thereby construct a normative definition of "ethics" if that is at all needed after so much of painstaking.

Considering particularly the cases of those who have not gone through this portion in Volume I (The Primer), and those who do not have a first hand information of the subject, a brief but clear statement of tasks of ethics is needed. First is to mention that ethics is a branch of study within the broad field of philosophy, which attempts at doing many important things barring constructing prohibitions of sex and do's and dont's in personal life. Rather, ethics is concerned with three major tasks of analyzing the *truth and meaning* of *ethical language*, which is to clarify the sense of important ethical notions like good, right, justice, duty, obligation etc. We call this the *metaethical* task of ethics, which also includes the discussion about *justifications* of ethical standards or norms. With this, we come to *normative* ethics or discussion about the *bases* of *ethical judgments* of intentional human actions having far-reaching social and personal impacts. The explanations of the standards, their various implications and use all come in this purview. The case of having a norm is also related to arguments for not having one or any such norms as is with the former task of ethics where we question the meaningfulness and truth of ethical language. In this context we often talk about *ethical skepticism*, its logic and its limits. Such skepticism is about norms, language and the practical application of ethical norms as well. Defying skepticism we now say that apart from the normative and the metaethical tasks, *practical or applied* task of ethics is equally important in which, we find out ways to resolve value-laden ethical problems of practical life. Apart from these three tasks, we have some other tasks of ethics. One among them, which is arguably old but lost in the controversy of philosophy or psychology, is to analyze the *psychological bases* of ethics. It is believed that without proper clarification of the psychological sphere in ethics one cannot realize what it is to speak about 'motives', 'intentions', 'free will', voluntariness and non-voluntariness in actions, and also about moral development and moral evolution in human beings. Closely related to this is another old but forgotten task called the virtue oriented discussion or simply, virtue-ethics, which is concerned with the nature, kinds and inculcation of basic ethical virtues in place of being theory-laden. Moreover, we now speak of occupational ethics and professional ethics within ethics because we realize that both the forms of human life need serious consideration about the ethically just, right and good as well as facing the value-crises in decision-making leading to pertinent resolution and formulations for guidance in social life. We will shortly realize that *environmental ethics* is also within ethics or that among many tasks that ethics performs, one is to deal with ethical concern in environmental studies (and also ecological studies). Ethics has become so wonderfully vibrant a subject within philosophy that it has perhaps become as unpredictable as the sciences to crystallize in exact terms what it could do any more. The performatives of ethics being so dynamic that it would be a little foolish to put a bracket here and say that these are the tasks and if you have understood them, you have known what ethics is. Yet, one might be satisfied enough to know what a general statement of "ethics", "ethical" and "ethicality" means.

We should keep in mind that an ethical discussion is "what pertains to ethics" and not simply moral churning of some moralists as moral philosophy or ethics is neither moral science of fashionable ethicists nor mere casuistry of Church fathers. Nor is it handmaiden of moralists who have taken the sacred task of building great individual and national characters by dint of formulating what to do and what not to do. We need to remember that any misgiving about environmental ethics might arise out of not understanding what ethical philosophy does and what it does not attempt at all. Another question may be that ethics is not truly ethics if it has nothing to do with metaphysics and religion. We shall remind here that ethics, metaphysics and religion are different categories altogether for their nature and tasks are different. Hence, they are not one and the same but that does not mean that they are completely unrelated. In fact, history of ethics is a testimony to the relation of ethics, metaphysics and religion, but the question is what purpose does the relation serve? We shall consider this question in details in the context of environmental ethics in a separate chapter. For the time being, we should understand that there is no room for confusing ethics and religion, and ethics and metaphysics. Truly speaking, there are metaphysical ethics and religious ethics as also metaphysical and/or religious environmental ethics, but we have to peruse into what they do or strive to do, and the way they contribute or they do not contribute in any way to the growth of ethics and environmental ethics.

It is proper now that we correlate "environment" and "ethics" in "environmental ethics" to reveal what is its nature and what are its functions. We should also discuss what the significance of putting together "ecology" is and "ethics" in "ecological ethics". Precisely now we should state that for all practical purposes we have resorted to the use of "environmental ethics" rather than "ecological ethics" in this book because we know that both the conceptions are quite different but closely related, and that a *study* of environment in its extremely *broader significance* does cover a study of ecology as a science to a great extent. However, the unique 'biological' overtone of ecological study, which is not predominantly oriented towards social, economic, political and other factors, does not rule out or even neglect to smallest extent the ethical orientation. Hence, it would be equally important to speak of "ecological ethics" separately but the unique 'biological' concerns in ecology that permeates ethical scrutiny is adequately taken care of by environmental ethical concerns. A separate biologically oriented ethical discussion may not be redundant or even superfluous but well covered under a much broader discussion, which we take as "environmental ethics" where it simply means any *ethical discussion pertaining to environmental matters or matters concerning the biotic world, the inanimate world, the components covering this animate and inanimate world and the manipulated and modified world of the human beings*.

It is here that we have to realize what the meaning of "pertaining to ethics" is. This is now not at all a difficult matter for us to understand because what pertains to ethical discussion has been unequivocally stated before. The main problem is, does it make any sense if we now place the environmental study in this context. Or, we may ask, is it sensible to juxtapose environmental concerns and ethical concerns? It is sensible no doubt, for the reasons are fairly simple. First, we have to accept the argument that any discussion within the field of ethics need not be so broad as to cover all the tasks of ethics. Nobody would accuse us of a gravity of mistake if we discuss metaphysics and epistemology and yet justify that such discussion 'pertains to philosophy'. Similarly, no one accuses us of not doing something 'pertaining to ethics', if we merely discuss about norms and practicability of norms. Going by this reasoning we ought to realize that unique way of discussing ethics within the fold of environmental studies (and also ecological studies) should not be conceived as undertaking all the tasks of ethics which we have enlisted earlier on. Second part is that what pertains to ethics in environmental studies is to find out how relevant is the question of *ethical values* in environmental studies. More precisely, do some environmental practical problems raise problems of ethical value as well? It is here we have to realize that *ethically* good, right, just, obligatory, dutiful are some crucial value notions, which does emerge prominently while we consider some environmental problems of concern in modern times. At this point one is reminded of an important problem to be resolved, that is, about the human centred understanding of the ethical values. Do we have to speak of right, good etc. only in terms of human good, right and so on or does it make sense to understand them in a broader perspective of life? However, in so far as intentional acts of man and evaluations are concerned, we do speak of ethical values of such actions only but those decisions and actions as influencing other animate and inanimate beings and things, raises question about the validity of the good and the right of these beings and things. Similarly, the question of value of human beings vis-à-vis the value of other beings and things make some prominence, which is understandably unavoidable. We have to consider this problem as well at a proper place. Third, we should realize that *metaethical* concern like the clarification of the *sense* of "environmental ethics" and the *justification* of or *possibility* of an environmental ethics does make a sense or else what we have been doing thus far would be labelled as sheer nonsense. Fourth, an inquiry about the possibility of ecological norms (environmental so to say) or ecological principles along with well framed ecological codes as forming an independent and important part of the body of ethical norms or principles should be thoroughly taken care of. There is however some controversy and to my mind some misconceptions about ecological principles and codes forming by themselves a body of ethical principle or ethical norm. Shortly, we will consider this problem. However, there is no doubt in my mind about ethical first principle like deontology and teleology and some secondary principles of morality like compassionate and empathetic acts have been found proactive in the context of the real life dynamics of applying ethics. At a proper place we shall discuss what such application of ethics is and what proactiveness of ethical principles are. Again, in this context it would be a good case to analyse whether or not ecological principles and codes as so-called independent bases of morality are themselves proactive. Here we have to inquire about the soundness of calling environmental ethics as a branch of applying ethics. Or, should we treat it as an independent branch in ethical studies as we treat professional and occupational ethics as distinct from applied or practical ethics? Finally, we have the possibility of discussing about basic virtues forming a part of human character (by virtue of being acquired dispositions or traits) which are deemed essential for the resolution of any ethically charged environmental problem. It is a pertinent question to ask whether or not without being a virtuous person first, someone can at all apply ethical knowledge of principles and codes to resolve value crises in environmental studies. Such discussion about virtues and virtuosity avoids to a great extent a reference to ethical norms but it is academically interesting to find out whether or not such an attempt to avoid certain things are possible.

In any case, what I was trying to justify is clear by now. In my mind there is absolutely no doubt about the involvement of ethics as a branch of philosophy dealing with the important tasks it performs, in the field of environmental studies and ecology. To be precise, many issues in environmental and ecological studies are not merely scientific and practical they also raise questions of ethical values (of goodness and rightness etc.), norms, application of norms and virtues. We should remember one important thing in this context that many matters in environmental studies are in fact *interdisciplinary* in nature as it gets related to not only ecology, geography, botany, forestry and so on but with sociology, economics, politics and also philosophy and ethics. Hence, it is not difficult to justify that the problem of "environmental refugees" and their rehabilitation owing to construction of mega multipurpose river projects remains unresolved if we do not take care of the ethical value questions apart from many other questions. We have to remember that the big question of ethical ought or should or, what we call ethically *desirable* should be properly addressed to because practical problems of urgency does not get resolved without an ethical face. If we boast to be civilized, it is not merely material and mental, it is something more, it asks about whether or not we are ethically right, just and good, that is, whether or not we take care of ethical values as well. We should also remember that practical decision-making in the fields of environment and ecology is a queer phenomenon for such decision-making, apart from scientific feedbacks, some concern about what is desirable or ought to be done is most important. From this, we find that there are at least three important problems regarding the "justification" of environmental ethics as an independent field of study. Thus far we have cursorily justified the possibility of having an ethical discussion in environmental studies, which however, needs some more clarification. We shall take it elaborately in the next chapter.

Before moving on to consider some extremely important issues in environmental ethics, we would like to provide some *reasons* behind the evolution of ethics-talk in environmental studies. This will give us a glimpse of what concerns ethicists and environmentalists alike to come together for an active resolution for the welfare of the existential. Indirectly, this discussion would reveal the value of the study of environmental ethics.

Jorgen Randers and Donella Meadows in their excellent essay on "The Carrying Capacity of Our Global Environment: A Look at the Alternatives" edited in *Western Man and Environmental Ethics* by J. G. Barbour goes like this: "Because our environment—the earth—is finite, growth of human population and industrialization cannot continue indefinitely...but its consequences pose an unprecedented challenge to mankind…In deciding on the ethical bases for making the trade-offs which will confront us in near future because in a limited world we cannot master everything for everyone"[4]. Randers and Meadows write these lines keeping in mind that if man is not treaded on earth, importantly there will be perhaps no one to cry about environment and environmental ethics. Yet, we should be equally concerned about the way man is going in this limited space, with limited resources at hand. The authors are also concerned about earth ethics in particular and the future of the next progeny in terms of their welfare. The authors have thus a strong basis to justify that environmental ethics (earth ethics in particular), is possible if and only if the finiteness of earth environment is a truism and our urge to grow, develop and be happy for generations to come not so limited. Hence, one grave ethical concern is the way we should at present encounter and use our limited environment for welfare. The authors have meticulously shown in their essay that agricultural land, forests, and minerals everything is limited and our means for material growth has been utterly exploitative. The question is, should we continue this way? Should we not take care for the future generations? Interestingly, here are bare remarks about the welfare of animals, plants and inanimate objects, though one can understand that ethics is basically a human affair and its different tasks need not avoid the questions concerning human beings.

If we are not immediately so fussy about such human centredness, we may accept the truth the much ethical concern about desirability of decisions taken by man in limited earth resources boils down to the following things:

(a) Our continuous current *growth practices* would certainly lead us to an environmental crunch including a cultural and economic denomination of the human race;
(b) Our objective is to *maximize* the physical *benefits* of the *people alive today* would lead us to reject concerns for people infesting the earth in future, and

(c) Our present actions influencing the physical conditions of earth tend to a "global non-equilibrium" tending to a global catastrophe.

From these we have fairly clear idea about three natural human aspirations. One cannot easily denounce them as unethical on face value because there is no point in "non-growth", "non-maximization of benefits" and "individual and local no-growth". What is unethical is that "short-term objectives" of human beings are *only desirable* and "global equilibrium" is a *mere utopia*. It is, therefore, that ethics-talk or the talk of the "desirable", the "just" and the "good" of "long-term objectives" and "global equilibrium" comes to the fore. Equally true is that the so-called 'unethical' things as stated above needs to be justified as 'unethical' in clear terms or else our convictions cannot be taken to be justified on face value. The unethicality of the short-term objectives of human beings depends on the *criteria of unethicality*. In environmental ethics the broad criteria fixed for calling short-term benefits of man on earth at the present time 'unethical' is general welfare or simply "welfarism". Though "welfarism" appears to be an economical concept and despite the close relation between ecology and economy (the oikos terms), it should be understood in the context of ethical criterion as *the maximization of good over evil in the overall existential situations*. This would cover the situation or the environment in which we live and in which other animate and inanimate beings and things exists as being enveloped by the essential elements around them to a great limit beyond earth, which sustains or help sustain life and things on earth. But this is an extremely broad conception of 'welfare'. People prefer to avoid vagueness in the conception by replacing 'good over evil' in more concrete terms as we find in 'well-being'. However, what is 'well-being' is often blurred by several misconceptions ranging from physical prosperity to solace of soul. It cannot be comfortably replaced by older concepts like "happiness" unless a more blunt "benefit" is not used as an alternative. So it would stand as "what is beneficial" to the existence as whole, notwithstanding human needs and aspirations. Some people clarify it more by saying it should primarily keep in mind the need of the most needy among human beings notwithstanding human urge to grow and develop along with other beings on earth.

Whatever may be the vagueness in 'welfare', the 'beneficial' (if it is a clearer notion), need not be taken merely as either economically beneficial or physically beneficial or materially beneficial. Most of the modern environmental ethicists tend to transcend gross conception of "benefit" or "well-being" and without being necessarily metaphysical, holds that some material advantages might be downplayed under the idea of 'quality of life'. Is this a clear conception of well-being?

Many people harping on 'quality-of-life' as the *mantra* of well-being think that a non-metaphysical conception is altogether impossible because its main proponents have exactly in mind the metaphysically oriented life as 'qualitative'. Some would hold that first it is important to conceive the whole existence as becoming Pure Being, which is essentially 'One', and as such, the existence in itself is complete, harmonious and good. If human life is led in accordance with the "sustenance" of the "pureness", "harmony", "goodness" and "Completeness of Being", it is "qualitative", therefore, "well-being" of the person is guaranteed. Such "well-being" is thus inextricably bound with many *non-metaphysical* conceptions as well says the metaphysicians. Others who are more soaked in such philosophy of life advocate divinisation of the existence as such and avoid such veiled assertions like Absolute Being or Pure Being because 'Being and being' duality of the Heideggerian legacy had its own non-metaphysical (even anti-metaphysical) import. A clear divinisation would tend to almost the same idea of "sustenance" of the perennial "one". Yet, the non-metaphysical and anti-metaphysical ways of interpreting "quality-of-life" may have a different conception of "sustenance" altogether. The claimants of "scientific sustenance" of the whole of existence would speak in another relatively understandable language. They would speak about the growth and development of human beings with other beings in a fine-tuned equilibrium or a state from where a conception of nature in entirety tending to a thermodynamic equilibrium or entropic state is not conceivable in far future thus assuring the *long-term objective* of *liveable life for future*. It has to be mentioned here that the objective of a "liveable life for future" or what is a scientific way of qualitative life or well-being cannot and need not be refuted by the metaphysicians. The metaphysician and the proponents of divinisation base their arguments differently as their presuppositions about man and nature are radically different from the material conception of man and nature.

In any case the metaphysicians, the scientists and the anti-metaphysicians as well derive several "qualitative" ways to live a life on earth. It is believed that *unnecessary* infliction of pain or harm on animate and inanimate existence is adverse to 'qualitative life' on earth. Though one animal does consume another animal and one animal does consume other life-forms, which need not be reversed by whims. In any case, hermits' ethics of rigorism does not find a suitable place on earth, as that might be too good for us to live. It is equally believed by the metaphysicians, non-metaphysicians and anti-metaphysicians that all people who live today on earth are as valuable as all people to infest earth henceforth (and hopefully for another millennium), therefore, we ought to use and reuse the limited resources on earth and around it in a manner so that growth and development of the human race in future is not hindered or gets poor in any way.

Keeping these things in mind, we are now in a position to say that in environmental ethics, we do not conceive of one basis of or one criterion of ethicality and unethicality, we conceive of many criteria. Though "welfarism" or the theory of "well-being" appears to be the criterion, it becomes a cluster of criteria in reality if the metaphysical and the non-metaphysical sense of well-being are kept in mind, though in itself, none is absolutely true. To be sweeping, welfaristic criteria are two, not one, the metaphysical and the non-metaphysical, which commonly speaks about *sustenance of existence among milieu of growth and development avoiding unnecessary suffering and a care for life now and then*. This says many things. The ethical bases we find in environmental studies are as plural as they could be and appears to be different from the deontological and the teleological norms taken from our traditional ethical basket. However, we have to delve into the reason for such claim.

For now, it is enough to say that ethical questions in environmental studies are of some worth at least because without it, we live a life but do not live an ethically charged life. Such life is desirable in itself or else tries to imagine its corollary if that suits a being called 'man'. Here it is important to point out that if relativists argue about the non-fixity of an ethically charged life because ethicality is subject to change, I do not believe that they will accept that at a point of time unethical life is better than ethical life to sound like sophist Thrasymachus of Plato's "Republic" fame.

With this we close this chapter by saying that environmental ethics has its unique value to compel us to realize that if human beings are not so capable as to sustain their life all alone and if life on earth is valuable for times to come, our deliberate act that goes against sustenance of the whole of environment in general, should be questioned for 'desirability' or scrutiny of ethical goodness. We are now at hand a fairly clear statement of the nature of "environmental ethics" in terms of the tasks it performs. To be precise, it has a task of clarification of important value concepts as well as justification of environmental ethics as within ethics but independent of applied ethics. It has also the task of justifying the ethical norm or norms to be applied in resolution of value-laden environmental problems. It is also important to see whether such norms, like "welfarism" bear different sense than the one we are acquainted in traditional normative ethics. It is thus important to find out whether or not traditional normative ethics is sufficient for norm related environmental ethical discussion. Finally, it deals with the way value-laden decision-making rules are formulated.

NOTES

1. Rasmussen, Larry L., *Earth Community and Earth Ethics*, Orbis Books, N. Y., 1996, p. 90.
2. *Ibid.*, p. 91.
3. *Ibid.*, p. 93.
4. Randers, J. and Meadows, D. in J. G. Barbour edited, *Western Man and Environmental Ethics: Attitude Towards Nature and Technology*, Addison-Wesley Publishing Company, MA, 1973, p. 253.

2

Justification of Environmental Ethics

A careful reader must be aware of the fact that analysis of "environment", "ethics", "environmental ethics" and other related notions provide us some idea about the "nature" of environmental ethics. Knowing about the different 'tasks' of environmental ethics also help us in unfolding the "nature" of environmental ethics. In order to know more about the 'nature' of environmental ethics we need to "justify" that environmental ethics make sense within ethics or else, it is something beyond ethics and still makes a sense. Not only this, in popular opinion environmental ethics is well justified as a branch within applied ethics, it does not need any justification as independent from it. We have to find out whether or not environmental ethics is at all a branch of applied ethics. Another gap has to be filled up. We have to find out the reasons behind the quantum shift from ethics-talk involving human beings to ethics-talk involving non-human existence. This is to fill up an important gap of justification of the unique ethics-talk we are so enthusiastic about. Furthermore, if ethics-talk makes some sense with regard to the cases of human and non-human existence, does normative ethical discussion along with the metaethical, the applicative and the discussions about virtues and values, in such contexts, contributes to the growth of ethics? If it does, justification of environmental ethics is made further possible as an important branch of study within ethics. Most important truth is that if these gaps are filled up appropriately, the "nature" of environmental ethics becomes pronouncedly clear or else what has been done thus far leaves a shade of ignorance, which is not desirable.

First, we shall provide reasons in favour of a shift from "human-centred talk" to "existence-centred talk" in environmental ethics in this chapter. First, we shall analyse the shift from human-centredness to existence-centredness based on *non-metaphysical* arguments.

Such arguments are most needed for "justification" of ethics-talk about environmental matters. I have mentioned in the previous chapter that the field of ethics is ever-growing because it is no more looked upon as a mere theoretical venture in philosophy, an armchair affair, cutting us from the actual concerns in life and society. The dynamics of ethics has projected fresh areas and justified them within the main field by showing that question of moral evaluation, norms, virtues, values and logic and language are closely related to the real dynamics. It is in this regard that we now accept without much fuss that professional and occupational ethics are within ethics as is business ethics, social and political ethics, and educational ethics and so on. However, so far we have been dealing with some such novel areas in ethics which are beyond doubt human-centred because human beings or moral agents have been considered as the focus of moral evaluation. In fact most of the branch in practical, professional and occupational ethics put human agents either as most important or only important agents whose intentional decision-making and actions are put to moral scrutiny whereas, the "other parties" who are influenced by human actions are put into the oblivion. It is not that we repent for the "other parties" for they were not treated on equal footing as human agents rather, they are not considered seriously for their values, virtues, rights and even beauty and symmetry, rather, to forget "other parties" whether humans or non-humans in moral scrutiny would be a great mistake. I have a feeling that with the advent of engineering ethics, bio-technological ethics, environmental ethics and agricultural ethics, the "other party" concern gained a momentum because we came to know that in an ethical interface, humans are not necessarily and sufficiently related to human beings, they are related to machines, microbes, molluscs, vegetation, robots, the non-living objects and even insane and vegetative people. We come to know at this point that 'human-centredness' in ethics is therefore a trend in ethics in which, the metaethical, the normative and the applicative concerns, three basics concerns in ethics, and also the concerns about virtues, values and psychological basis of ethics keeps in mind "human to human relationships" and judges or evaluates the right, good, just, duty, obligation etc. of intentional human actions. It has to be noticed that 'non-human centred' does not amount to the same as treating "non-humans" (including living and non-living entities) as human beings and thus evaluate their functions ('acts' will be a laughing term). To be precise, 'non-human centredness' in ethics is not to consider a stone or a waterbody or viruses as if they were as good as moral agents as human beings and thus their "functions" (mostly unintentional) should be judged for morality and immorality. Even if we consider active and dormant cases of consciousness, as in plant kingdom and non-human animal kingdom or cases of thoughtful actions of an orangutan or monkey, we do not say that they are as good moral agents as human beings are and their deeds and acts should be morally scrutinized. Hence, in ethical discourse, 'human-centredness' has a specific meaning in which crucial are "other parties" and "relationship with these parties through intentional human activities". Clearly, though "other parties" are important enough, these non-human other parties in particular do not make a justifiable case for moral evaluation though they make a justifiable case for *moral-talk*. What we get from this is that ethical discourse need not be necessarily and sufficiently looked upon as merely human-talk rather, justifiable non-human-talk. This is in fact a growth of the field of "ethics" itself, which is important to understand. We have to understand why environmental ethics makes a good case within ethics even if profuse "non-human-talk" permeates the field. Certain points need more clarification at this point. One is, how do we say that metaethical concern in ethics which has been human-centred and most important in ethics should now include non-human centredness? In traditional human-centred ethics even language concerns were human-centred because what is good, right, duty, obligation etc. have been analyzed for their meaning keeping in mind what human actions in human-human relations are good, right, just and so on. Even the theory of indefinability of the moral term "good" needs human intuition to explain its nature. Yet, history of meta-ethics is testimony to the fact that such ethical notions might not be clearly analysed for their meaning by completely overlooking the interesting uses of the ethical notions in *human-non-human relation contexts*. If we have insight to such truth, why should we think that language concern in ethics is solely human-centred? It would be foolish if we imagine that language concern in ethics is human-centred solely as here we are not considering monkey-talk. Do monkey's talk ethically? Who knows? Certainly if they do, we do not respond because we are neither Tarzan nor monkeys (now). So none on earth thinks that humans use human language, ethics is unique human language use and, therefore, it is human-centred. Let us eradicate this doubt from our mind that 'human-centred language use in ethics' is such use in which ethical notions are analyzed for meaning keeping in mind human-human relations merely whereas we want to stress that such notions are properly clarified for their meaning when human-non-human relations are covered as well or to be simple, when ethical notions are understood in terms of the ways in which human beings use them in order to *talk ethically about* "other parties" including non-humans. This would amount to acceptance of "non-human-centred meta-ethics". But this is not all, in meta-ethics we discuss about truth of ethical assertions and about the logic of normative and applied discourses. Do these things not transcend human-non-human centeredness? First, no language use and determination of truth transcend human-centeredness. Rather, it cannot transcend non-human centredness either. The language of *ethical* good, right, just, etc. to be meaningful needs care for the function or the different performatives of such language, which is uniquely human-centred but that does not rule out non-human centred uses, that is, *not* as *used by* non-humans rather, as *used about* non-humans in *ethical* talks. Hence, 'human-centredness' here has a significant import as we have seen in other cases above. The truth of ethical language does depend on the way we (humans) use such language in human-human and human-non-human relations in actual world. In fact, I have been harping for quite some time now that language use and its truth does not overlook empirical concerns in real life (see Volume I). The *logic* of normative ethics and normative ethics *per se* does not forget this simple truth either. To justify through formal and/or informal ways that teleology or deontology (any kind) holds good or does not hold good begs the statement of these norms (as basis of moral scrutiny) in fairly clear terms. Once the statement is made possible, logical scrutiny for validity of accepting norms or rejecting them is all very much human no doubt and, therefore, the whole affair is human-centred. And this is so commonsensical that it is painful to remind that human-centredness still does not rule out non-human-centredness in normative thinking and its logical qualification. What I mean to say is that the norms themselves does not speak of human-human relations, they speak more than that. Teleology and deontology whatever, does not speak of human teleos in relation to humans only rather, in relation to "other parties", and human "duty" is more pertinent in relation to "other parties" and not merely in relation to humans. Coming to logical justification, justifying that human "teleos" is such and such holds good is not merely logic rather, deeply associated to many *empirical considerations* involving humans and "other parties". Really speaking, any amount of rules of syllogism or rules of valid inferences or rules of induction do not justify teleology or deontology unless the human teleos and human duties are heuristically understood in relation to other human beings and non-human entities in real life. It is a matter of good mental exercise to peruse into why teleology and deontology are not fully justifiable if for instance we hold that our purpose in life is to maximize good over evil for us (humans) and that our duty in life ought to treat humanity as intrinsically valuable. A careful analysis will reveal that teleology and deontology as they appear above are not justifiable or unjustifiable by the formal logical tools we have. These tools may be useful only when some empirical facts are put together, for instance, when we further claim (on empirical basis) that maximization of goods for man has not turned out to be good *in reality* as "other parties" have been deprived in many ways. Or, when we claim that treating animate and inanimate entities as inherently valuable proves to be broader duties of mankind *in fact*. There is a danger in my position here as one might think that this way, I am ruling out formal logic as useless. Or may be, mine is a veiled *intuitive* argument. What I am stressing is the impossibility of *logical* justification of norms of ethics without first of all analyzing clearly what norms state and what empirical grounds constitute the basis of such statements. I am also referring to an impossibility that empirical basis of norms does not rule out concerns for non-human beings, and for that reason, human-centredness in logic and normativity though commonsensical, in a different way signifies that human-centredness does not discredit non-human-concerns in logic and normativity.

The practical or the applied task of ethics tells us the same story. If we remember what practical ethics is (Volume I), we can easily understand that an application of ethical knowledge of good, right, just etc. (not necessarily of norms), is again a human affair but not merely a human affair. This is further crucial in understanding that practical ethics is not human-centred in its traditional eloquence. Rather, an application of ethical knowledge of human beings covers resolution of value-laden practical problems pertaining to other human beings, corporate, other institutional entities and other non-human entities as well. Here again we realize that our so-called human-centredness in ethics should be overhauled. Human involvement in ethics does not imply that ethics involves only human beings. Human interest in practical ethical affairs involves non-humans as well.

Nothing is different when we consider the virtue ethical interest, the axiological ethical interest and the psychological ethical interest. Speaking about virtues and values has never been exclusively human affair. If we speak of virtue ethics or of ethical values, nothing changes much as despite humans are virtuous (or should be so) and humans have value (or care for values), they are not so when in relation to human beings only. This false human-centredness should be shunned. When we speak of human beings as virtuous and caring for values, we relate them to "other parties" and thus scrutinize for virtuosity and valuableness. What about psychological basis of ethics? Is the same thing applicable here as well? Undoubtedly yes, for one instance, if "purity of intentions" is so important in decision-making and acting ethically, can it be judged "pure" only if human-human relation is kept in mind? In no case will it be possible. Though intentions are uniquely human, we speak of purity of human intentions not when they are pure towards humans but towards "other parties" as well. Should a person's purely intended love for man allow him to plunder the natural objects where we live and thrive? We can think over it but for sure we shall find that purity of intention raises a question of relations human-human as well as human-non-human. Hence, the psychological basis of ethics considers human beings as most important, but it considers non-human-focussed questions regarding "purity" as well.

What do we get from the entire discussion about ethics as visualized as "human-centred"? Is it a misconception? No, because any discussion in the field of ethics involves human beings and relation of human beings with other beings. To be clear, only humans speak ethically, think ethically and write about ethics (also ethically), and only they speak, think and write unethically but that is not human-centredness as understood in ethics. Human-centredness is also not merely human beings' ethical concerns about other human beings. It is human beings' ethical concerns about human beings and non-human entities (and we know what these ethical concerns are).

Now the ground is ready to consider the basic question we have raised: Is transmutation of human-centredness to existence-centeredness in ethics essential for conceiving environmental ethics? I would like to reply in 'no' and 'yes'. The answer is 'no' with a qualification that ethics as such can never be imagined as transcending human beings because it is out and out concerned about the deliberated actions of human beings, which we evaluate for moral goodness. Moreover, human-centredness is unavoidable as we ought to speak of ethicality in complex human relations in societies, and human ethicality in relation to animals, other life forms and inanimate objects. It will be thus impossible to imagine an ethics where non-human agents are evaluated for morality of their actions. However, transcending human-centredness in ethics is possible in yet another qualified sense. It is nothing but transcending some such ethical paradigms (mostly archaic) that tend to invoke the idea that human beings are to be evaluated for morality of actions if and only if such actions are affecting human individuals, groups of individuals, a society, a state and humanity at large, nothing less, nothing more. What has been said in some of the passages before makes it amply clear why such ethical paradigms have serious flaws. Most simple being completely looking human agents as moral islanders, who on earth are morally secluded among complex environmental interrelations.

It is thus notions of "complexity of human relations" and "moral evaluation of actions of humans by humans", prepares the ground for transmutation of the so-called human-centred-ethics to what I prefer to call "existence-centred ethics". I would like to justify my preference at a proper place than leave it as mere predilection. However, let us continue with our 'shift' in ethics, which must have changed "ethics" itself to a great degree as has other historical 'shifts' in near or remote past. One wonders however that given the treasures of the Aristotelian ethics and the Christian ethics in the west and the richness of the ethics of India, (and they are too old), we are still lingering with doubt about this 'shift'. Was there really a 'shift' or was it ever under our nose, which we forgot for a theoretical hegemony of later times? Didn't existence-centredness in ethics exist for long, which we might have rediscovered? If such discovery is really a 'shift' from our modern and contemporary ethical hegemony, well, I have nothing to say.

First, take this conception of human-centredness against the backdrop of what I have said just now to corroborate our views or to have more revisions. Robert Elliot in his essay on "Environmental Ethics" proposes the following about human-centred ethics:

A. "Some think that environmental policies should be evaluated solely on the basis of how they affect humans (*cf.* Baxter, 1974 and Norton, 1988). This entails a human-centred environmental ethics… classical utilitarianism… is one example of a human-centred ethics."[1]

B. "A human-centred ethic could lead to substantial agreement with environmentalists about policy. This would depend on the facts about the effect on humans of changes to the natural environmental… reached by considering the interests of humans alone… this ethic treats only humans as morally considerable."[2]

C. "On human-centred ethic… the species as a whole nor the individual members of it are morally considerable: it is only the happiness and unhappiness of humans which is morally considerable and this might or might not be affected by what happens to the turtles."[3]

An analysis of Elliot's conception of human-centredness reveals in (A) that it evaluates environmental policies solely on how they affect humans; in (B) that it considers the effects of the changes in natural environments on humans alone, their interests only and that only they are morally considerable, and in (C) that the whole human species is morally considerable not any one individual, and that the happiness and unhappiness of the species is morally considerable. What Elliot finds in (A), (B) and (C) is hardly any different from what I have said, which truly reveals the 'archaic ethical paradigm' in which "only humans but not other parties" make any sense in so far as decision-making, interests and happiness are concerned. Read between the lines that the moral, good, right, and just of decision-making, interest fulfilment and happiness aimed at depends on the fact that only humans have been taken into consideration, though not one but all because only they are morally considerable. There is no confusion here that a 'shift' in our 'archaic' ethical thinking is a shift from such human-biasedness in "moral considerability". This needs some reflection as I have said that beyond doubt human agents are 'morally considerable' and that is true to the core for times to come that it is our decision-making and actions which needs to be morally evaluated, not those of the turtles. However, we should remember in further discussions that 'human biasedness' in 'moral considerability' should face the plain truth that only actions of human beings affecting other human beings are not subject to moral consideration, turtles should be included if they are affected as well. This hints a 'shift' at least to turtles and then to others or taken together, the *existence* as such. We shall come to this soon but not before something more has been said about human-centredness and its inherent problems.

Something interesting is said by L. Westra in "Environment and Environmental Ethics" edited in *The Blackwell Encyclopaedic Dictionary of Business Ethics* by Werhane and Freeman:

D. "We must go beyond anthropocentric paradigm to establish who or what might possess moral standing (*cf.* Van De Veer). Environmental Ethics is broader, more inclusive than other practical ethics."[4]

The human biasedness is what is now cryptically termed as "anthropocentrism" which is clearly a bias about "moral standing", and most interestingly, environmental ethics is far different from other interests in practical ethics which focuses human moral standing exclusively. Westra further points out in the essay that such anthropomorphic view of ethics might surface in two forms: "weak" and "strong". The former view is supported by Norton and others who speak of human interests, welfare, happiness etc. over and above those of animals and others but does recognize environmental values in the face of exploitation of nature. The stronger view is supported by Feminists, and Carl Cohen who speaks of rights in moral community, which is merely human. However, Westra reminds that Peter Singer, Tom Regan, Paul Taylor and Arthur Leopold argue that all natural entities are valuable and merit inclusion in 'moral community' and, therefore the 'shift' we are speaking about is possible. Still one is reminded of arguments provided by some thinkers against non-humans having any claim to be members of a 'moral community'. Either this misconception is eradicated or else there is no 'shift' in ethics.

Though it has to be admitted that the reasons and actions of human agents are scrutinized for moral value and that something is bad in acting in favour of their well-being by disregarding completely that of others, there is thus in the real sense no point in thinking that only their well-being and interests are considerable for they are 'moral beings'. In reality, moral agents consider the well-being and interests of other people of different age, sex, groups and health status, and many of them are neither merely egoistic nor altruistic. Most importantly, many such people do not favour benefits to their species alone because they are selflessly caring about particular non-human animals and other living and non-living things. There are people in this world who do not care for interests of these non-human entities for maximization of their benefits and, therefore, are not only self-centred and human-centred.

One so-called strong argument in favour of anthropocentrism is that human agents are the conscious evaluators of objects and acts as morally good or bad, right or wrong, and for that reason they belong to a 'moral community' not those non-humans who are bereft of such qualities. One wonders if this is a credible form of argument. What does it aim to prove? Nothing perhaps because it sates a stark truth that human beings (to add, only mentally able and mature ones), value certain human actions as morally good and bad. It will be a little overdoing if they do so in case of animate and inanimate objects and in case of insane people or incapacitated people or too young children but does that prove that those animals, inanimate objects and insane people do not call for moral considerations? Entire confusion here is due to confusion about *moral vocabulary*. The confusion is not in the vocabulary or in its use but in its *understanding*. Undoubtedly, sometimes the confusion may be due to wrong *use* as well. What we have in mind when we use 'moral community' in this context is that 'one who is capable of moral action, moral evaluation and also an object of moral evaluation,. This is the case with human community none others whereas the truth is that 'moral community' refers to an 'agglomeration of humans and non-humans in space and time who demands some degree of care for moral considerations, that is, they call our attention to consider whether or not such and such intentional human decisions and actions were morally right, just and good or simply, morally desirable. Also whether or not such and such moral evaluations in terms of good and right etc. is pertinent enough, that is, based on valid grounds and whether or not such and such actions at all make a fit case to claim it either moral or immoral'. If we keep this in mind, though humans alone morally evaluate human actions, they are not the only members of moral community for along with their decisions and actions towards other human beings, those numerous decisions and actions affecting and involving non-human animals, plants and trees, and inanimate objects make proper cases for moral evaluations of which valid grounds should be there. Hence, a 'moral community' in this context does not refer to humans as a moral subject and a moral object or both but those *existentials* calling our attention for moral evaluation for they have been some way or the other influenced and affected by human intentional decisions and actions. Hence, even if animals, plants and other things are not moral evaluators, they belong to a community calling for moral attention of people. And as we have some ways to avoid these "other parties" in moral considerations, our avoidance itself is an object of moral judgement to prove that a natural shift from human-centredness to existence-centredness is possible. One point may be extremely important at this point. The moment we ask, why animals should be dragged to moral considerations for they are essentially non-moral? We have pages filled with arguments for or against, they are worth considering but in the midst of the controversy did we forget that as usual we are not pressing for judging non-moral actions in ethics, we did not lose our good habit, there is no category mistake either, we are only extending evaluation-possibilities of human actions, which were usually done in ethics and not in any alien subject lest we do not suddenly jump the gun and say that ethics is meta-ethics, we only need to analyse language and nothing else. But that is a different stance, I have replied to it in Volume I. What we are reminding ourselves about such habit is that now in the name of applied ethics, professional ethics, occupational ethics and environmental ethics, we are staging revival of ethics of extended human moral accountability. If we remember this, what great things would counter arguments to include non-human animals or objects of moral consideration prove is not known to me. Non-humans are not moral beings is no argument at all, all this we know from the history of ethics but the same history tells us other story to include "other beings" for moral considerations and shift from fixity of human oriented ethics. Did we care to read the history of ethics? If we did, let us now quickly shift from the ethical-hegemony of recent times to disregard other parties in ethics.

Another popular way to counter the inclusion of non-human beings and things in moral scrutiny is that though such beings and things have some value, mainly extrinsic or value as means, only so far as humans are there to evaluate them as such or else, such beings and things do not have any value in themselves. This is another good example of directionless argument. What does it aim at? Among many values, intrinsic and extrinsic are well known to us and for the sake of argument let us say that non-humans are not valuable in themselves rather, assume extrinsic value owing to human evaluation. So, who on earth would refute it and why? We would all sanely agree that we evaluate dogs as dutiful, plants as benevolent and rivers as great and we agree that most of the time we say that dog's dutifulness helps us to protect our life and property, plants help us to inhale oxygen for life sustenance and they take poisonous carbon dioxide and rivers are great because they flourish human civilization. Even if we do not hopelessly dangle with their inherent values as without this life on earth or earth itself will be not as it is, what would the hyped external value, the value we impute, prove? In fact, lack of imputing value or lack of power of judgement of value as inherent or external etc. is not a disadvantage, it is not valuelessness either, to be precise, it only proves that something is not like humans but not that they are not to be seen as objects in nature whose toil in the face of human actions is not fit to raise questions of moral desirability. Non-humans as naturally devoid of doing morally valuable actions and moral value imputation proves nothing about "moral community" and moral questions because the entire moral vocabulary has been grossly understood. Moreover, we forget that our moral evaluation won't exhaust all values of natural objects for moral values are not all values. Plants and dogs do have many other values than moral ones, one is surely an inherent value to constitute the biosphere in its marvellous life sustaining form we see. Even if some subjects like us or one in the alien land values plants and dogs as morally good or bad or nothing at all, they would not lose their several values including the basic one as mentioned above. There is another point to be closely followed. Natural entities conceived as imputing moral values is nonsensical, but their having no value is equally nonsensical. Most nonsensical is to confuse about their 'having values' and 'having moral values'. Let us remember our first lesson for ethicists that confusing value categories by vagueness in usages of names is not good at all. The "goodness" of dog does not imply that the dogs like human beings, think and behave well or are goody-goody guys for they protected our properties. We *call* them by this 'name' does *not* guarantee their inherent moral property. Their goodness may be completely *non-moral* goodness but when we protect our dogs from wanton pain or if we inflict pain for reasons known to be against the dog, "goodness" and "badness" does have moral sense. Yet, when we wash our dogs with rose-water, it may be good or not so good in a *non-moral sense*. All this is said to clarify that more often than not value imputations in non-human cases are clumsily confused for moral category of values and we tend to say that the dog is morally valuable when it is really non-morally valuable. But all confusions of these kinds do not allow us to keep dogs and plants out of the 'realm of moral evaluation of human actions'. Moral evaluation of human actions does include dogs and plants and others and for that reason the notion of "moral community" is used broadly to include these beings and things which should not be misunderstood for meaning.

This account also focuses an inherent problem in a 'subjectivist' account of valuation for it allows some value to non-human entities by human beings proving the point that *anthropomorphism* or what Richard Sylvan calls "Sole value assumption" is completely mistaken because at least one subject on earth at some time would impute some value, may be extrinsic value or whatever to dogs etc. and thus refute the idea that only humans are valuable. Now everything in the argument being good, we should still be careful about category confusion in values and valuation. We must be clear in mind that if all subjective imputations of values to dogs are 'moral value imputations', then the subject imputing such values does not know the basic theory of moral values and their imputation. In fact, here in this argument we are speaking of *some non-moral value* of dogs as judged by person, which does prove that only persons do not have 'some' non-moral values, dogs do have them and, therefore, on earth, only human beings are *not* valuable and only their interests and well-being need not be taken care of. It also shows that for certain reasons only humans do not have rights, animals also have them, and it is immoral on *our* part to forget that animals possess *some values.* Now if that be the case then anthropomorphism amounts to the case as we have been harping. It is a view that overlooks effects of human actions on non-humans and that it disregards "some values", at least non-moral ones which non-humans have, and thus puts moral obligation on man to take care of their interests and well-being. If such obligation is cared of, moral evaluation of man is made possible in another sense. All this enable us to justify that a 'shift' from human-centredness in ethics is possible but the question is, to which position do we shift?

A commonplace 'shift' is to the "animal-centred ethics" because as Elliot make it simple that it "is a view of ethics which counts not only humans as morally considerable but non-human animals as well: it includes all animals in its scope. Many of the things which we do to the natural environment do affect non-humans adversely and this, it is suggested must be taken into account"[5]. In this statement, the expressions, "morally considerable" and "adversely" make clear suggestions for the 'shift', which is clear from our earlier analysis as well. However, to eradicate some confusion Elliot clarifies that "An animal-centred ethic enjoins the moral consideration of individual animals not of species: what happens to species is only of indirect concern in so far as it affects individual animals"[6]. Hence, animal-centredness rules out "speciesism" for that might amount to some sort of parochial treatment. But there are important questions in animal-centredness which we shall consider while we go along with some supportive views first.

In his essay "Rights across species", Tom Regan has systematically refuted anthropocentrism in clear terms, though by this it has not been justified that he is anti-people. He favours speaking about "rights" of animals or non-human things and beings at large which, apart from helping to justify the favoured 'shift' from human-centred ethics, also helps us to analyze the meaning of value term "right". Regan with his belligerent zeal and language in the essay puts forward the following ideas:

(a) To be in favour of animals' protection against cruelty caused by human beings unnecessarily or unjustifiably and to be in favour of their welfare does not prove anything, it may be a mere emotive move forgetting that there is a danger of accepting the view that if cruelty is necessary and justifiable, it is desirable. Still worse, when is something unnecessary and unjustifiably cruel and when is welfare of animals done are not pronouncedly clear. These are all vague ideas, mostly having no sound basis.

(b) In fact, fallen in their trap, the anti-cruelty to animals group and animal welfarists often swallow the bitter pill of acceptance that significant pain (sometimes at least) is justified mostly on greater welfare of a number of other animals and human beings. The *animal rights movement* or their supporters do not adhere to such views as they are glossed versions of anthropocentrism and egoistic hedonism, and thus immoral. Though it is again not immediately clear why egoistic hedonism or simply egoism or anthropocentric notion is "immoral". Rightly people have grounds for such declaration, which Regan clarifies.

(c) Regan analyzes the animal rights' position to show that they are by far supporting their zeal on the basis of universal egalitarianism, which is further not the proper basis of animal rights or as Regan says, it should be better understood as against complete animal liberation. The reason being the impossibility and illogic inherent in "equating the interests and welfare of all". Reagan asks: How can slaves' and slave-masters' interests be equated if both are to stay in the world and not alienated? Would slaves' equitability of rights be ever determined and maintained in an egalitarian method? He does not think so. Similar is the case with animals because an egalitarian philosophy would never be able to show how they could be equated to human beings for their rights.

(d) A renewed understanding of animal's rights and thus animal liberation should be based on "systematic justice" or avoidance of "systematic injustice", and if we think deeper, it turns out to stop violation of "basic moral rights" of slaves and/or animals. Therefore, he says, "Animal Liberation is goal for which the philosophy of animal rights is the philosophy. The two—Animal Liberation and Animal Rights—go together, like a hand in a glove".[7] Now Regan has no doubts about "systematic justice" "being fairly clear basis for his 'a hand in a glove' but what about "moral rights"? Is the notion fairly clear and justifiable?

(e) Though feminists championing "individualism" objects to the idea of "moral rights" of non-human animals as does Cart Cohen on grounds completely unacceptable to Regan, "moral rights" does not amount to anything abstract or queer. It is nothing but "unjust oppression" of animals in many forms, one being using them as means to another end. If this be the case, only then do we have any ground to say that non-human animals (and even humans) have "moral rights" and thus "animal rights", or else, they do not have any.

In an eloquent reason to justify the shift to animal-centredness, Tom Regan in "The case for animal rights" holds that beings possessing an *inherent value* enjoy "rights". These values, as different from the external values are good-in-themselves or good irrespective of what they do to others, or the way they are useful. In his words, only "subjects-of-a-life" has inherent value, that is, all mutually normal mammals of a year or more are "subjects-of-a-life" in so far as they have consciousness, desires, wants, some deliberations and actions have inherent values and enjoy their rights to be projected. Though these rights enjoyed are not 'legal rights' for such rights are products of law variable in different societies. Moral rights are universal rights of subjects-of-a-life to be treated with dignity as an individual member of a community harping inherent value, which are *equal* irrespective of species. These inherent values of animals are neither lost nor regained by acting virtuously or in an evil manner. We should proceed to another support in favour of animal-centred ethic now though Regan's view has its own problems.

Undoubtedly, Regan's view is implicitly egalitarian one which he tries to conceal. It is also one respecting individual value as equal. Another supportive view that of the *utilitarians* like Peter Singer does not rely much on the equal value of all beings. The reason may be inclusion of man as subjects-of-a-life along with others would not be equally treated so far as practical solutions of conflict of values are concerned. For instance, the one mentioned by Regan himself in the above essay, is the case of throwing off one among five consisting of four men and one dog on a drowning boat. Utilitarians have a different opinion than the egalitarian Regan regarding this matter. Regan thinks that equality of 'inherent value of beings' being so important, does not still make a case that, therefore, all beings, men and animals demand equal treatment. This is so because some loss to life may be *prima facie* a huge one or irreparable one in comparison to other. In this sense, Regan holds that a thousand dogs, on sinking boat, needs dumping to save only four human lives. Now this greater *'prima facie'* loss and harm is not the same as losing one who discovers medicines or writes plays and so on, rather, one whose loss of life is *incomparable* to that of the dog or such animals.

Peter Singer and many utilitarians alike would refute this "inequality in treatment" of harm and loss of inherently equally valuable life. Utilitarians hold that they would like to underplay that very idea of 'equality of treatment' of dogs and men as such for they are equally valuable because more important point is to consider equally the abilities to live life in the world and ability to suffer in life. Hence, in a given situation of equal interests of all beings affected by an action, they must be considered equally. As such, harm and loss to valuable life and sufferings in life are all equal in a situation of a drowning boat. There is no point in such a situation as the drowning boat to overlook the suffering of dog by an arbitrary declaration of human life as comparably more important and thus needs to be continued on earth.

However, Singer and utilitarians like him do not hurriedly come to the conclusion that, therefore, any one person should be dropped (and that sounds practical as well for his or her weight may be more than the dog). In fact, utilitarians remind us that in such complex situations several factors contribute to a decision-making though horribly against a life at least. It is not as simple as weight because putting on weight might soon turn out to be a curse on earth. Rather, we have to consider different life-situations of all four people on board, their complex relations, families and friends, and consider their pain and suffering as well. More trifling situation will be the one if all relations, families, friends on board have been killed in the catastrophe. As such none stands as a referent to judged pain. The utilitarians coming to their non-anthropocentrism concludes that then one man ought to peacefully die (perhaps by administering lethal injection), and thrown but the dog remains on board because going by their first principle of morality (utility), the presently least happy and not likely to remain happy in life in future should be dumped. This way the ideal of maximization of pleasure would be least hampered by minimization of the same prudentially.

We assume now what controversy will be raised when using animals in experimentation comes to focus because it is also a case of deciding to take a toll on life of animals and human beings as well. Regan would argue (as he does in the same essay), that any trivial and unwise treatment, which is unjust to animals for the sake of good of people is to be morally shunned and stopped. So it is a matter of rights and equality of treatment that decides against animal experimentation. Singer on the other hand holds in "Animal Liberation" (1976) that experimentations contributing significantly to the maximization of human beings are justifiable if and only if not lesser animals but equally positioned persons in a *vegetative condition* be put to the same experiment because he would be the least one whose elimination causes minimum problem for maximization principle. Singer's non-anthropocentrism does not however, amount to a view that all vegetative people henceforth be put to deadly experiments rather than animals, but that unreasonable preference for one's own species over others for saving life is immoral, which is called "speciesism".

A careful reader would not be dismayed at anti-anthropocentrism and the more moral 'shift' to "animal-centrism" because we have still not taken up their inherent defects for a revision. Animal-centredness is not the right answer to be accepted in any way without further scrutiny. Let us first analyze some ordinary criticisms against animal-centredness for they may not be proper at all before we come to more pertinent criticisms. First is provided by Michael A. Fox in his book *The Case for Animal Experimentation* (1986) in which he tries to justify that animals (non-humans) are not members of "moral community" and for that reason, no human being has a moral accountability towards any animal. According to Fox, 'a moral community' is constituted of some such beings having free will and moral autonomy that socially interact with each other. In such a group individuals understand moral matters and moral concepts, and they recognize the autonomy and individuality of members of the group. The most important concept in this regard according to Fox is "autonomy" of people, which is to be critically self aware, understanding and working with complex concepts, interacting in sophisticated language and ability to morally deliberate, choose and act. As animals are devoid of all these properties, particularly the last one, though they might have other unique properties to constitute a community of their own, they do not have as qualitatively superior community as humans have and thus do not have qualitatively higher life. This is enough ground for Fox to conclude that the members of moral community do have sufficient reasons to sustain their superior life form by using such lonely life forms as means towards fulfilling their ends and thus do not have any moral binding or accountability towards animal life forms. For us, it is important to notice that the 'shift' we thought so natural, is not so because human-centred ideas have their so-called strong reasons. Fox in fact, does not have strong reasons because whatever he says of human beings as having unique dispositions including moral autonomy is nothing new to us. It is also not new that we have been trying hard to distinguish us and also relate us to animals since a long time on the basis of morality and soul but does it prove in any way that in our moral quest in life, we do not have moral obligation towards other animals? Fox completely misses the point that we do have moral concerns but for him we should not have such things as those lower beings, so degraded as not to have moral capacities and thus at best means to our goods, does not make a good case of moral obligation. Interestingly in the glossy cover of human autonomy Fox forgot that human beings are autonomously immoral when they inflict unnecessary pain, suffering and cause disasters in wars. Do we have accounts of world wars among animals? Or, they devastated the environment by whims? Fox is terribly mistaken in thinking that moral dispositions of man give unforeseen powers to them to use animals only as means to their end. It is this idea that bereft animals of inherent values we have been speaking about and their abilities to sustain the ecological community, which is so very important for any moral community to thrive. It ought to be said that one who forgets or neglects the truth that moral autonomy is apart from other things, a disposition to deliberate on the good, right and just of their intentional actions affecting animals, adversely or favourably, is a terrible mistake. It is this plain truth that enables us to understand why a moral shift from human-centeredness to animal-centredness is possible. Other attacks on Fox are equally important, one being Frey's logic in his work "Interests and Rights". In so far as we interact in language, holds Frey, we are disposed to understand several things like each other's attitudes, emotions, beliefs and intentions. Human being's ability to conceptualize moral matters, deliberate and act certainly gives moral status to them as does their unique language use. Animals lacking both these cannot be grounds for their lack of value or worth, they still possess inherent and instrumental values. One interesting thing is when people use people as means towards end, what do we think? Are they immoral (as goes Kant) or, do they have superior qualitative life over others (as goes Fox)? What about people who are not able to know about means and end, choose such things and act? What do we say about very young children, spastics, mentally sick and people in vegetative conditions or those suffering from Alzheimer disease? Do they form a part of Fox's moral community? Certainly, they do not as a dog doesn't when it bites a cat or a man. Going by Fox's theory, we should use these 'hopeless' people as means for our ends and that will be good enough, morally speaking. But Fox was not headless, he soon realized those problems and reverted his stance in "Animal experimentation: A philosopher's changing views", an essay in *Between the Species* to hold that "Our basic moral obligation to avoid causing harm to other people should be extended to animals... it would (likewise) be wrong for us to benefit from the suffering of animals"[8].

Though Fox returned graciously to the 'shift' in ethics one of his ideas, not so well developed in his essay has been projected and argued well by R.G. Frey in *Between the Species* (No. 3, Summer, 1988, pp. 191-201) bearing the title: "Moral standing the value of lives and speciesism", Frey holds that though animals do constitute a moral community for we have certain moral obligations towards them, nonetheless, a healthy grown up animal life is incomparable to that of a sane, grown up person for the person possesses moral disposition and thus a qualitatively better life. It is due to this that persons in a vegetative condition or those too young does not have qualitatively as good a life as those who are not in such conditions. Frey, therefore, comes to the conclusion that using animals rather than using vegetable—persons are never justifiable in experimentation but he has no doubt that they should be used by human beings for their well-being when it becomes necessary. Does it really enable us to make a 'shift'? No perhaps, for Frey has smuggled the quality-of-life on moral grounds, an old idea, to come to conclusion that sane people are really moral beings because they would risk lower lives at the cost of their well-being deliberately making things impossible for a 'shift' in ethics. In fact, Frey stands where our 'sinking ship' was; still, the utilitarians and the non-utilitarians championing animal liberation does not speak much of the moral community in which moral obligation stems from moral dispositions of human beings to form a qualitatively better life of such beings. Hence, Fox, Frey and even Regan and Singer have perennial problems regarding 'taking life' for well-being of human beings. However, concern of these people for human species particularly when it comes to distinguish normal from the abnormal, vegetative and too young, raises several interesting questions, which cannot be resolved now before we have discussed other things.

It is important to find out why discriminating animals from the realm of the morally oriented humans be a necessarily evil when differences are so very prominent. Yet, such interesting things when addressed to should not overlook the truth that within the field of moral discussions, only humans do not count, it has really shown up its broadness. Isn't this good enough to have 'shifts' than fixity in ethics? If the answer is in the positive, one thing is clear, 'shifts' themselves are not *fixed* because a lingering point haunt us: the existence as it is, should not be reduced merely to human beings and animals, it is something more. If we take this point seriously, human moral accountability need not stop with a shift to non-human animals. This may be taken as one ground to show, where animal rightists, liberationists and even Frey went wrong. To add to this we may point out to Mr. Regan that equality of all, the basis of an egalitarian view is really difficult to maintain without contradiction when 'sinking boat' comes to the fore because it would be queer to hold that at least in one case among equals, some are more equal than others. Another problem with Regan and many others is that the value of human beings cannot be a shade better than animals by virtue of the former using others as means because human beings are used profusely for meeting ends of others and thus instrumentally valuable as well. Why on earth such a value is derogatory to human quality of life is not known to me. If an intelligence officer for the sake of duty uses me as a "decoy", am I less valuable than the great officer in true Kantian mould? More troublesome would be a case where a "decoy" dies or is sure to die or is used to die. A frantic search for a "vegetative decoy" or a "medicated insane decoy" would be a better option perhaps? Mr. Singer's case is not free from many other problems.

Mr. Singer's 'utilitarians' should be extremely enlightened for he or she has to judge who on earth are able to contribute to the great ideal of maximization of happiness. How such computation is possible is not known to us because taking 'all factors' in view is too enigmatic. Singer's "preference utilitarianism" which argues about specific preferences for sustained existence on earth by self-conscious, rational human beings goes against throwing human beings on a sinking boat. Hence, if Regan throws dogs or Singer throws a man, either equality or preferentiality in utility has to break down. What completely breaks down is the myopic view that beyond animals and humans nothing exists for human beings in the world for care and well-being.

Ethicists have been not so myopic in this regard. We did 'shift' from animal-centredness. Let us discuss it now. Whenever we speak of the class of "living things" we are immediately remembered of plants, algae, fungi, single-celled organisms perhaps bacteria and viruses. When we are in view of a wide panorama of "life" why should we be myopic about humans and animals? And if everything has now boiled down to moral accountability or obligation for deciding what it means mainly to discuss morality in cases other than humans (my "other parties"), one should consider these "parties" for moral considerations. For this however, one should first decide, what is living? The point is to remember that at least "living cases merit some moral consideration". Fairly simply stated, living systems are 'self-regulating' ones which do not necessarily aim consciously at certain goals and strives to achieve them. Though consciously fixing goals, striving for them and in doing so affecting many favourably or unfavourably to be judged for respective values or moral values, is what is specific to living beings who are moral but we have time and again said that this is not needed for moral considerability of all living beings. A life-centred ethic thus considers all living things as morally considerable or that an ethic which takes human beings as morally obliged to all living things. In another way, life-centred ethic takes intentional human actions affecting all living things, not only humans and animals, for moral scrutiny. The main reason may be that some intentional actions might be against the basic right of sustaining life of all things low and high though they may or may not be conscious of doing certain things. However, we may differ over the morality of differentiating some forms of life over others for their sustenance and/or growth.

In the recent times with the flourishing of environmental ethics and the constant shift, therefore, from humans to animals and then to all life forms, ethicists are believed to be speaking "deep", or they have left the "sole" or "shallow" value assumptions in ethics to move into a "deep" value assumption, sometimes referred to as "greater value assumption" (*cf.* Richard Sylvan). However, people like Sylvan holds that a greater value assumption might amount to a view that though humans and all other living things are of great value no doubt, all these parties cannot match human interests and for that reason moral considerations of different types. This view is supported by Robin Attfield as an "Intermediate" view of "shallow" and "deep" types.

The dichotomy of "shallow" and "deep" ecology and ethical concerns in ecology has been focused by Arne Naess in his seminal essay, "The Shallow and the Deep, Long-Range Ecology Movement: A Summary" in *Inquiry* (1973) and late in 1984 in "A Defence of the Deep Ecology Movement" in *Environmental Ethics*. Naess argued in favour of "biospherical egalitarianism" not the type of egalitarianism which we have just discussed (*cf.* Regan) rather, the one in which all life forms having *equal* rights to live and flourish. With this Naess has been taken as supporting 'equal value' to *all* life forms which are a revision over greater value assumption to adopt an egalitarian value-theory, which is still 'deeper'. In one sense Naess has been looked upon as a radical life centred ethic supporter for not only are all living things morally considerable but those having equal moral significance. What is 'radical' in Naess is that he is less interested in academic ethics than articulating our experience of the existence itself because such articulation would lead us to a new philosophy of life, which he calls a *philosophy of world experience or "Ecosophy T"*. How this could be much away from moral view is difficult to say, what is more clear is Naess' vision of multifarious "intrinsic value" of the biosphere itself, which is basic to his biosphere-equality. Such 'equality' relate human beings inseparably to the entire existentials and tries to block the hierarchical thinking and biased treatment of human desires and aspirations. This is radical 'shift' of anthropomorphic ethic to biospherical ethic. Naess thought that this shift proposes a new human wisdom about oikos or the planetary "dwelling place", and added "T" for "Tvergastein" to stress that ecosophies needs to be evaluated from a specific locale, with which the wise philosopher is acquainted with or has personal relationship as has Naess with his Norwegian locale called "Tvergastein".

Naess does not rule out an "intuitive experience" basic to such world view, which helps in recognizing the "autotelic value" of all life forms. For this reason, therefore, one has a "deep" ecological vision that no anthropocentric thought may be treated as valid for it champions humans as ends-in-themselves. The truth is that all living things are ends-in-themselves. Naess' intuition leads him further to some deeper 'Self-realization" inherent not only in man but in nature as a whole. Capital "S" of "Self" is worth noting because Naess does not rule out that Ecosophy T comes closer to some sort of religious philosophy or metaphysical stance about which we shall discuss at a proper place. Apart from this primary drive in nature Naess conceives of twelve norms, mostly of imperative nature linked to fifteen hypotheses. Apart from Self realization we thus have Self realization for all living beings, diversity of life, complexity, symbiosis, local self-sufficiency and co-operation, local autonomy, no centralization, no exploitation, no subjection, having equal rights to Self realization, no class societies and Self-determination. Here we have to be careful about use of "Self" at many places, not necessarily all of them having religious-metaphysical import. If we take a closer look, Naess suggests that the first norm does have some *ontological* significance, whereas others in the imperative form are *deontic* in nature. Taken together, they suggest "intrinsic value" of all things around us and thus "biospherical egalitarianism" is well understood in so far as their well-being and rights are concerned. Naess also stresses on local importance in environmental decision-making and a non-violent, classless social structure. Though Ecosophy T is clearly non-anthropocentric, human beings are not subordinate to the non-human things. Naess is not suggesting that altruism is the answer, he suggests a "shift" from ego-centred philosophy of life to biosphere-centred or eco-centred one. This 'shift' is the basis to ethical concern as well. But some confusion arises owing to several unique terms used by Naess. One of course is "Self" whereas the other is "biocentrism" which enabled some interpreters to say that Naess is nothing but a "Leopoldian naturalist" or "environmental holist" who in reality takes quantum shift to land and landforms or if simple term is needed, the whole of existence (my "existential" on the whole). But this is not wanted as Naess had no intention to be a naturalist as he was not at all eloquent about 'non-living' objects as important in philosophy of life and ethics. So, it would be good enough to call it "biocentric" or "ecocentric" or "biospheric egalitarianism". However, this does not mean that Naess did not have some problems. In fact in some of his later remarks, we find many problems.

The antagonists of sole-value or shallow ecology grew in number and intent with Warwick Fox, Bill Devall and George Sessions in particular joining Naess' "deep" vision movement. They tried to overcome some of the unclarities of Naess or may be some of the misgivings to form what we call, "platform" for the deep ecology movement, which is deeper. Fox speaks of "transpersonal ecology" in his work "Towards a Transpersonal Ecology: Developing New Foundations for Environmentalism" (1990) for the simple reason that ecological discussions, particularly the ethical ones have been long dominated by *personal* and *political* views which is an outcome of the anthropocentric world view. This "personal" outlook considers the natural world along with its various species as useful things or resources for human good and, therefore, is merely good as means or that they have 'external value'. A natural outcome of such thinking is thus optimum use or in a sense plundering these valuable resources, which are nonetheless inherently valuable for many reasons. In opposition to such a view Fox called for a "transpersonal world view". During 1980s Fox, Naess, Session and others made a "platform" for realizing and propagating new value conception of nature as a whole or a new axiology, which would transcend the ontology-deontology of the norms propounded by Naess. The eight platforms are to recognize the equal interest value of all things in nature, to recognize certain intrinsic values as multiplicity, diversity and complexity, to allow human use, killing of living things and environmental use only to meet vital needs of humans, to decrease human population, to accept the truth that humans are at present hindering vital life processes, to counteract those human actions and decisions which are ecologically disastrous, which we owe mainly to unacceptable socio-economic policies and philosophical-cultural perception calling for urgent changes, to acknowledge the value of "quality of life" over "standard of life" and putting all these platform principles into action for they are moral obligations. Several interesting things crop out of these principles. One is to allow some use of natural things at least for human sustenance, change modern mindset, more focus on quality of life and most importantly, not to think but to act according to these principles as if we are "morally obliged" to do so. However, one apparently gets an idea that if these "platform principles" are followed literally, a natural confrontation with anthropocentric mindset and policymaking is but natural. Most acute may be how to weigh which policy would be most important to be implemented at this point of time keeping in mind the environmental disasters. Naess, therefore, proposes further guidelines or rather proposals of 'priority' and "nearness" to meet these crucial problems in practical life. Clearly, he prefers the consideration of more vital interests over the less vital ones, where "interests" of all parties, not of humans alone are involved. Not only this, the more proximate interests are important than the remote ones in space, time, culture and species. It is due to these that Naess would suggest considering Narmada oustees over a heightened dam or, Sariska tigers over poachers. Nevertheless, there are questions about Naess' and others' position in this regard to which we shall turn at a proper time.

Whatever may be the responses to deep ecology, it has contributed significantly in understanding the concept of "rights", which was erstwhile considered as making sense only to 'human rights'. As all living things have inherent value, they have a right to live. This enables us to realize that only human beings do not have certain rights for they are moral beings, rational and perform obligatory actions because it is an odd idea to correlate right and certain types of human dispositions. If we take a careful look at several other animals and living beings, we find some unique properties and dispositions, which by themselves might be extremely important for the particular being and the biosphere as a whole but one species or an individual of the species for that reason is not a right-claimer. Right to live is therefore one right which may be claimed for all species absolutely but to neglect them or violate them in case of one species is absolutely wrong says the deep ecologists. This enables us to speak of "rights of trees" among many valuable life forms.

Deep ecology movement thus gained variation and status with many people joining them, arguing separately for the ethic of protecting the rights of trees and other living beings. The arguments have been many, each considerable for their worth. To mention some of them we have Christopher Stone's work: *Should Trees have Standing? Toward Legal Rights for Natural objects*. Also Roderick Frazier Nash's work *Focusing rights of nature* and John Tallmadge's mystical *I-You* identification in the Bubberian vein. Peter Wenz in his book *Environmental Justice* also supports Naess' basic arguments through justification of "principle of anticipatory co-operation" even at the cost of some personal losses and transformation of fixed cultural notions and ways of living. Taken together, Naess' protagonists talk about universal rights of the whole of wilderness with some reservations about the imperatives to be applied in casuistic fashion. It is not that the entire movement does not have some gappiness; a 'shift' very emphatic is noticeable in ethics.

Still the shift is not complete as the existence as a whole is not about "living beings and things", it is something more. Do we have to speak of ethical obligations of man towards inanimate objects? We shall come to this, perhaps the final shift for an emphatic environmental ethics to be shaped but it would be a philosophical blunder to skip some of the problems which the earlier 'shift' faces. The life centred ethic we have just seen requires us to consider all possible impact of our actions on every living thing affected by them so as to determine how we should act. Among many affected beings, humans are important, and even Naess suggests material well-being of all human beings notwithstanding some necessary steps to fell trees or kill some life for basic subsistence. This is really difficult to do. How do we evaluate what things should be necessarily damaged to what degree to meet the basic demands of human beings universally and always? To take a 'radical' stance would make the problem more difficult for us because such egalitarian radicalism would literally disallow human interventions in the natural environment in any form on moral grounds. Interestingly, the moral "grounds" are either "ontological-deontological" or fixing "platforms" for actions in casuistic fashion, part or whole of which may be put to further scrutiny for validity. In fact, a number of imperatives and suggestions thought to be universal in nature are not so keeping in mind the cultural, political, social, economic difference in changing times. The movement asks for a dynamic and radical change in our ways of thinking itself which transcends such relativistic factors but that is a difficult task in asking when our ways of thinking and living themselves are invariably bound to the regional and the local, socio-political and economic set-ups. Perhaps this has put significant stress on the egalitarians themselves as changes in their views are noticeable. Though "preferential" decision-making and treatment has been preferred by the "movement", it leaves room for differences in opinion among those who care for nature. Among the tribes, for instance, or, among their sympathizers, it may create a furore if tigers are preferred over them or their cattle when it comes to dwindling national reserve forest and the human refugees in and around it.

Taking this problem still deeper to the level of "value" considerations, we are at great sea. Did Naess and others propose an axiological theory of value in the name of "equal value of all species"? If that be the case, their "preferential" compulsion in practical cases is not understandable. Pressed hard by this perhaps, Naess, Fox and Sessions argue that biospherical egalitarianism was more of an expression of an attitude of antagonism to anthropocentrism rather than stating a new value-theory. In fact, these people dislike entering into an esoteric coterie of value-theorists. This is however, not a welcome theoretical move for Naess and others for their built-up egalitarian structure itself gets a jolt. It does not appear that they were not serious about equal values, rights and justice rather there was an 'outrage'. 'Outrages' does not matter as a matter of fact in philosophy. Warwick Fox and his sympathizers thus warn that the rejection of "shallow" view is not all in "deep" view. It follows that if egalitarianism has to hold then preferentiality ought to go, and for that reason non-anthropocentrism is a mere excuse for 'deep' thinkers. Critics, therefore, hold that among groups of creatures their basic interests has to conflict in the face of the claim that still all interests should be equally cared, which demands preferential treatment than not. But how to prefer is the bone of contention, that is, what ought to be basis of such treatment? Can presence of "consciousness" be the basis? If yes, many problems creep in cases of those beings having fade or no consciousness and for those having high consciousness. Do we speak like Darwin to hold that conscious beings grow, sustain and multiply in the struggle for existence, and for that reason ought to be preferred? Or, do we have to say that the most conscious ones have turned out to be the worst marauder of existence and, therefore, it is better that they do not win by sheer whims? Another problem is, should we not include apart from interest and consciousness, the capacities to grow, self-maintenance and to multiply in case of non-conscious living beings for preferential treatment? If we do broaden our view in that manner, do we extend it up to bacteria and viruses? Or, there has to be some 'no-preference' in some cases? Naess and company do not face one problem, the more we unfold it, and the "deep" vision is exposed for deeper difficulties.

To save the case, Richard Sylvan in some of his essays published in *Radical Philosophy* (1986-1987) argues that Naess and his company is good enough in regarding living beings as equal provided they remember another important principle, which he calls, "biospecies impartiality". This is a natural move to counter the attack on "preferentiality". This principle rejects favour or preference of one species over the other on the basis of the species itself. It is a counter to "speciesism" only comparable to "sexism" and "racism". There is no way but to accept the truth that one species does differ from another species on several counts of properties but just one member of the species on the basis of being a member of that species cannot be differentiated from others. So differentiations which are so natural should be on grounds similar to both but not 'arbitrarily'. Hence, human beings and turtles differ on several counts inherent to respective species, which remains the same and unchangeable, though just as human beings are in a particular species on such and such grounds should not be arbitrarily differentiated or preferred or favoured over turtles for being in that species. Conclusively, there remains "biospecies impartiality" and as such egalitarianism holds good notwithstanding certain *unavoidable preferences*. Sylvan tries to justify that Naess was through and through trying to focus this in his writings but was misunderstood.

Misunderstanding is not a healthy habit in philosophy lest we misunderstood. What we understood clearly about Naess and Sylvan is that even if 'biospecies impartiality' holds well, it is so for all living beings and not a few. What about viruses? They do differ on many counts from us and it would be grossly unethical to say that being included in human species for such and such 'properties', we need some favour or preference over viruses. As such, we would like to eradicate them as undesirable. But where does biospecies impartiality vanish? This cannot be clumsily managed by the said 'unavoidable preference'. Sylvan and others forget that species are favoured or preferred not merely on the basis of properties but on the basis of their "capacities" as well. Attfield in his essay, *The Comprehensive Ecology Movement* mentions that even the recent writings of Naess accept this truth[9]. Undoubtedly, human beings have the 'capacity' to ponder about how they die, how viruses let them die and how viruses die and how viruses should meet their end etc. and this may be added to the species differentiating properties for preferential treatment. Now, if biospecies impartiality does not accept "qualities" alone but accepts "capacities" as well, it will tend to break down "impartiality" in case of human beings for their "capacities" are numerous and strong enough to outwit others and, therefore, human favours would outwit other species, though on some "capacities" among non-humans, one outwits others for preference as well. This would thus prevail in the entire life systems. Do we then accept that being a member of a species and including unique capacities, we should be treated preferentially? Some would add here that our "capability" of ethics-talk should be a good ground for preference treatment. Everything thus boils to the stark truth of 'inequality' and 'preferentiality' along with some urges to treat others 'equally' despite adversities. Perhaps Naess and Sylvan does not satisfy us though they took a good 'shift' in ethics.

Richard Sylvan still holds his ground for he has stronger grounds for biospecies impartiality in the face of "preferential treatment" for human beings. In his essay, (1985) Sylvan holds that though non-human animals, plants and others those of "normal" status are bearers of inherent values, yet they are never as valuable or as important as human beings. So even if impartial treatment of species for just being human having different properties does not hold good, Sylvan thinks that they should nevertheless be treated as most important beings. This sounds much like Peter Singer in *Animal Liberation*, sometimes termed "environmental apartheid". Tragically, a shift to non-anthropocentrism has resulted into such a position. However, what we understood from Naess' writings, deep movement amounts to the loss of human beings at times when "other parties" come to serious ethical consideration. Sylvan, however, did not do a Singer because as compared to Sylvan, Singer preferred non-human interests over and above lesser human interests. Even if we compare Holmes Rolston's views[10] it is not comparable to that of Sylvan because for Rolston all humans are in all cases superior to all animals in all cases, and demand preferences over non-human animals. Attfield argues though that Rolston does forget several non-human "capacities" compared to humans as superior thus demanding preferences at times. This is more prominent when we compare non-humans with vegetative humans or young children. Why should Sylvan sound confusing as either Singer or Rolston, when he is not both? Attfield analyzes to conclude that "Sylvan in fact regards as intermediate any position which accepts any of the three formulations… as long as no one is misled into supposing that the relatively mild kind of differential treatment which is not itself based on species-membership, amounts to the speciesism".[11]

Attfield thus finds that in the midst of the controversy of "shallow" and "deep" ecology, when preferentiality faces equality and impartiality, it does not create a muddle if an "intermediate" position is maintained. We thus have another way to move away from anthropocentrism of Naess and others or even the earlier position of Sylvan. The question what does this "intermediate" position amount to? Attfield speaks of three basic tenets for resolving the conflict of fulfilment of the interests of all living beings and impartiality in the distribution of interests when all living beings are inherently valuable. The tenets are: "(a) That like interests should be accorded like consideration; (b) that some interests are greater than others and their satisfaction of greater value and (c) that sometimes (but not always) such greater interests are actually at stake in conflicts between human interests and non-human interests".[12]

The first tenet is nothing new, Attfield accepts that Peter Singer advocated it in *Animal Liberation* (1976) and *Practical Ethics* (1979), which Sylvan appears to agree as well. However, Rolston has some reservations about it because he persists that looking at the 'whole' environmental flux no human interest is like that of the animals, that is, their interests are never at par thus demanding preferential treatment for human beings. Attfield does not disagree that at times if you look in isolation or in association, human beings in the biotic community are greater than other beings, and that there are conflicts in distribution of interests and advantages to them. But to say that Singer always takes the matter of interest isolatedly and in all cases human interests are outwitting, are palpable factual falsity. This may be the basis of moral wrongness as well. Attfield argues that a comatose person at an irreversible state cannot be at par with gorillas or normal dolphins. However Rolston is adamant for human preferentiality by arguing that domesticated animals under human care as different from wild animals is the basis to determine how to take care of their like interests. Moreover this focuses why humans need to be preferred for welfare for they are caretakers of animals (of wild animals as well), and thus take on greater moral obligations and have greater value therefore. Attfield does not refute this position rather, mentions that this need not be the sole principle in preferential treatment though this is a good case of human beings for certain favours when they take on greater responsibilities. It points out equal importance of favours extended to non-human animals as well. It may be interesting to notice that Sylvan and Attfield does not stress much on the present socio-economic and political scenario, changing times and locale or else they would have noticed that human value to take care of domestic animals need not necessarily ask for greater human value than taking care of the wild animals. In fact different locals caring domestic animals may be "economy oriented" whereas caring the wild animals is "life oriented", the choice of greater value is perhaps clear.

As Rolston fails to refute the first tenet, two others for Attfield are fairly easy to justify because he has shown (and also our analysis) that some interests ought to be satisfied over others for greater value and though not always, human interest fulfilment have priority. From all this, Attfield comes to the conclusion that Sylvan's deep vision may be covered for unnecessary attacks by saying that in normal cases of humans and non-humans or, generally speaking, the interests and welfare of both species are necessary for the existence as a whole though there are instances of preferential treatment of humans over animals, which is not to be emotively discarded rather, sanely judged on the basis of numerous factors, like changing times and locals. In fact, one is no less human if sometimes they lose of if they gain; gain and loss are all relative perspectives.

Everything said about the deep, the shallow and the intermediate positions, we are still wondering about whether or not a shift from life-centredness is possible. It appears to be possible for we have just seen that human beings, non-human animals and trees, algae, fungi and numerous living things are in a complex interrelation in an abode which is itself material. Even the intermediate position appears to have not considered this point seriously leaving a question of value wide open. Are human beings valuable if they do not care for non-living things? Does earth community as a whole constitute a case of consideration of morals and values? I believe ethics discussed thus far has not taken as great a "shift" as to answer these questions. However, by now there is a significant "shift", which has been made possible by several contributing ethical movements owing largely to the "deep" movement. Coming to this "shift" one is reminded of the recent reactions against anthropocentrism or "shallow" ecology, which however, is unlike those discussed early on, and for that reason, moving from the "deep" vision. The general argument to bring back "human being" at the helm of affairs is what we call "ecological humanism", supported by Ann Arbor, Henryk Skolimowski, Thomas Derr and others. Such humanism, unlike scientific humanism is *metaphysically oriented.* It is thus an ontological ecology as Naess envisioned. The difference is that it does not merely speak of Self-realization rather, "self-transcending". Human beings ought to realize their nature of self-transcendence to permeate with the broader natural self. This is the uniqueness of human beings which claims for its supremacy over a virus and a mosquito. This self-transcendence was however turned by Bryan Norton into "weak anthropomorphism" and Thomas Derr renewed it to "Christian humanism". Though these humanists are against the hierarchical attitudes of some of the anthropocentrists, they have their own problems in so far as their ontological basis is concerned. Many "deep" thinkers and others who would not like to be ontologically based, would still conform to a basic contention that human beings should transcend their egoistic and hedonic self to permeate with a larger moral community. We shall come to this problem in the next chapter in more details.

Though "shallow" view has not been preferred by many contemporary ecologists, a group thinks that what is most unlikable is not anthropocentrism rather, "androcentrism" because the real chauvinism in ethics is male-centred so far as ecological decision-making and actions are concerned. The dominant male attitude has affected the entire environmental set-up just for the sake of benefit of man. There has been against it a wave of "eco-feminism" which speaks of more caring and loving relation with the material nature and natural things. Its ethic is thus far moved from androcentric ego-phobia and hedonic overtone. However, a close look at the feminist trend reveals that the earth as a whole with its creatures is not looked upon as merely material and organic because some kind of "nature-mysticism" deny nature as treated in that manner. "Eco-feminism", has been supported by French thinker Francoise d' Eaubonne, Karen Warren, Carolyn Merchant and Sallie McFague. McFague along with Rosemary Radford Ruether are theologians as well. Many poets and writers also contributed to this movement. We will skip the theological viewpoints and other contributions here. However, an excellent synposis of views may be seen in "What are they saying about environmental ethics?" by Sr. Pamela Smith.

Eaubonne and Warren argue that women have the potential to initiate an ecological revolution and to challenge the logic of male domination in nature related affairs. What is most abusive in such domination is that environment has been abused to as great an extent comparable to racism, sexism and classicism. Eco-feminism tries to fight such oppressions of males in nature not by imposing rights but caring personal relationships. Sallie thinks that eco-feminism is much akin to the deep principles of biological egalitarianism, symbiosis and classic societies but it is not the same as deep-movement for it is technical and academic. These thinkers call for a complete liberation from the mechanistic, scientific and the masculine-macho explanation of the earth which was once viewed as more feminine, that is, an epitome of nurture and nourishment, which would prevent us from its deliberate plunder lest we have this inner vision. This vision of mother, the nourisher and carer soon enters spiritualism and mysticism with the metaphysical "Body of God" and "Gaia-God" explanations, which we shall leave for now. To my mind, eco-feminism also brings in the ideas of earth as a living organism and the intuitive ethic of protecting the sacred organism.

Eco-feminism tends to break away from the "deep's platform" on many accounts. It appears to be the vision of a "technological ethics" for it advocates harmonious, organic, flourishing of all living systems. This way the well-being of every individual entity is made possible so that living system itself is taken care of. Every decision and action that threats disharmony, damages and death to this system, is ethically undesirable. The most important feature of eco-feminism, which is a step forward to the final "shift" we have been speaking about is both its "biocentric" and "ecocentric" orientations. Reuther in her work "Gaia and God" speaks of the whole of earth as a dynamic, living thing. The pebbles and the rocks are not non-living. This is the basis of eco-feminism's bio-centrism. As it also holds that the parts of this whole have intrinsic value but completely related in the mesh of existence, eco-feminism uphold ecocentrism as well. However, they have been vocal about the "feminization" of the "whole" thereby enabling Warwick Fox to charge "gynocentrism" in place of "androcentrism". Most interesting is, its "feminine" bias makes it "anthropocentric" and not any mystic or metaphysical view as we think it to be.

Donald Davies in his paper "Ecosophy: The Seduction of Sophia" in *Environmental Ethics* (Vol. 8, 1986) tries to patch up eco-feminists' gynophobia by claiming that we should better have an "androgynous non-hierarchical thinking" but ultimately says that such thinking enables us to be empathetic in nature, which is in fact "female". How can a teleological "androgynous" approach be "female" is not known to us. This enabled Warwick Fox to claim that rather than being androcentric or gynocentric, if we include both being transpersonal and transforming our thinking to zoophilic, herbophilic and geophilic all together, it would be more appropriate. This constant reference of "geophilic" or the "Gaian" has enabled us to further conceive that in eco-feminism and in some of its criticisms, the "shift" is almost clear. We would like to conserve this for further discussion but here we have some more afterthoughts by eco-feminists that their view amounts to "anthropocentrism". Thinker like Deborah Slicer argues in *Environmental Ethics* (Vol. 17, 1995) in her paper "Is there an Eco-feminism—Deep Ecology debate?" that critics like Fox are really macho-thinkers as they subject eco-feminism to gynocentrism on the basis of androcentric writers like Michael Zimmerman and Jim Cheney but not based on eco-feminist themselves. Deborah argues that eco-feminism squarely rejects anthropocentrism in all ways, though they are not anti-humans. Their human-centred views are quite unlike the logic of "human-centredness" visited by us. Deborah shows that the basic logic of eco-feminism is huge expansion of the life forms to merge as one self, the "oneness" not any way graspable by "transpersonality" of Fox. The simmering metaphysicalism has to be focused elsewhere.

Interestingly Zimmerman himself is not in favour of a head-on collision course with eco-feminists. He thinks that eco-feminism is compatible to deep ecology and a charge and counter-charge would prove nothing substantial. Both these visions about environment generally speaking, are against particular attitude and damages caused to nature. However, Marti Kheel thinks that there are problems in eco-feministic "holism", some of them psychological as well. We will not go into the psychological intricacies though it has always been tough imagining a shift from egoism to altruism or expansion of human self to the whole or to oneness at the mental level of man. Greater problems arise when we come to the interest of the individuals in the whole, the distribution of advantages to them and using as means one individual by another. The problem of rights and obligations of humans towards animals are also trifling. A lingering problem is that how would one define a "woman" who voluntarily contributes to a masculine decision-making and corresponding action that plunders the nature? Truly, eco-feminism is not a female dominated movement, it has to still answer this question without traditional gossips that such exploiting "women" are themselves "male dominated" or have such "racial residues" in their minds. In fact, if we move from gossips to history of mechanization, natural exploitation and natural plundering they need not be viewed as exclusive male affairs. It may be argued that eco-feminists speak of a "male-like" attitude and not of males or females in such cases. The "male-like" bullish and plundering attitude is what is unethical whether the contributors are male or female. Truly, to "plunder" is wrong thing to do. Let us now all become "female" in attitude, very caring, sensitive to all life forms, even sand-dunes are cared but equally true is that we have to exist, and for that we have to use many things and beings. What are the measures and intuitions that enable us to know when 'using' is turned to 'plundering'? Human beings as omnivorous as he or she is, has to use up several things for basic subsistence. How much more or how much less would determine a feminine psyche in us? Or, will it be a hermit-like desisting from everything? Extend this to the questions of living, wearing, educating, acculturation and civilization, I believe many questions have to be answered, almost all boiling down to how do we commingle in the whole when each of us and others have aspirations, interests and asking for 'using up'. The natural answer would be eco-feminists or whatever, an experience may be intuitive of 'where to-stop-while-we-subsist' is crucial. A caring person on earth thinks about such things and thus a moral being is basically "feminine" in attitude (irrespective of sex). Does it amount to human-centredness? Perhaps no, because the "stopping" in the mind of humans is a limit put to such centredness. Is it life-centred? Or, do we reach "holism" this way? Can "holism" justify environmental ethics as a final shift? Perhaps yes, but how these concerns can be far away from the "existentials" not so well represented till now, the non-living ones so to say. So let us start enquiring whether or not non-living things are morally considerable.

Starting from anthropocentric turn in ethic to the life-centered ethic through the animal-centred ethic, one finds much interest in the discrete individuals, either humans or animals or trees etc. separately. There is a noticeable change in perception nevertheless as in some way with the existence of the kinds of individuals to an agglomeration of all living beings, we gradually came to know that ethics-talk has taken quantum leaps. It is tending now to a climax of including inanimate objects in moral discussion, which may be thus far conceived as the phenomenal leap forward. Here at the outset we must remember that inanimate objects by funniest and most fanciful imagination does not cry for morality but we do strive to be morally responsible, dutiful and obliged to these things, and raise serious moral questions of desirability, good and right of our decisions involving inanimate objects. We shall gradually come to know that this is not an intellectual overdoing, and we have clear idea in our mind that here we are not speaking of legal duties and legal rights in case of such object as we did not bring in legality in our earlier discourse. Though legal discussion regarding these matters are closely related to ethical discussion, which is not to be taken up at this point. It is also not our task at this point to enter into the trifles of imputing 'life' into non-living things or to give it a 'metaphysical' twist by calling non-living an expression of the Absolute or Pure Consciousness or Self or things like that. However, we can come to these things elsewhere. For now, what concerns us most is the stark "naturalistic" explanation bearing ethical outcomes. We are here concerned with not merely non-conscious objects because there are many such things claimed nevertheless to be 'living' rather, here we are concerned with non-conscious objects not claimed to be 'living' due to the lack of rudimentary biological organization but involve us in moral discussion. Robert Elliot prefers to call if "everything ethic". We may consider here whether marvellous white sand on sea beaches containing materials for enriching it to prepare nuclear power or to devastate rocks to size them for flooring material or to use celestial bodies for star wars are ethically justifiable. Hence, we are not aiming at doing anything in 'everything ethics' for it would be over enthusiastic to call something an ethic if it takes up the case of digging a hole deep in ground to plant a sapling in garden as if by that 'mother earth' will 'bleed' inside writhing in pain. We consider some such cases of deliberate human actions involving inanimate objects, which have far-reaching ecological effects, pressing us for moral scrutiny. On the other hand, a concern like this one about sand, pebbles, rocks, minerals, celestial bodies, empty space above, air, water involves a conception of "rights" of these objects, which we are obliged to protect. In fact, the objects understandably do not claim rights; we do that and are subject to analysis of worth of such claims. Correspondingly, more often than not, we speak of our duties and obligations to protect the rights they have and in the ways this should be done. Everything ethic, as usual, may take many forms according to the degree of moral significance allowed to natural objects, and the way such objects are seen in relation to many living beings, including human beings. One important question is about allowing egalitarianism in such cases or not. Another is the way things may or may not be used as means even if they have intrinsic value or else we may ask if they have any intrinsic value or any value whatsoever.

In this sense then environmental ethics in a general sense amounts to human ethical concerns about almost everything, every "existential" if we prefer to use this term. Most narrowly understood, it would amount to moral consideration about humans, non-human animals and other life forms. In this sense, 'life-centred ethic' is more of an environmental ethic than 'animal-centred ethic' and 'human-centred ethic' (which is however the crudest notion of environmental ethics). Hence, environmental ethics is made possible with "shifts" in our ethical thinking about us and beyond it up to the whole of existence. Some thinkers like J. Baird Callicott (1979) uses the term "ecological holism" for it considers two things: the biosphere as a whole and the large ecosystems that constitutes it for moral discourse. However, there is a tendency in such "holism" to overlook "individualistic" concerns in moral considerations; these individuals matter only in so far as they are parts of the whole. The "whole", in fact, is morally considerable.

The "naturalists" as the group is known championing biospherical holism owe much to Aldo Leopold who had initiated the movement in Yale University's forestry department through his works such as *Game Management* (1930), *Conservation Ethic* (1933), *A Sand County Almanac and Sketches Here and There* (1949). He has been ably supported and propagated by Susan Flader, J. Baird Callicott, Peter Wenz, Roderick Nash, Holmes Rolston III and many others. Leopold's ethics is also known as "land ethic". The range of such "ethic" has been nicely portrayed by Sr. Pamela Smith in *What are they saying about Environmental Ethics*? in these words: "Such themes as ecosystematic interdependence, reverence for all creation, the perception of the interactive mix of intrinsic and instrumental values in non-human beings, an active pro-biodiversity stance, a broad extension of "moral considerability", a commitment to sustainable living, a concern for earth restoration, a pursuit of the "natural", and an esteem for the "wild" mark the ethical reflection of these writers".[13] It is easily understandable what environmental ethics amounts to when it is finally grasped at its broadest level. Sr. Smith provides us a good account of the same.

This broad view emerged slowly with the varied experience of Leopold as a forester. He appears to be enthusiastic about nature as a "living thing" and even as "divine", though, for the forester he saw much of an interest to care for the *whole gamut* of interrelated species thriving on land and, therefore, the land itself, but not merely the individual species just for the fact that they are dwindling or extinct. His land ethics also cares much for "land", which without much divinization in his later writings, has been looked upon as not merely soil rather an epitome of "energy" which is conserved naturally in many ways and needs careful conservation for more life to thrive. Hence, the "biotic pyramid" a flow of energy pyramidically from land to complex organism is possible by human care and co-operation. Clearly, any human act that is either adverse to the land system or is lackadaisical for management would amount to ethics-talk. It is also clear that land by itself has some value which has to be retained by human beings who profusely use it for other needs. Thus human beings do have certain moral obligations and duties concerning "land". This further enabled Leopold to envision land as a "community" in which human beings exist and use it variously. This "community" should, therefore, be loved and respected. The viewing the land as a "community" is an important human perception and involves a matter of how we approach at it. In fact, it is important to understand that "soils, water, plants and animals" form a "society" in which we are a member not conqueror. Leopold thinks that "land ethic" does not overlook the matter of "economic expediency" either but it does not overplay with it. The ethical dimension in "land ethics" is brought out quite clearly by Leopold in these words (and I quote from Sr. Pamela's book): "A thing is right when it tends to preserve the integrity, stability and beauty of the biotic community. It is wrong when it tends otherwise".[14]

The ethical vision of Leopold has been variously extended by holists Flader and Callicott in particular. Callicott thinks that Leopold has been arguing in a "deontological" fashion because he imagines the intrinsic values of land be restored by humans as duty. Most interesting feature of Callicott's interpretation of Leopold's writings is that in case of "preferential treatment" he believes that Leopold prefers human beings in the integrated community over and above the animals. In this way he focuses Leopold as an antagonist of "biocentric egalitarianism" as found in deep ecology. This must be the main basis of Leopold and other holists being non-attentive to species care or care of individuals as primary. What Leopold suggests in holism is that perceptions about environment ought to be integrative as a community of non-human organisms, species and ecosystems, which separately are morally significant entities. They have certain interests, urges, want some benefits and strive for well-being. The "highest good" in the community for human beings is to have good life that is, thriving in unison in multiple existentials as if the whole system is undisturbed for its basic structure and harmony.

Leopold's uniqueness in holism is that his apathy to individualism did not allow him to take interest in Indian or African or Tom or Dick, he did speak of human beings in general, wherever they are, should be prior in receiving benefits as they are the chief proponents of ethics and caretakers of the natural world. Hence, some degree of land clearing and alterations may be allowed for human welfare, which is clear abeyance of egalitarianism, speciesism and the Gaia trends. Yet, Leopold advocated minimum disturbances in the natural environment and there is no point in thinking that viruses (a point I raised earlier) need not survive for the simple reason that they do not match the "complexity" of human interests and aspirations. There is however, a tendency among most *ecocentrists* (as we may call them now) not to make absolute statements about environmental matters. It is for this reason that intrinsic value is not absolute in a being, neither is a species absolutely valuable, though instrumental value of many animals and things is reserved for human beings. The basic idea of "preserving" the *diverse* value in the whole is central to "land ethic".

Much has been said about environmental ethics when we come to its justification in the first part. We have seen that such an ethic is possible within the vast realm of ethics in so far as ethically viable talk makes sense when there are gradual "shifts" in our perception of environment as such. No doubt, the shifts in our perception about environment and the ethic of environment have been too smart but there are gaps. Holism taken as it is, does not remain without these gaps, the most disturbing among them being the interests of humans, non-humans and non-living things and their distribution. Holism has least repentance for using, killing, transforming and altering the nature for human beings for the whole of ethics-talk is human in nature and they have to survive. It appears that in the whole, a minimum disturbance may be caused for maximum holistic well-being but do we so finely compute such "minimum" in nature? Is it not true that the "minimum" is relative to the locale, which is very difficult to ascertain? Moreover, how can the whole be cared for well-being without caring individuals, is an open problem for us to decide in real-life management? Most interesting feature of ecological holism is that when it comes to actual decision-making it does not significantly differ from biocentrism or life-centred ethic. It is possible that holistic ethic of this kind finds a place in almost all ethic we have analyzed thus far, with a distinct feature that it does not isolate human welfare at the cost of other welfares but prefers it no doubt without much disturbances. I have said that this 'much disturbances' may be an eluding thing for decision-makers or it goes to human minds that basically does not know the "much" for they are in an integrated whole of which they "perceive" almost nothing; not even the locale in its completeness.

Still environmental ethics is well justified for the simple reason that some inherent defects in and within the "shifts" does not prove that the efforts are pointless rather, it is fairly logical to believe that human beings qua moral beings should not restrict ethics-talk to human action, the effort should be more extensive. In fact an ethical philosophy which is itself ethical needs to be so broad as to talk about human responsibilities and duties towards non-human beings and towards the existence as a whole and as parts. Another interesting feature of environmental ethics is that it becomes possible (or at least sensible) when we forget an obsession with "hermit ethics". To talk about environmental matters does not demand that human beings overnight become "hermit" who lives in caves or in deep forests, all alone, they do not need a society for their basic subsistence. Even they will not take a fruit from a tree unless it ripens and drops for picking. They would not dare use animals any way, not even drink milk. The semi-naked people will use leaves and other things (perhaps when they wither away). They may go to the extreme of putting a natural filter on nose to prevent inhaling tiny living entities as lungs would destroy them. These people do not need electricity for they do not have house and a cot to sleep and as such do not need coal, water, dams, mining, mortars, boulders, anything which disturbs the nature greatly or minimally. If we have such people in mind, coming from pages of Indian mystical writings, environmental ethics is never possible. It is thus a misconception that environment is never effected or need not be affected for the members therein. Another notion is that human beings are at times rogues in reality for it is they who disturb the nature not others. Truth is that many beings in nature affect the natural settings, things and beings significantly, not for mere subsistence but for growth, development and multiplication. They must be habitually or instinctively doing so, not intentionally so to say but that does dot make the matter any different. The nature as a whole has its own ways, even non-living forces do have the capacity to disturb nature phenomenally though not intentionally. An "intentional" disturbance in nature is not in any way disturbing, the environmental ethicists would argue because changes, manipulations, modifications, alterations, using-ups and discarding are within nature constantly going on which does not make humans a shade of a rogue. When all these intentional actions go to an extent of destroying deliberately animate and inanimate existentials for the sake of human desire satisfaction or affluence to an optimum or tending to an optimum level, do we feel like environmental ethical talk more sensible. The 'optimum' is not non-computable, we may have a probabilistic scientific record of it. Even common sensically, the entropic situations are realizable (which need not be taken as mystical as self realizing). Any rundown condition need not be hidden, it needs to be revealed why such things ought to be stopped and how we ought to make up the loss. Another interesting feature is that the rundown and the make-up for loss are for the existentials "now" in view of the every "then". And "now" the world we live in has rundown greatly, all over almost, so an ethic-talk need to start here and now. It would be a little too much to expect the possibility of environmental ethical talk only in the context of the "will be". There is perhaps a self-satisfying, humbug hypocrisy when the environmental ethical talk is only about the "will be". The present "is" is crucial for all of us, and important in environmental ethics as well for enabling decision-making with a moral face. As such, those who justify environmental ethics in its broadest sense, are not worried about lofty idealism, which at times appears to be unrealistic. Perhaps some people on earth are trying hard to achieve a moral ideal but there can be no such achievement if stark ground realities, habits, customs, habitats, beliefs, faiths, desires, interests, welfare are not in mind of man in the flux of time within unique locales.

Yet, ethics involves certain universal elements, when we keep this in mind we find that environmental ethics is justifiable if we consider some values intrinsic to beings and things in nature, which does not vary with time and tide. Likewise, some instrumental values are also unavoidable in so far as beings and things exist. These considerable values are crucial in the 'shifts" in ethic. Perhaps our realization of such values, particularly the inherent, universal values are responsible to talk more about trees, small and big animals, rocks and celestial bodies. I have a feeling that we suddenly start realizing certain things, which were right below our nose. Environmental consciousness, environmental values and environmental studies are so old in our rich literatures but of late we are pressing hard that environmental ethics is a justified ethical talk. Is it true that when humanity gets an extinction threat, it starts speaking morality? Or, is it a spurt of education which nourishes our civilization and has widened itself by making space to new disciples closer to quality-of-life? It is not only broadening of views, it has also been the ability of human beings to study the "organizational complexity" of beings and things which has enabled "shifts" in ethics. The complexity when commingles with "beauty", "symmetry", "balance" and "harmony" also causes some moral concern even for the deadly beasts and lifeless sand dunes. Not only this, our growing moral concerns in environmental matters is justified for the value of that which we have not created or cannot create. We take care of the non-technical marvels just for the reason that what we can never make, should not be ruined either. It is due to this that the self-managing, self-regulating natural systems need some human response if they are adversely affected. Let us not forget that when something wrong is done to these systems, 'law' is not the only way out; a law without a moral face may be equally undesirable. Killing, ruining, devastating all call for moral scrutiny—if it comes to the natural system, nothing less is needed. Often killing and ruining comes up with new organizations and arrangements as if to cover up the natural with the artificial which are equally attractive to our eyes and also believed to be useful. Yet, we do not leave them unanswered for moral questions—environmental ethics may enter to challenge the man-made beauties and goodies. We care our natural systems so much for we know that they are richly diverse and the diversity ought to be restored for numerous reasons we have seen. It is due to all this that I started speaking about "ethics-talk" with regard to "existential" or with regard to the whole of existence.

NOTES

1. Elliot, Robert, "Environmental Ethics" in Peter Singer edited., *A Companion to Ethics*, Blackwell, p. 285.
2. *Ibid.*, p. 286.
3. *Ibid.*, p. 286.
4. Westra, L., "Environment and Environmental Ethics" in P. H. Werhane and R. E. Freeman (eds.), *The Blackwell Encyclopaedic Dictionary of Business Ethics*, Volume XI, Blackwell, 1997, p. 205.
5. Elliot, Robert, *op.cit.*, p. 286
6. *Ibid.*, p. 286.
7. Regan, Tom, "Right across Species", in Edgar Morscher, et al. (eds.), *Applied Ethics in a Traditional World,* Kluwer, London, 1998, p. 49.
8. Fox, M., *Animal Experimentation: A Philosopher's Changing Views in between the Species*, No. 3, 1987, pp. 55-60.
9. Attfield, R., "The Comprehensive Ecology Movement", in Edgar Morscher *et al.* editors, *Applied Ethics in a Troubled World*, Kluwer, London, 1998, p. 13.
10. Rolston, Holmes, *Environmental Ethics: Duties to and the values in the Natural World,* Temple University Press, Philadelphia, 1988.
11. Attfield, R., *op. cit.*, p. 15.
12. *Ibid.*, p. 16.
13. Smith, Pamela Sr., *What are they saying about Environmental Ethics*?, Paulist Press, N.Y., 1997, p. 47.
14. *Ibid.*, p. 50.

3

Some Theological Justifications

Our analysis in the previous chapter has revealed that to speak of environmental ethics is not enough, it needs some justifications. We have not one but many justifications based on ethics as such. However, these justifications do not necessarily and sufficiently conceive of earth and the universe containing many beings and things as divine or creations of the divine. In fact, the justifications do not venture into the vague presuppositions of the Absolute, the Pure Consciousness, and the like, which though not always referring directly to God, does have strong metaphysical fervour. However, we came to know that Ecosophy and Ecofeminism have some inclination towards explaining the existence on metaphysical lines. Hence, we have thus far analyzed the possibility of ethical-talk in environmental matters on the basis of some such arguments which are essentially non-metaphysical and do not transcend the field of ethics as such. We have mainly seen how differently different ethical thinkers justify that it is possible to talk about ethical good, right, just, duty, obligation, desirable and ought about the human intentional actions affecting the natural or the physical beings and things around us and permeating our existence. In this regard, we have seen that how environmental ethical-talk has "shifted" from a basic or shallow level to an elevated or deep level or, may be to a very deep level. With these "shifts" we came to realize that environmental ethical-talk is not properly so-called if it covers desirability analysis of human acts affecting humans only, it is something more. Hence, the "shifts" more or less revealed what a proper environmental ethical-talk is. What came to the fore is that all deliberate human acts affecting either *whole* or the *part* of natural existence in space and time awaits "moral evaluation" and, therefore, a moral-talk is sensible in case of humans, other animals, vegetations and inanimate things. Yet, most interesting feature has been a non-committal attitude in this chapter about what these affected beings and things are or what is the proper nature of the natural existence on the whole. This non-commitment is natural because we were primarily engaged in analyzing the 'nature' of ethics and environmental ethics but not the metaphysic of nature or the fundamental nature of the natural existence. It was, therefore, not at all desirable qua ethicists to analyze those views including our own views about whether or not "existence" as such is non-material or non-physical, and that its peculiar non-material nature owe its existence to someone who is much beyond the physical. Interestingly, thus far we had no business to reveal views and form our own about whether or not there is one who is beyond, possesses a unique nature like this or like that. Hence, ethics-talk about environment as it is in itself, does *not* necessarily and sufficiently strive for nature clarification of existence as based on divine reality. Nevertheless, we have to accept that ethics, when taken closer to environmental sciences for a "talk", is neither similar to the environmental science nor similar to theology and/or metaphysics because ethics has different *tasks* from environmental science, metaphysics and theology. Though one may argue that ethics may come closer to these disciplines, which is enough to show that they are all closely "related". A theological environmental ethics is, therefore, not a misnomer. There is little to contest here because ethics may be imagined as closely "related" to ecology, as in 'ecological ethics' but that is not the same as saying that 'ethics is ecology'. No one prevents us from imagining the possibility of *theologically based ethics* either. Still, we have shown how logically environmental or ecological ethics is possible on different non-metaphysical and non-religious grounds, and we are expected to be logical if theologically based ethical-talk in ecology is shown to be possible. Clearly, once we move on to fresh set of justifications (nothing less than logical but not strictly based on formal logic), no doubt they are *not* the same as those provided before. Hence, there is no logic in saying that theological environmental ethics is environmental ethics, as there is no logic in saying that theology is ethics. We are, therefore, at hand an attempt to justify ethics-talk in environmental matters by moving beyond ethics on the ground that though ethics is essentially different from theology and metaphysics, their close relationship justify that an ethics-talk in environment is possible as based on theology and metaphysics.

However, one wonders that though there is little doubt about ethics coming closer to ecology as a science there is no doubt about theological ethics coming closer to ecological matters! Many people argue that on the face value the state of affairs in the world and the universe does not appear to involve ethics as such. On face value, the same appears about theology and metaphysics. However, the truth is that differences between ethics as a normative science and ecology as a positive science does not prevent close relationship in order to widen the realm of knowledge; the same may be the case with theology and ecology though they are categorically different in nature. But the problem boils down ultimately to the "ground" or the basic "logic" as mentioned before to show that metaphysical ethics is closely related to environmental affairs. The problem is graver when we speak of the relationship among ethics, theology, metaphysics and ecological or environmental matters. Yet, for many thinkers the graveness is apparent rather, things are very simple because theologians' "God" or "Allah" or "Brahma" permeates both ethical and environmental matters without any doubt. It is fairly simple to show that the "whole of existence" and "existential" we have been speaking about when ethics was involved in environmental matters, is not what it physically appears to be, it is in fact, Divine or Godly. Similarly, the moral agents (human beings), being Divine in essence, *should* strive for Divinity through actions because such aim is what is "truly moral". Hence, with the double-edged Divinization of existence and ethical strivings of man, will there be any doubt for a logically grounded "theological environmental ethics"? The ground for counter arguments and replies are ripe if you ponder carefully but for all that, we have to first find out what theologians hold to be good enough. In this chapter, we have to undertake this task first.

It appears that only theological justifications are good enough. The truth is that there are many "non-theological-metaphysical" justifications in favour of "metaphysical environmental ethics" as one finds in the Bauddha and the Jaina trends. Here we are not considering these "non-theological" justifications though enough may be written about Bauddha and Jaina environmental ethics. We are considering two extremely strong theological claims only for all "practical purposes". We are here considering the theological or the divine environmental ethics of the *Christians* and the *Hindus* to analyze a number of problems raised by us. Again, I have to apologize for not considering the Islamic and other theological views as their features may be well covered by the two views in consideration though their unique features still waiting for a more scholarly exposition of a well-versed person. However, here are glimpses of some theological and metaphysical views regarding this matter.

The metaphysical foundation of environmental ethics is too old. Environmental values and environmental concerns have been well appreciated by the wise people of the east and the west since a long time. The appreciations are documented (many lost and undocumented) in historical, spiritual, religious works of the past. Hence, it is not uncommon to have accounts of ethical concerns about environmental matters, a concern which is deeply soaked in metaphysical—religious presuppositions because a strong belief in a metaphysical ethics as 'ethics truly so-called' was prevalent. With the advent of the two most strong religious philosophies, the Christian and the Hindu, the growth of a number of well documented appreciations of environmental matters based on ethics (truly so-called), we came to know about religious and/or metaphysical environmental ethics. However, apart from the Hindu and the Christian concerns, we are also acquainted with the Buddhist, the Jainist, the Islamic, the Sikh, the Zoroastrian, the Taoist and many others ethico-religious concerns regarding the same matter. Of late, a number of books have been written by eminent thinkers who analyze and re-analyze these ethico-religious concerns in environmental matters, with a new turn, the turn of contemporary environmental crises, ways to resolve them and the role the ethico-religious or metaphysical—ethical systems might play or should play in this regard. One significant expression of the contemporary ethico-religious thinkers in this matter is the Declaration of the World's Religions in 1993 in Chicago. It is not difficult to assess that some non-metaphysical and non-religious ethical justifications of environmental ethics have been deeply influenced by the growth of ethico-religious philosophies of east and west. This is evident in the writings of Naess, who clearly hint towards "ontology" in his Ecosophy T, though his "Self-realization" is taken differently by most ecologists, not necessarily on a particular religious philosophy or a metaphysical line of thought. However, we shall soon come to see that many non-metaphysical justifications we have discussed are ably supported by some thinkers who could not successfully veil their ethico-religious inclinations.

Coming to the Declaration of 1993, it is truly religious in spirit as no one religion finds prominence. It may be largely due to conflicting views about the nature of ultimate reality. Many theologians, therefore, have different views about the nature of the existence as a whole from the non-theologians like the Buddhists and the Jainists. This is however crucial in many ways; it will be sufficient to tell here that even if many religious people do not presuppose nature as divine, would still come together to common Declaration about how environmental crises should be met today. Significantly, many secular people of 'scientific spirit', that is, essentially non-religious and non-metaphysical, might do the same (but for this, we have to wait till Chapter Five). Among religious people, theologians might logically argue that bereft of God, Allah, Brahma, or whatever, nature does not remain sacred. The other way round, non-religious people, would logically argue that "sacredness" of nature is an emotive notion or, a class-ideology imposed on us (as is the notion of God etc. imposed on us) for several exploitations, and is so powerful that its care demands our attention. However, within the non-religious groups, their Gods might beg either economical or political or social or some other excuses for showing the "value" of nature. A complete fire-fight of words is at the mercy of these people, and it would be philosophically disgusting, particularly, if this is called "ethics". May be, ethics does not need both the religious and the non-religious war of words; it may be independent enough to shape as "environmental ethics", who knows?

However, the last possibility need not be taken seriously now. We may carry on with some religious justifications about environmental ethics. Though Jainism and Buddhism, the two non-theological views are averse to divinization of nature as a whole, they differently prove that existence as it is, should be treated as an epitome of "good as such", which needs to be looked upon reverentially. If nature is looked upon as a mere flux of evil as no doubt evil pervades it, there would be wanton mischief all around. The existence, with its entities, the human and the non-human, should thus be free for the realization of the good itself, the fullness of being. Almost similar line of thinking emanates from the Tao and the Confucian ethico-religious philosophies though they are different in many ways. The whole of nature in both these philosophies is regarded as epitome of intrinsic worth or value, and for that reason humans should not look upon it merely as means or an extrinsic value for their benefits. Though humans should depend on nature, use it, the interference should be as minimal as possible so that a balanced relationship exists between them. Theological Islam bases its environmental ethic on Allah's creative design and human acts in the creation. Though in Islam human beings enjoy some primacy over and above other creatures in Allah's creative domain, human beings cannot be the "dominant" party. Rather, human beings should be "grateful" towards Allah and strive for the complete sustenance and flourishing of all creatures. As nature-abuse by human beings is looked upon as abuse of the Divine Will, it is unethical. These human beings are expected to act ethically towards the environment as an exploitation of the Will of God. It is not difficult to find many such instances of Divinization of nature in the world religions. Interestingly, many primitive and tribal religions conceive an all pervading *spiritual force* in nature, which should not be transgressed by human intentional acts as the hindered force would further make the nature unliveable. Much natural tragedy is, therefore, due to the cruel unethical deeds of human beings. What is most interesting to notice is that almost all religions come to *non-anthropocentric worldview*, which is the first "shift" towards a comprehensive ecological ethics.

Once we reach the Christian ethical views regarding environmental matters we meet an exceptional concern of humanity that transcends the human interests, benefits and welfare. Non-anthropocentrism, the cornerstone of eco-ethics is prevalent in the Christian writings. I am greatly indebted to Rasmussen's reflections on this matter. To speak of the Christian views, one is not necessarily restricted to what has been jotted in the Bible or its different interpretation, one is reminded of the numerous thinkers, scientists, philosophers, writers, contributing to an understanding of what could be a sensible view of those participating in a Christian way of life, says Rasmussen. 'The Christian way of life is one of faith, not merely a faith on the transcendental but an *earth faith*. This is a truism as Christianity is a radical incarnational faith owing much to Judaism, whose founder and foundation was believed to be *earthly incarnation of God*. Hence, from the Judaic unfailing faith did Christianity derive the truth that both creation and redemption are matters of earth, matters that grow and fizzle out in space and time, with it a fleeting history is formed. It is but natural that in Judaism, Christianity and also Islam, God's "faithful" are "strangers" on earth and their religious motif is alienating from the earth. Yet, the strangers are not escapists, they are not skeptics as well, their religious conviction is that the created earth is good as work of God, which needs its fullest flourishing or redemption in totality. This idea must have influenced many non-anthropocentrists of the west in contemporary times. The earth (in its broader connotation, universe), being 'work of God' is in essence epitome of *love and justice* and truly speaking, when ordinary folks, the unfaithful 'alienate' from love and justice they do not really "alienate" from the follies of the world, they fall in it. Hence, Christian call for "alienation" is from earth's ordinaries, not from its foundational 'love and justice'. The faithful's deepest urge is to rid the earth of its distress, not to be distressed in earth and distress the earth by follies.'

'The "earth community", therefore, needs careful upkeep. In the words of Apostle Paul, "the mutual upbuilding of community"—the continuous upbuilding of the oikos as a whole. This turns out to be global citizenship and earth patriotism, which as said in Deuteronomy 30:19, duties of "choosing life" and "living in accordance with choice". Naturally, such choice is a moral one, keeping an eye on rehabilitating for those damages suffered and inflicted upon. However, global citizenship does not mean that the benefits, well-being and aspirations of the particular spheres ought to be neglected. The first Christian, therefore, was concerned about the sustenance of a moral culture on our home-earth so that a "new age" could take shape'. No doubt the Christian zeal to lead a life of *moral* citizen transcends the locales but does not neglect it (remember Naess?). For the Christian ethicists, therefore, rather than "systems ecologies", a global thinking is possible.

With a simple background of the Divinization of the universe and our deep found faith to find the Truth, we can move on to what the "faith" could bear on ecological crises. Let us first accept that the phenomenal spurt in Christian ecological ethics started in the modern times with a negative note for historian Lyn White in 1967 argued in his paper: "The historical roots of our ecologic crises" that the Judaic-Christian tradition's emphasis on God's transcendence, the human alienation, a pilgrim's progress to an alien world for refuge from pains and human domination on earth as the powerful creation of God has devastating effect on environment. The main reason for devastation being conceiving the existence on the whole as "alien" or "other" and, therefore, is at best means for our good or if no use for redemption to a higher life. Not only this, the sceptical challenges have been faced head on by Gregory Bateson (*Steps to an ecology of mind*, 1972), John Cobb (*Is it too late? A Theology of ecology*), Jorgen Randers (*To create a different future: religious hope and technological planning,* 1972) and others like Mesarovic and Eduard Pestal (1974). In 1990, a team of scientists led by Carl Sagan and Hans Berthe issued an open letter charging crimes against creation.

Professor White's charges however had positive effects on the Christian thinkers and scientists in so far as they started understanding the Judaic-Christian texts *de novo* in the backdrop of causes of ecological disaster. Some notable efforts in this regard have been by Eric Katz, Larry L. Rasmussen and many others. These thinkers come to the conclusion that Christian concepts of alienation, pilgrim's progress to a Promised Land and human supremacy have been grossly interpreted by White and others. They argue that the Jewish tradition teaches that humanity is just the *trustee* to the existence as such and not its owner because God is the sovereign owner of all that is. Most importantly human ethicality lies in the ever-growing job of perfecting the existence and not derogating it. These thinkers point towards the Judaic environment friendliness because in it man has been pictured as a protector of the environment or, to be precise, that should be man's *moral aim*. In fact, Rasmussen has detailed the reasons for proving that human "alienation" has a deep religious significance, along with "the pilgrim's progress to an alien promised land. These are in no way skeptical or anti-environmental concepts. There is no doubt that much misgiving about Christian conception about nature is due to its several improper interpretations of man's relation with the nature and the natural things and beings, as well as understanding of disdain for the body, the earth, a desire for transcendence, spiritualization, resurrection, creation and cosmic destiny and dismaying earthy images and metaphors. Paul Santmire in *The Travail of Nature: The Ambiguous Ecological Promise of Christian Theology* writes that "the best way to get out of many ambiguous understanding of Judaic-Christian texts is to adopt as Christians an ecological reading of biblical faith".[1] The "faith" calls for shunning inherent dualism that treats man and earth as "one" and the "other". Hence, a plethora of Christian thinkers, Catholics and Protestants strive de *novo* to restore and revise the biblical motifs positive and negative, understanding human relations with God, other human beings and all of existence as such. This might converge into what Sr. Pamela Smith calls, a "concentric" view of the Christians about ecology, which avoids the problems of anthropocentrism and ecocentrism. Rasmussen in his typical oratory vocabulary has stressed that the capacity of the theologians and/or the religious persons to face adversities like the one above is to find out new religious cosmologies because man, the "story-teller" has been since time immemorial finding out the ultimate in cosmic story. He says: "This habit of letting grandiose meaning-stories is as ancient as the oldest human records and apparently unique to Homo sapiens. We conceive but not only conceive, we draw the imposing maps we call cosmologies".[2] Truly, the religious person, as the Christian is, has unique capacity to find such maps: "religious cosmologies display our thirst for exhibiting in symbols, and explaining in ritual, rite and reflections, nothing less than the totality of things.[3]

Nevertheless, the ability of "story-telling" has not been the same even with the Christian thinkers. Rasmussen thinks that the Protestant theologians from 1930s to 1970s have been remarkably deficient of theology based ecological ethics because during this period many theologians placed human beings in the cosmos as distinctly alienated as the "other" one from the nature as a whole. Nature has been largely portrayed as inert object at the mercy of creative genius of human beings. Thus theology was heavily bogged down by the modern material sciences and assumed that nature is for our use, we are makers of a world to live in and we are powerful as to model and remodel nature, society in umpteen ways. The cleaved nature-man relationship changed the meaning of life and perspective of life within the Christian tradition. But this is tragic for the tradition which has been early on forcefully articulated by Francis Bacon who stressed that the human vulgarity to dominate and boss on nature, people and societies has been crucial for ecological crises. Hence, all these things comes to the way man acts or behaves in the milieu of nature though the way man perceives nature is most important. However, as Langdon Warner rightly remarks that "nature" does not mind human "subjugation" (for nature knows what will be the fate?). In the wake of human domination and scientific onslaught over the nature we had the *existentialist theology* of Bultmann and Kierkegaard who placed the true human, the one who possesses "freedom", at the helm of worldly affairs so that he or she could shape his or her personal world and established meaning with everything else. The nature as the "other" has been thus looked upon as the mute object whereas human beings played with it by virtue of freedom. The free human beings, who become queer "abstractions" in nature do religious activities as they create values, make meanings, produce symbols but "nature remains disjointed in every affair as it wanted the "free" human to "give it a meaning" and make it of "use". This Christian ethics on the basis of neighbour-love and justice did everything to a few people but neglected many. Great thinkers of this line, Karl Barth and Reinhold Niebuhr proud of "systematic theology" were thus never interested in "ecological crises" or meeting the crises with an ethic, which is theologically oriented. The result has been telling in human culture as the "faith" they nourished has been always as "outside" nature, a "faith" that sets nature against us and, therefore, makes it religiously and morally morbid. Religion, love, justice and peace along with nature moved around human welfare—the moral realm has shrunk to a great limit.

But, the Christian faith has been moving; it found a way out of this dull anthropocentrism and embraced gradually theological eco-ethics of different type that is not romantic but religious. In this respect, we can conceive of a religious man in nature in many forms. Let us have a look at it. One way, which is theoretically speaking hated by contemporary Christian theologians for its revered anthropocentrism is the "dominion" status of man in nature as has been championed by Christian theology of the past. The present, however, has not up pushed it in the limbo for it has a compelling "practical" advantage over other more glossed one. Cardinal O'Connor as late as 1990 and more recently in 1995, in the official papal teaching "The Gospel of life" propagates *dominion status* for the religious man in nature. The reasons are simple. God created everything including man but the "life". He created is quite different from those of other creatures. Though human life is created from no special material, it enjoys a sublime status owing to its unique bond to God as *God's image*. This is textually undeniable truth in Christianity, so how can we overlook it? The Book of Genesis affirms the truth Pope John Paul II utters: "when… it places man at the summit of God's creative activity, as its crown, at the culmination of a process which breeds from indifferent chaos to the most perfect of creation. Everything in creation is ordered to man and everything is made subject to him".[4] He concludes: "We see here a clear affirmation of the primacy of man over things; these are made subject to him and entrusted to his responsible care, whereas for no reason can he be made subject to other man and almost reduced to the level of a thing".[5] The "things" however receive the faint glimpse of divine glory and thus "different". But lingering questions are raised: Who has given the dominion to man? Is He, an earthly God? Is He Himself creation of our imagination? It would be difficult to avoid more nagging anti-anthropic questions if we press this dominion status of religious man much as moral chauvinism then, it may finally explode out of shell, which would not be liked by the Christians of our times (or, may be, was not liked in the past).

Does it give way to another model of religious-man in nature? The model of the *steward man* is sometimes advocated by the Christian thinkers as the basis of a religious eco-ethics, which has an inherent voice of "dominion" nevertheless because the "steward" is often taken as the master, the controller and the custodian of nature. Rasmussen thinks that 'the "steward" does *not* refer to such overpowering person rather, understood as *trustees* of the nature as a whole. This understanding is mainly owing to human-nature relationship as understood in more rational terms and a renewed understanding of the nature of human beings. The nature of human beings need not be understood always in hackneyed terms—the material, the grossly physical, the mentally egoist and the overpowering marauder of things around for benefits unlimited. Human beings are also powerless beings, at nature's mercy, always a little short at perfection and beauty and though possessor of power of knowledge, the knowledge that may lead him to the ultimate truth, is nevertheless in want of it, ever ignorant. Yet, in the situation of the positives and the negative awfully rolled in one simple creature in the 'image of God' (it is only an 'image' but not God), the "moral" aim in human life is clear enough. It is a just and sustained use of power of wisdom and knowledge so that certain greater (or wider) *responsibilities and duties* are taken and postponed in nature. Hence, in Genesis (1:28) and elsewhere the ideal morality for man is to permeate, dominate and multiply so that most humbly and restraintfully human beings live with all living beings and many things of use. *Stewardship* is thus not dominion; it is most rational and ethical way of life in nature. It only refers to a relative prominence in Ecological Community where all live and dwell in and with God. It is truly "Theo-centric" and neither anthropocentric nor biocentric'. Our previous discussion (in Chapter 2) now reveals the germs of such Christian thoughts as present in many views of thinkers having a Christian way of life. Hence, the "dominion" status of man has been vastly changed as "guardians of earth". However, the Genesis (2:15) suggests that when the custodians are morally wrong, nature certainly would react to keep them on track. What has been conveyed in the Hebrew writings is far more clear than our English usage because the "Shomrei" truly speaking is the one "who serves and preserves", of course, the existence on the whole. Going by the Hebrew text, Adama (or Adam as we use it), refers to "earth" or specifically the "topsoil", from where (the "dust"), human beings originate and return to after death. This "dust" he owes much and is its custodian for life. From this, we find no great antagonism between the "dominion" and the "steward" models in Christianity lest both have been understood in more "deeper" ways. Only then perhaps, the most tough and nagging questions of the anti-anthropomorphist might be answered to a great extent. Nevertheless, anthropocentric charges against "stewardship" has not faded till now because of its links with the "dominion" model and, therefore, the real "shift" in ecological ethics is still a little tentative feels many. The main reasons are that the Hebrew legacy and the modern official Roman Catholic teachings have still obsession with Homo sapiens at the centre of "moral discussions" relating to ecology.

The question is who else has to come to the centre for only the man is moral or immoral? Truly, man is moral or immoral, and man takes on "greater" responsibilities but the "custodian" image (as in the 'image of Lord's servant'), does not rule out some overdoing for their survival, multiplication, happiness and welfare. The "use of the other" for all "practical reasons" may be religiously glossed for a morality that suits us. Further, such imaging of man on earth as steward prevents us from imaging morality into nature itself and God's redemption of the nature as a whole. It may hurt the views of the tribal people who live in and with nature who believe that in no way are they "stewards" who subjugate nature. Nevertheless, all criticisms and doubts notwithstanding, Biblical ethics still takes the earth as "Lord's" and that humanity "own" really *nothing*. In this sense, the "caretaker" of nature is a trustee of Lord's possessions; what may be distasteful in it if we go by the main thesis that "all is Lord's"?

It may well be that we have another way to put things right. We may speak of man as a "partner" so that the homocentric obsession lingering in the dominion and the steward models are revised to place man as a softer and humbler being in nature. Man in nature is sometimes taken as "partner" of beings and things, not one who is supreme, should govern and control rather, one who *participates and empathizes* in and with nature. Such caring and empathizing is the crux of "morality" therefore. St. Francis refers to the earth in particular as an abode of "holy democracy" for here all small and big beings and things enjoy equitable share of nature's bounties for sustained living, and in such "democracy", the wise man is a "patron saint". The picture of a holy earth is thus of interdependence and interconnectedness among beings and things so that a shared life of peace and harmony is possible. Hence, wise men should be able to follow earth's dictates of harmony and act ethically. It is here that a more "decentralized" view of earth (cf. Rasmussen, p. 236) comes to the fore in which human beings are not looked upon as the only important members of the holy universe. The universe in complexity and wholeness is a "moral" universe because all living entities have a "quality of life" and strive to fulfil their life's potential to the fullest. Rasmussen in his book (p. 236) refers to W. Godfrey Smith who calls upon the moral man to conceive "other" living beings as not merely of stock value or "silo", value as an object of harming or "laboratory" value, value as recreation or "gymnasium" value rather, as possessing "cathedral" value or ethico-religious and aesthetic value. Man being an important member in nature does have all these values and it would be equally immoral to consider them as silo or recreational etc. Here, however, much hue and cry is still about "living beings" (only obliquely could one refer to things), and also about "human beings" as most important because the ideal of a "partner" in a moral universe gives moral responsibilities to human beings and rejects the idea that many other beings do share great (or greater) responsibilities in nature for sustained living. It is a myopic moral thinking that other beings being not like us by virtue of wisdom, knowledge and higher consciousness does not have great moral responsibilities. The "decentralization" is thus not as complete as we think. Thinkers like Jay B. McDaniel (1989) say that "Rasmussen rightly remarks that the "partner" is only relocating the important "steward"—it is a mere variation of that thesis. There is no denying the fact that the steward naturally by virtue of umpteen powers dominate over other things but the nature may have the last laugh, which, however, does not refrain us from being a powerful partner. Rasmussen remarks: "The recentred human partner recognizes that, whatever power we wield as a species, we do not legislate the laws of an encompassing nature… justice and peace aren't human goals and states only, but all creation's terms. Creation's integrity sets the terms and requires 'moral considerability' of other kind in human decisions".[6] It is due to this that the Franciscan ascetic man fall short of "creation-loving ascetic". He fails to realize that in ecological ethics there is now a strong need to "shift" from anthropocentrism to *life-centrism and beyond* as humans is wondrous "microcosms of the macrocosms" (*cf.* Chun Hyun Kyung in Rasmussen's book, p. 237).

Nonetheless, theologically based eco-ethics has another model of moral man, where man is a *priest*. This is the *sacramental* conception of man in nature, which is believed to move away from homocentrism to at least "life-centrism" by shunning the "dominion" and the "steward" conception along with the variation of the thinkers of the pre-Augustinian Christianity and Celtic Christianity and finds modern voice in Kosuke Koyama (1992) and in a publication titled "Orthodoxy and the Ecological crises by the Ecumenical Patriarchate and the World Wide Fund for Nature International" (1990). The Sacramental clarion call does not ask for human beings to put at the centre of all creation but ask him to see as "priests of creation" who are *representatives* of creation participating in the praise of God, the Creator. This however, does not mean that "others" in nature are dumb, rather the whole of earth and heaven sing in the glory of God. Humans pray in psalms in one way whereas heavens have other ways to praise. This "cosmic liturgy" is unique and simile of "song" refers to the various dynamics and transfigurations in nature for self-fulfilment. In one way, speaking humanly, our gratitude apart, our moral acts for the perfection of self are ways to praise in the glory of the moral governor. Most interesting in the episode is that we are pictured as having a life in Ecological Community and when we praise, we do so on behalf of *all* because our dynamics and transfigurations in moral life are not in isolation rather in unison. The "images" of God are thus *all* in praise of whose image they are. This sacramental view sometimes referred to as Hasidic Jewish and also pantheistic, does *not* in any way turn entire nature to God. Hence, nature and man is not God themselves, in this case, nature is not divinized and, therefore, not object of worship. Sacraments are the *symbols* and *signs* participating in Reality, they are finite in which the Infinite glory is immanent. Nature is thus not God but of God—God is present where nature is present. The ethical part for man is clear, 'man should act in the glory of the Infinite because the Infinite is present in him and He *ought* to be realized fully'. The moral status of man is thus not one of mastery, control, caring at a distance rather immanent in and with nature. He is a priest who mingles and empathize us and with nature as he does realize that nature *reciprocates* equally morally, too. There is thus in nature a "spiritual unity"—the physical is apparently so, it is "spiritual" in the immanent glory of God. It is here that slowly the idea of an earth community, the wholeness of existence and the humble moral sense of man that we owe much to *all* fellow creatures and thus have greater responsibilities percolates. It is here that we are "shifting" to an eco-ethics, which has freed itself from the shackles of anthropocentrism and even life-centrism. The idea gets again and again reflected and appreciated in the ethic of indigenous people over the world who would not mind daily practices like hunting and farming etc. as anti-nature or unethical. It is all about learning to live the best life not with the sense of "use" but "sacrament". (Refer Leopoldian naturalists)

All said but not done may be the real tragedy for the sacramentalists for the glory of God and the spirituality of the nature does not make a complete ethic if we are bereft of concrete actions. In the face of "mother" nature being so thoroughly defaced and our urge for fulfilment, what *should be done*? In fact, living a harmonious community life is an ideal, we have left it far away *now*, what course of action than praise ought to be done? What strong political decision has to take place? It is not difficult to find out sacramentalist ethical corruption through the ages when people are in feudal overtones sing God's glory but not the glory of the poor, the wretched and the devastated. Some kind of devotional "healing" that is a "*status quo*" in devastated nature is in place of the ethic of action for the troubled. So in nature (as we are) not at all beautiful, moral and fully realized, our revelation of God is promised—a distant one though. Here the nature does not work because the present state of affairs and the dreams do not match. It is due to this perhaps that the voice of indigenous eco-ethics along with eco-feminism (see Chapter 2) brings together earth-consciousness, earth-sufferings and specific socio-political actions for the battered present community we live in. There is thus no doubt that the Process theology, the torch bearer of natures' sacramentalism with eco-feminism's liberationism might strengthen what has been so far left vacant by the old sacramentalists but the inherent spiritualism and metaphysics needs to be punctuated by social and political analysis and action. Yet, the Christians are not averse to such analysis and action lest we do not forget that through analysis and action, the moral goal in vision is not the drab physical, it is inherently divine.

Keeping in mind the 'now' of the ecological community and to some extent not overplaying with either sound metaphysics or ethics, come forward the "ecoprophetic" thinkers providing us a *prophetic* or *covenant* theistic model for ecological ethics. Ecoprophetism is in fact an urgent ethical call for improving the cosmic conditions because such conditions are unacceptable to a religious mind. The prophetic mind being essentially dynamic for the restoration of what has been derogated in nature, it first asks for a massive "conversion" as the prophets does—conversions to the Divine Life, to the faith of God and thus away from destructive ways in life. The prophets next assume the status of a "teachers" pointing out the basic causes of ecological disasters. In fact, neither *status quo* of the priest nor theoretical churnings of the intellectuals are enough at the present state, we need first a religious consciousness and zeal to make up for the loss. The prophetic model is thus essentially creed-free, it does not belong to any religion—it is basically "religious" in spirit, though the truth is that such model finds its voice in the works and deeds of the Christian theologians. These *evangelists* in particular take ecological resurrection as their *covenantal responsibility*, herein lay the ethic of ecological matters. Rasmussen remarks that the masters of the covenantal order remember what God said in Genesis (9:12-13), "This is the sign of the covenant that I make between me and you and every living creature that is with you for all future generations: I have set by now in the clouds, and it shall be a sign of the covenant between me and the earth". The ethic is prominent, it all depends on performing a covenantal duty because doing so we live well and live long. This is the unique as the divine covenant has the power to remind people now and again that no false gods help; it is only the covenant that helps. All said well but did we really move away from the idea that only human beings, the prophets have a covenantal duty? Isn't the Genesis Verse implying a greater truth? Isn't it true to take that the whole nature has come to a covenant in the Divine order and *all* tend to take that duty selflessly? Somehow, one feels that in the Christian models of ecological ethics, we are still not grasping two things: one is the importance of the existence as a whole and the other is the role the existential plays in order to live a life for them and for us.

The Christian theologians were in no sense ethical myopic not to have noticed such important things. Along with the basic Christian monotheism, thinkers have been coming forward to move away from anthropocentrism and even life-based ethics to some kind of holism, which is not adverse to the monotheism. This is again nicely clarified by Sr. Smith. One instance is Sallie McFague's metaphysical explanation of the "body of God" in her "The Body of God: An Ecological Theology" (1993), which is one important writing in eco-feminist trend about which we have been saying important things. She calls for a more pertinent post-modern metaphysical understanding of nature-God oneness not leaving any component of existence, therefore, so that the old sacramental organic view is successfully replaced thereby refusing "human" oriented metaphors. McFague thus proposes, "The metaphor of the world as God's body".[7] Sr. Smith thinks that the metaphor tries to focus the truth that 'God is immanent or in embodied form in planet and cosmos on the whole. This metaphor is well accept the Christian thinkers in so far as it implies that all components of creation are intrinsically valuable. Being intrinsically valuable, they enjoy unique rights for self-fulfilment. This model overcomes the patriarchal dualism and the tendency to look natural objects as matters of exploitation and use. This also helps ecologists to envision the truth that things and beings in the cosmos are not merely organically related. It also stresses on the need of the renewed ethical understanding of human behaviour towards nature as a whole. The nature's recovery is not only a matter of scrappy make-ups and for the short-term. McFague thus comes to believe that taking nature as "the body of God", we can find a good combination of the ethical, the ecological and the theological. Interestingly, it is believed that such understanding is for us not for others for we must have lost such wisdom or the knowledge of the sacredness of the universe as such. The model is not averse to the Christian conception of Divine Transcendence rather, it is preserved and redefined in another way. The transcendence of God is restored in the idea of the spirit of the cosmos, the *élan vital* that puts things to life and motion. The conception of an immortal Christian is restored as well for the Jesus is *cosmic Christ* who incarnate to universe of existence in many distinct ways. The Christian attribute of omnipotence and omnipresence of God has been also restored nicely looking at existence in this way, our notion about "bodies" also change radically. They are not grossly physical, mute and of use rather, we have to love and revere it as divinely valuable. There is thus no point in helping it or degrade it as we have done before because we love only female bodies and none else. There is thus no point in the loving and loathing that are merely "organic", the bodies ought to be seen as divinely rich. McFague thus move ahead of biocentrism in her theologically based ecoethics for which she thinks that an attention "epistemology" is most needed—the attention is to find in the diversity, *oneness*, which evokes in us solidarity for existence as such. McFague may not be as vocal and prescriptive in the Prophetic tone, she nevertheless thinks that it is important to decentralize and also relocate human beings in the helm of ecoethical matters.'

McFague has been further extended for a Christian ecoethical understanding by Rosemary Radford Ruether in her Gaia-God thesis. Sr. Smith explains that Ruether in her work "Gaia and God" (1992) holds that our ethical relation with the environment is grounded on the basic fact that we are one species among many who have the ability to surpass the moral limits and thus become one with the whole of creation through our intrinsic kindness. However, Ruether comments that this surpassing of an organically and morally limited person is the essential "ambiguity" which is to be accepted. Yet another human "ambiguity" is that within the accepted "finite limits" of nature and natural beings and things human attempt at shaping and reshaping the nature, that is, attempt at its various transformations. If these ambiguities are unavoidable for human beings, one cannot live in confusion. Human beings in reality do not nurture such confusion because they are wise enough by the grace of God. They realize sooner or later that the universe is *sacred* and God is the ultimate retriever. Once we realize this, says Ruether, we come to know that the earth is *Gaia*, that is, a *living system*, which is dynamic, growing and creating organism. It is as if an all embracing whole, the living whole, much like the Greek earth goddess, "Gaia". For Sr. Smith Gaia as 'earth-god', the "matrix of life" merges "small selves". The earth-vision, a spiritual one is not *status quo*, it is formative nevertheless as self-surrender and oneness of selves itself is not a mere vision or a metaphorical imagery—it needs action. But Ruether does not prescribe its details. In its place, Ruether speaks of "right relations", the *covenantal* one where human beings, lesser beings and things respect for equity, justice and dignified subsistence. It is, therefore, a culture of "compassionate solidarity" much in abeyance to the patriarchal hierarchy. The gist is that for Ruether ecoethical behaviour must have a *spiritual* basis. For her Gaia, the goddess is both transcendental and immanent, much like Christianity's attempt at recognizing the infinite in the finite. In the sacred universe Gaia is seen in many forms, all striving a life of own, each growing for fulfilment, each trying (like humans) to surpass the limits. Here is the ethical aspect of the whole of existence. 'Ruether has a tendency to move farther, she perhaps does not want to be a stylized Christian as she wants to combine Christian and indigenous African cosmology as well so that the human-Christ relation is more focussed and we come to know that in the name of feminism this theological basis of ecoethics does not relegate God to sex ratios, rather, like the indigenous wisdom the Gaia is "Pachamama", "as much mother as father",'[8] remarks Sr. Smith.

It has been analysed thoroughly to find that the ecofeminists', including Ruether's views leads to "liberation ecotheology", which has been more explained by Boff, McDonagh and many others as well. Liberation ecotheology is a wonderful effort at combining the spirits of Ecosophy T, animal liberation, ecofeminism and naturalism in one fold. It is thus a balanced effort to become *holists* without having much disrepute as obsessed towards truth of anthropocentrism and other such opinions. Sr. Smith tells us that this philosophy is for the first thing about life and light but not death and darkness because Gospel of Matthew asks us to cherish the "promised life", "peacemaking" and most importantly "shalom", that is, a state of wholeness and integrity—"a condition of life that is in harmony with God, other people and nature".[9] This effort of the theologians is essentially "liberating" as it is for the liberation of the battered and the poor, and the Christians have an ethical task to give some options for the poor to live up. However, it is not shallow ecology, for the "poor" is not "poor people" on earth but for the existence as a whole because for Boff every existential has "intrinsic value" on the grounds of "religious faith". A social reconfiguration ought to be based on a "spirituality of integration" and thereby liberating the poor from the damages done. However, the "integration" so important owes to basic "synergy" or interrelation and interaction among species, systems and societies for well-being.

Much like Boff, McDonagh in his important works such as *Passion for the Earth* (1994) holds that the Genesis narrative and Christian Panentheism enable us to speak of the "dignity and intrinsic value" of everything around us and, therefore, we ought to have a non-anthropic world-view. McDonagh also speaks profusely about "sustainability" as does Rasmussen because it is the key concept for a human direction to activity for the liberation of the battered nature. McDonagh sounds much like Ruether when the idea of "pastoral ministry of sustainability" that integrates tribal ways of life and Christian mysticism.

All said and done, what we have is the strong concept of an "ecological community" in Christianity, a community that is essentially Divine. Without a basic spiritual vision as this, "The integrity of ecosystem communities and their many citizens is not recognized".[10] And "Indeed the moral formation of society and world is a critical action of its own".[11] But is everything fine with this ecoethics deeply soaked in theology? We shall not analyze this now as the Hindu theological view is in the offing. However, we have to be careful here to find that after so much of churning with religious theology and Christianity, we are at hand a "justification" of ecological ethics, which is like what we have found in Chapter Two of the book, permeated with "shifts" from the homocentric to ecocentric and to the holistic one, though the "shifts" were possible on different non-scientific grounds despite "ecology" being itself a science. Does ecological science in connection with "ethics" enable us to bring in the inevitable "non-scientific" in the "scientific"? We shall come to it but here we go for the Hindu theological ecoethics.

After so much of churning about "environment" and "ecology", we hardly ask more about their nature and significance but when we come to "Hindu ecology", we face some problems for clarity of meaning as "Hindu" has many things to convey. Taking a relatively simple way to clarify its meaning, we can speak of "a way of life" particularly prevalent in the Indian subcontinent. It also refers to a "religious way of life" distinctive of its *monotheism* though references to gods and demigods are not to be forgotten; yet, it does not amount to pluralism and henotheism. It also refers to several religious practices, sacrifices and worshipping methods. Broadly, many anti-theists and non-theists participate in this "way of life." Hindu religion in its theistic forms and beliefs speaks of One

Supreme Reality as the basis of multifarious existence. All sorts of practices, rituals and worships of minor gods and demigods notwithstanding, the unique "religious way of life" is not disjoint of "ethical way of life" as Hindu religion does not keep off from the practicability of life on earth.

A Hindu environmental or ecological philosophy and a Hindu ecological ethics essentially refer to the criss-crossing theistic, anti-theistic and non-theistic philosophies and ethics concerning ecological matters. It is due to this that broadly speaking people participating in a Hindu way of life have different ethical opinions pertaining to environmental matters. Hindu ecological ethics would thus amount to a broader anti-theistic and non-theistic ethic of ecology and a narrow, theistic ethic of ecology. In this work, I will focus on the "theistic Hindu ethic of ecology" or "Theistic ethical ecology of the Hindus". This would give us an idea about how a large chunk of the theistic Indian Diaspora and theistic Hindus abroad think about the *possibility* of an ecological ethics. Hence, we should proceed with the "justification" of ecological ethics as propounded by the theistic Hindus.

In order to perform this apparently simple but tough task, the best thing to be done is to first deconstruct the Vedic-Upaniṣadic concepts of *Satya* and *çta*, and also refer to *çõa* and *yajña* in the context of ecological ethics. The next is to deconstruct the ecological ethics of *Bhagavadgitā*, which is a good back up thesis of what has been propounded in the *Upaniùadas*. We can further move on to refer to some relevant aspects in the great epics based on the theistic philosophy. We may also refer to some modern theistic thinkers to get an idea of ecological ethics of the Hindus. At the outset however, we have to remember that numerous references to the *devis* and *devatàs*, who are often referred to as goddesses and gods need not be confused with the God of theism we are concerned with because here we simply have in mind One Supreme Reality or the Absolute or the Brahma as the basis of existence as such. We do not here recognize any pluralism and naturalism, as is being traditionally understood owing to the lack of knowledge of the theistic Hindu traditions.

The conceptions of the *çta* and *Satya* are Vedic in origin and according to me, without knowing this, appreciation of the theistic Hindu ecological ethics is not justifiable. Most interesting feature in this respect is that unlike many religious and non-religious ethical traditions, we are here not justifying eco-ethics by "shifting" gradually from anthropocentrism to holism, rather, the Hindu view is essentially "holistic", it right away starts with the exposition of the existence as a whole and conceives it *spiritually*. It is natural that to realize what Hindu theistic ecoethics is, we have to keep in mind that several truths about human, animal and nonliving material beings and things are found in the Whole Reality. According to this view the existential in the world and beyond or the whole existence *per se* is in the form of *"Satya"* or "Truth" or the "Absolute". In the Vedic *mantras* along with *"Satya"* there is a reference of *çta* as well which speaks of an "eternal spiritual order" or an "eternal system" in and with the so-called worldly or universal order. In *è g Veda* for instance, it has been said: *"çtam satyam"* and *"çtam satyācārah"*. In due course however, in the theistic interpretation of ecological ethics *"çta"* acquired much prominence than the more profound metaphysical spiritual concept of *"Satya"*. May be that some "deep" and "abstract" significance of *"Satya"* was replaced by a less abstract one in the name of *"çta"*, though *"çta"* need not be hastily taken as mere ecosystem or of organic nature.

"Satya" or "Absolute Truth" is the fundamental metaphysical Reality or "Being" whose multifarious concrete expressions as order of beings and things or what we call becoming is known to us. It is interesting to notice here that *"çta"* originates from the root *"ç"* having double significance. One is "that which is moving or dynamic" and the other is, "that which is not moving or static or unchanging". Together they convey the meaning of *"çta"* which is the same as, "that which is unchanging in all changes". Now we can understand that in an abstract sense *"çta"* is the Principle of Truth that is Being or Unchanging Reality, whereas in concrete sense, it is the principle of change or becoming. In this sense, therefore, we say, *"çtam Satyam"*. It is here we come to realize that *çta* is not merely organic world order, it is not material or physical ecological system or ecological order—in essence the so-called systems or order is "spiritual", an epitome of the Absolute. Another significant feature of *çta* is that it refers to the "moral order" as pervading in the universe as a whole because as an expression of *Satya,* it should be taken as that which expresses the Ultimate Good and Ultimate Value. Hence, the becoming is as it is an ecological order of change,

it is also a moral order of eternity. This order should not be disturbed, as it is a vice in itself. So both the ecological order and the moral order are closely related as both are in and with the Absolute Order. Now if we look at "existential" of the existence, nothing is different as "existential" exists in and with the existence and for that reason are expressions of the Absolute One, the *Satya*. Being of a spiritual nature essentially, existential themselves order—natural and moral. Take an individual man, he is as a natural being the expression of the Absolute and as a moral being good in itself. An individual is thus in and with the order of the *çta*—showing signs of ever-changing actions and modifications but in itself remaining unchanging as Absolute Good. The picture is true for any existential, high or low, animate or inanimate. The holism of the Hindus is thus bang on target. Going farther, there are certain Laws or Rules of the moral order in the system of *çta*, which are spiritual in nature. We have to abide by these Eternal Laws as doing so we realize the unchanging within multifarious change and there is unique harmony, balance and beauty. If one wants to go against the Laws, the natural and the moral order crumbles; so the simple ethical guideline is not to go against the fundamental Laws of natural and moral order. In the system of the *çta* therefore *all* of us have equal rights to live, grow and liberate to the higher realization, which determines the way our moral life should be led. Though possessing greater wisdom and knowledge is a virtue, by virtue of these men can no doubt try to control and guide the system of *çta* but cannot disparage its Laws—its harmony and balance. An ignorance of this leads to *"ançta"* or ignorant actions of man leading to bondages in life and utter sufferings, physical, moral and spiritual.

To elaborate the nature of natural and moral order, *çta* is metaphorically explained as the "stream of water engulfing earth"—*"çtam rõãtmakam mūlāpa jalam"*, and as *"viùõurūpa"* or in the Form of Lord *Viùõu* or the Supreme. Hence, there should be no doubt about the *eternity* of *çta* in the Form of *Satya* or more clearly now, as, *Viùõu* (*çtam Viùõuh*). *Mahābhārata*, the great Indian epic thus explains *"çta"* as the Supreme God who saves the world in distress by taking as many worldly forms (or incarnations) as does water in many containers. Equally true is that the world-order is the Absolute order who has given world its dynamicity and its sustenance. Interestingly, *çta* has been explained as the "sun" in the *Satapatha Brahmaõa—"çtah såryah"* and further in the *Yajur Veda* it has been seen as "fire"—*çtah āgneyah çùbhedah"*. This is not naturalism or henotheism or pluralism but many concrete ways to understand the Truth, which in essence is not in any concrete shape and form. Moreover, it reminds us that air, water, fire, sun and the whole of existence is in and with the eternal order—it is in and with God, the Absolute Truth. Hence, as Being, the existence is seen as *"Brahmā"*. Not unlikely therefore that *çta* has been explained as *"Brahmā"*. In *Mahābhārata* we find: *"śāśvatam Brahmā parmamam dhrubam jyoti sanātanam"*—the existential order is as eternal as the pole star. Also in the *Satapatha Brahmana* it has been said that *"çtah parama Brahmah"*. The saying about *çta* has been so emphatic because the becoming need not be wrongly understood as organic or physical in essence, rather *"Brahma rūpa"* or essentially as Absolute God. With this, the actual meaning of *çta* becomes clear. We come to know why in the Hindu theological ecoethics, *çta* and *Satya* go together as inseparable concepts. In fact, to my mind, any Hindu theistic exposition as we find following the *Vedas*, holds that an understanding of *Satya* and *çta* are crucial for the realization of an ethically oriented talk about ecology. As *çta* also conveys the sense of "moral order", it is but natural that the ecological existence is wholly pervaded by the virtuosity and the goodness of God, which order has to be maintained as sacredly harmonious not because it is for our benefit or for mere well-being of existential, rather, it leads us unfailingly to Oneness with the Ultimate.

Closely associated to these concepts and crucial in an understanding of theistic justification of ecological ethics are the notions of *çõa* and *yajña*. In some sense, these derived concepts are more sensible in the context of human morality though it may be argued that a broader explanation may be possible. Clearly stated, if we are wise enough as moral agents in this moral order to realize that essentially the whole of existence including us is sacred, then the wise should feel *obliged* towards the whole of existence (and the existential). Naturally, if "moral islander" is an impossible notion to be believed, we are in deep interrelation in nature and we owe much to one another. May be we come to conceive here that each of the existentials owe much to the other for liberation to a higher form of life. The whole moral order is thus one of interrelatedness and there are no moral 'islanders'. The wise with the sense of obligation and deep felt duty thus act morally. This sense of *obligation* is expressed in the notion of *'çõa'* and real heartfelt selfless *performance of duties* to pay back what we owe or what is due, is expressed in the notion of *"yajña"*. *'Yajña'* is however grossly translated and understood as 'religious sacrificial rites', which does not explain the real significance of "sacrifice" in terms of "performance of disinterested duties", a much appreciated concept in the *Gitā*. What is crucial for our understanding here is that a religious realization of what we "owe" and how to "pay back" are important in the ethic of the Hindus. As we start speaking about the types of obligations and duties, we find more interesting things relevant for ecological inquiries. In the *Taittirīya* and the *Baudhāyana Dharmasūtras*, we speak of what human beings in particular owe and what they should do. They owe to the several natural gods, the elevated wise people and to parents and forefathers. Accordingly, they have duties in the sense of *'yajña'* towards the gods, the wise men and parents and forefathers. In the epic *Mahābhārata*, we have references of obligations towards humanity at large and in *Srimad Bhagavadgitā* we have the reference of *"bhūtādiyajña"* or what we owe to the "ecological community" at large because our existence is unimaginable without several inanimate objects, the five basic material elements and other living beings. In fact, the whole of ecological existence as it is an expression of the spiritual Being, whether it be basic materials, the building blocks or several living and non-living beings and things, sustain human beings in many ways. Human beings in turn are expected to sustain them or at least are *morally* expected to do so. Hence, a deep-felt gratitude to *"bhūtas"* lead us to *"bhūtādiyajña"* or sacrificial duties to the ecological existence as such. In the *śāstras*, a *gçhi* or a householder has to perform such duties apart from many others. He is expected to care for his small garden, the trees, a brook or a river nearby, the small and big animals, try to keep some domestic animals and even take care of the so-called lifeless stones and pebbles. Interestingly, it is not the henotheistic or the naturalist Hinduism rather, the theist Hindu who worship stones, trees and animals and performs sacrificial rites for a deeper symbolic way to express gratitude. All these things are also important for an understanding of the ecological ethics of the Hindus.

In this context now, we can go a little deeper than "deep" ecology without much hodgepodge about anthropocentrism and so on to holism, which focuses the value of nature as a whole in the theistic *śāstras* by exploring into the discussion of five basic elements and sacredness of nature. Our environment and the conceivable universe is said to have been constituted by *five great elements* or *pañcamahābhūtas*, which are the embodied energies. Every element has its unique existence, properties and *élan vital*—they are thus having definite forms, locations, directions and umpteen forces to enact the cosmic drama of formations and transformations. Moreover, these elements are not kept isolated and act at a distance in the vacuum but are interrelated and interdependent in a whole. These primordial elements are space, air, fire, water and earth (*ākāśa, vāyu, agni, jala and pçthvā*). The said classification is not only according to the atheist *Sāmkhya* but also according to *Nārāyaõa Upaniùada.* Unlike the *Sāmkhya* the *Upaniùada* relates these building blocks of the universe to *Brahman* and therefore, the conception of a sacred whole of which each part is looked upon as sacrosanct is natural. The result is that the nature, constitution and functions of the said physical cosmos cannot be understood by itself, it needs reference to a higher Reality. The *Brahman* is in fact the causative ground of the universe. It creates the universe as a whole by means of the basic elements, enters into the whole of existence as *life force* and may be better expressed as all pervading *consciousness*. The human beings being construction of these elements, the causative *Brahma—Caitanya* (Supreme consciousness) pervades them as well. Nevertheless, the expressiveness of the consciousness is not one and the same as many things and beings like humans do not perform conscious moral acts in the same degree and in the same way. Hence, the whole of nature is not outside us, we are in and with the nature. It is also clear that the *animate* and the *inanimate* worlds are interrelated and interdependent as they are with manifestations of the Ultimate. According to the *Bhāgavata Purāõa* God is thus conceived at two levels: at the abstract level it is all pervading consciousness, the spiritual basis of all existence, and at the concrete level, it is the world of formations and transformations. However, in the

Bhagavadgitā (Chapter VII, Verse One) the nature is said to have been constructed out of eight elements, the five said above and three others such as mind, intellect and ego-sense: *"Bhumir apo nalo vāyuh kham mano buddhir eva ca ahamkāra iti yam me bhinnā prakçtir astadhā".*[12] However, it has been categorically stated by the Lord in the following verse of the same chapter that the *astadhā* or the eightfold elements constitutes only the lower Being of the Higher Self. The Higher Being is the true Self by which the world is upheld is the *spiritual* Self or the *Brahma* itself—the Supreme Consciousness is thus the pervading substratum of the lower existence of the Self: *"apare yam itas tv anyām prakçtim viddhi me parām jivabhūtam mābāhoyayedam dhāryate jagat".* Both these verses are crucial in the understanding of the theistic ecological ethics of the Hindus. God does not create from something, it manifests its Pure Conscious nature into the multifarious lower nature where emerges the animate, the inanimate and person of mind, understanding, ego-senses. The Truth is unity of the Divine Self; multiplicity is lower expression of it but not an illusion. The foolish dualism or pluralism that is read in these lines should be avoided. God is one, expressions are many and real—the nature is God at its lower appearance, the nature is not morbid matter of the *Cārvāka* or the physicist.

Interestingly, the perception of nature as Divine itself has important impact on ethical decisions and actions. Another point to remember is that much awe about the purity, balance and harmony in the cosmic system is answered once the sacred aspect is realized properly. The Reality in itself is epitome of Truth, Beauty and Good, there is no disparity and disharmony in the nature of the *Reality*. Hence, its manifested aspects, whatsoever 'lower' it may be, are pure, harmonious, beautiful and good inherently. The unique law pervades the entities, balances it, proportionate it and thus constructs the balanced nature as a whole. It is a matter of common sense that if human beings, who are essentially pure and unblemished, by dint of misdeeds break the harmony by polluting any part of the nature, it must be treated as *unethical*. The crux of the environmental ethical action thus consists in the unblemished love of the nature and its protection from damages. It has been mentioned now and again that our misdeeds adversely affect our gross body and the subtle mind, it leaves effects not immediately know to us. The ill effects carry their potential of devastation to a long period from which we have to suffer one day or the other. The ethic of environmental protection is thus well realized if and only if the whole of nature is sacrilized—sacrilization changes human perception. It has to be noticed that nature does beg sacrilization—it is needed for human perception.

Let us now come to *Srimad Bhagavadgitā* for some deep realization of theistic ecological ethics. *Srimad Bhagavadgitā* is not just another book, it is a guide to realization and liberation thinks many wise men of India and elsewhere. We have said something about it before. The *Gitā* nicely goes along the faith that the Ultimate Truth of Reality lies in the existence of a transcendent Supreme Being on whom the whole of universe depend for existence and sustenance. The transcendent Self or *Ātman* pervades all of us, it dwells in every human being as well. *Gitā* asks us to realize this simple truth of existence and thus get eternal freedom from ignorance, greed and desire to be born again and again. Like many Indian ethical philosophies, the *Gitā* propounds the idea that the ethical life of human beings in particular may be conceived at two levels: the worldly and the ultimate. At the cosmic level, our ethicality demands infatuationless actions, that is, performance of social duties as per *stages* of life and training and temperament so that there is no clinging to the fruits of actions done. This is the social ethical ideal of *niùkāma karma*, which in turn, may be means towards achieving the higher teleos in social life, that is, to attain the knowledge of the Truth or the Supreme Self *(Ātman)* and achieve liberation or *mokùa* from the bondages of life. Socially speaking, we are obliged to maintain the world order and universal welfare which is expressed in the ideal of *lokasamgraha*.

In fact, *lokasamgraha* or the relentless function for universal welfare does not refer to human or animal welfare; rather, *"loka"* includes the whole of existence where we sustain life. Hence, the *Gitā* calls upon human beings to uphold their *"dharma"* or to do their duties qua human beings. In this way, they would also be able to safeguard their own *"dharma"* or, the ethical-spiritual nature inherent to a human being. It is evident that *Gitā's* call to wage a war if necessary for the sake of the protection of *"dharma"* is backed up by the idea that this is done for *"lokasamgraha"*. Hence, without having the knowledge of the sacred universe, the knowledge that God pervades the whole of existence, there are no welfare possible. The ethic of *Gitā* is thus founded on its spiritual philosophy of attaining the state of *Brahma* through desireless performance of duties. The protection of life and things around us is ethically desirable, and what is ethically good is justified by the higher spiritual teleos. It is, therefore, important to read some of the passages in the *Gitā* that clarifies the spiritual sanctity of nature and our duty towards its protection. In chapters VII and VIII we have many references to the sacred nature as:

In Verse Six: *"etadyonīnibhūtāni sarvānityupadhāraya aham kçùõasya jagatah prabhavah pralayas tathā".*[13] It has been said that all beings have their birth in the divine self. I (God) am the origin of this entire world and its dissolution as well. God is thus thought of as including the universe within Him, emanates it from and resumes it within Himself. The universe is thus His nature itself.

In Verses Seven and Nineteen in the same chapter, it has been affirmed that God or *Vāsudeva* is all in all—*"vāsudevah sarvam"*[14] and *"mayi sarvam".*[15] More interesting is to notice that the reference to *"bhūtani"* or beings in the earlier passages does not relegate *Gitā's* ecological ethics to ecocentrism or life centred because in Verse Eight we find God uttering: *"rasoham apsu kaunteya prabhāsmi śaśisūryayah praõavah sarvavedeùu śabdah khe pauruùam nçsu".* It is said that I (God) am the taste in *waters*, in the *light* in the *moon* and the *sun*. I (God) am the syllable *Aum* in all the *Vedas*; I am the *sound* in ether and manhood in men.[16] I think the holistic ecological utterance which is essentially divine cannot be more pronounced. It is not that the ecological community consisting of men and animals are divine in origin, everything, the sun, the moon, the light, water and even the ether filled with sounds and syllables are divine in essence. The problem is where does ethics come here? The *Gitā* has clear answer to this in the doctrine of *"lokasamgraha"* and *"sarvabhūtahita".* We have referred to these early on but we shall go for more details shortly. We already know about the broad outline of *Gitā's* double level ethics. Covering both, the social ethics of disinterested duties and the spiritual ethics of the realization of *Brahma* as the One and all. It has been stated in chapter VI, Verse twenty-nine that: *"sarvabhūtastham ātmānam sarvabhūtāni cātmāni ikùata yogayuktātma sarvatra samadarśanah".* It has been said that the *yogi*, whose self is in perfect harmony, sees the Supreme Self as pervading in all beings and all beings in the Self, everywhere he sees the same.[17] Interesting is to point out that Arne Naess in his "deep" ecology, quotes this passage to enable him to stress on the ontology of ecoethics as he is sure of the existentials trying to fulfil their self as most important moral journey. However, in this passage, we have also references to *act* like a perfect yogi, whose vision and action is of harmony as he has detached himself from all external things, fruits etc. and has devoted himself to performance of disinterested duty. Hence, the ecological vision of worth of all beings, with the spiritual vision of *oneness* has been nicely backed up by welfaristic as well as stewardly duties of an ascetic. The theistic ecological ethic of the *Gitā* is extremely clear. There may be a charge here that *Gitā* is still not free from Naess-type of anthropocentrism as the theistic and the ethical views both moves around human beings but not least has been said about the *"bhūtas"*. It is not enough to say that *"bhūtas"* enjoy equal rights but more important is to know whether the entire non-human existence realize the Self by doing something welfaristic. This sounds a queer question because from the *human point of view* nothing unconscious or semiconscious or poorly conscious can ever protect their rights, do their duties and realize Self. Is it so impossible? So laughing? Or, so queerly metaphorical? If it is laughing, then anthropocentrism has its ground. It is not difficult to find that some commentators of the *Gitā* has disliked twisting the above verse to suit to ecological ethical demands. However, it has been said by Samkara and Rāmānuja (who appears to hold a Gaia type thesis), that God is essentially a transcendental Being and having no doubt that the Being pervades all existence, the person having the vision of the *"yogi"* and who serves like a *"karma yogi"*, realizes the above Truth if and only if their selves are detached from the material existence or when they are freed from the shortcomings of association with *matter*. Modern commentators like Jacobson argue like this as well. Otherwise, many people have mentioned that the unified vision of the *"yogi"* is impossible. Truly, many people, who claim to be wise in reality act differently as they dualistically or hierarchically put one existential over another or do not behave with equanimity with lower selves or the lower castes. The *Gitā* does propagates the qualitative oneness of the whole of existence but does not take away the credit from nature that though clinging to matter, it does have the intrinsic *Brahmic* disposition to free itself to attain *cidātmarūpa* or the Absolute Self. The whole of existence may be seen as constantly striving for the consciousness of the God or to liberate themselves. They may be seen in their different ways functioning most disinterestedly to fulfil their selves. These things have been found in the revaluation of Sri Aurobindo of the *Bhagvadagitā* as explained in his conception of *"sarvamukti"* or "wholesome liberation of selves". For him the divine *involution* to the cosmic states has to be backed by cosmic *evolution*. At the level of "evolution", the nature is not biologically *evaluating*—the nature being in essence spiritual, moves spiritually. Hence, the higher *teleos* in the *lives* of existentials, their unique ethical progress is to attain the Absolute State. We should remark here that Sri Aurobindo's revelation of the *Gitā* does not make him a greater theistic ecoethicist or that he has overcome the threatening dualism and hierarchicalism eminent in the *Gitā* or in the views of its several commentators. Sri Aurobindo puts *Gitā* as it is, non-dualistic and non-hierarchical. Equally true is that the *Gitā* is not inert about the cosmic order, its oneness and its protection lest the conception of *"lokasamgraha"* and *"sarvabhūta hita"* are not overlooked by us. Here again, in the discourse situation the prescriptions are meant for Arjuṇa the disciple, by the teacher, *Kçùõa* and, therefore, sounds to be true for human beings. Yet, if we go by the broader implications just clarified, much should be read between the lines—the message may be that the whole of existence appear to evolve to *Kçùõahood* by means of both welfare of the *"loka"*, the existence as such (not human welfare) and the welfare of all *"bhūta"* (not only human beings). When existential evolves in tandem to the *Kçùõahood* where lays dichotomy? And where is the human orientedness? Most importantly those who do not see anything in the *Gitā* for building a theistic ecoethics or at least an ecology, like Lance E. Nelson[18], we have the suggestion to follow properly its *non-dualism (sāmkhya tattva)* and then look into its several passages depicting God-World relation and their oneness (*yoga tattva*), the ideal of ethical life (*niṣkāma karma* and *Brahmisthitilābha*) and finally the concepts of *lokaìmgraha* and *sarvabhūtahita*. More interesting it may be if these thinkers read carefully some sections of the epic *Mahābhārata* and *Bhāgavata Mahāpuraõa*, they would find many interesting elements in support of what we say here. The ethical ideal of welfare of existential has been well depicted in the chapter III, Verse twenty and chapter XII, Verse four. Whereas the concept of well-being of the ecological community by the performance of one's duties steadfastly (as a *sthitaprajña*) has been well explained in chapter III, Verses twenty and twenty-one and chapter V, Verse twenty-five. If one reads carefully the concepts of *"loka"* and *"bhūta"* combined with *Gitā's* conceptions of *sāmkhya*, *yoga* and *niṣkāma karma* tells us a simple truth—'though the world is God's creation, it is not with another material, the creation is one with God, therefore, all have worth, which needs protection by a person of equanimity who performs his duties not for petty gains'.

As I have said now and at the initiation of the Hindu theistic discourse that the ecological ethic of the *Gitā* should be evaluated not merely on the traditional lines of Samkara and Rāmānuja rather, on the basis of our own equanimous findings, which we most importantly forget because the *Gitā* or any theistic doctrine is not a field of verbal battle or theoretical churnings where most players are fighting a lost battle, for there is little or no self realization. The Hindu texts are those of realized people, they need more careful treatment. Yet, the *synthetic* nature of the Hindu texts tells us that many unrealized but sympathetic readers like us who are not at the gloss of post-modern ecoethics or post-modern theistic ecoethic looking at the loopholes of dualism, naturalism and hierarchism of the *Gitā*, would still take help from other texts as we will now do. We will now deal with what the epics, particularly Mahābhārata (of which, *Gitā* forms a part), and some of the *Puraṇas*. We shall then come to Gāndhi and Aurobindo for modern understanding (though something of Aurobindo has been said in confronting the *bheda tattva* of Nelson).

In several verses of *Srimadbhāgavata Mahāpuraōa* Lord *Kçùõa* utters the following:

"sarvam āvçtya tisthati", that is, "He (God) dwells everywhere" (13:13) and most importantly clarifies in Part II of the book that ether, air etc. *Mahābhūtas* and also earth, *all planets*, *all creatures*, *all directions*, trees, rivers, seas, everything are *organs* of *God's body*; and the devotees are expected to revere all entities for they dwell in God. It is in this context that in the same part of the Treatise there is mention of *"vāsudeva sarvam iti"*, that is, 'Lord is

all in all'—He is all pervading. Now following this theistic notion of nature as a whole, the *Purāõa* clarifies its ethic in no unclear terms. In the same chapter there is definition of an ethically good person or a devotee who is the one having vision of unity, seeing all creations as the presence of God and thus empathizes with existence as such. It has been thus said that: *"sarvabhuteùu yah pasyed Bhāgavadhavamātmanah"*. In the same tone in *Bhagavadgitā* in Chapter V, Verse eighteen, a wise man is explained as having equanimous view of all living beings and even the outcast to show that one is humble towards God. Though Nelson does not like this passage of duality and hierarchy (!) and thus does not find any ecological ethic either. One interesting matter to notice is that none of the old texts or *śāstras* of world's living religions directly refer to environmental science, environmental ethics, environmental crises and resolution of crises because we did not have such things at that periods but they do have some such sayings which pertain to exactly these things we now speak of. So it is all a matter of realizing what *śāstras'* inner implications are. Much depends on *heuristic understanding* as well as knowing the correct significance of the several aphorisms which in nutshell convey so many things that have engaged wise people since a long time for so-called proper interpretation. If we take this seriously, theistic ecological ethics is one possibility, which is strong enough. In the same tonality, in the epic *Mahābhārata* (the chapter on *Mokùasādhanā*), reference is of *"Sarvabhūtātmbhūtastho"*, that is, God inheres in all souls (living beings) but "deep meaning" reveals that the whole of the universe and the objects therein are an abode of God because all has been created out of Divine nature. The things are for everyone's welfare and the *ethical course of action* is to learn such judicious use of things for sustaining life and for flourishing that the rights of no one species is hindered. Perhaps to bring some sense to the *Purāõic* stories, to save the fallen earth (don't read ecological crisis of our times), God himself incarnated as fish, a tortoise, a boar, a dwarf and even a combination of man and lion. Lord Rāma in *Rāmāyaõa* and Lord *Kçùõa* in *Mahābhārata* have been supported by and also protected by monkeys and herd of cows. If we come to the family of *Durgā* and *Shiva*, they all seem to be with lion, bullock, rat, owl, swan and peacock. Not only this, Lord Shiva is said to have swallowed the poison churned out in the search of a magic potion that would give eternity. Shiva also carries the poisonous snake, the Ganges flow from the locks of his hairs, snakes adorn his neck and He is offered all types of poisonous things in devotion. Even in *Durgā Pujā* for all nine days of devotional rites, the goddess is hailed and offered different types of plants like banana, turmeric and so on. Well, we do not read much into all this but we need not trash them either. In the cultural adoptions of several rites, hailing from the Divine inspirations, we find deep relation with natural things and beings—a sense of interrelation, dependence, stewardship and selfless love does flow from all these queer stories. To take some more instances, in *Viùõu Puràõa* it is said that the God loves one who does not inflict unnecessary harm or destroys helpless creatures, and time and again it has been said that cycles of birth and death bind persons of unethical characters to lead them into a terrible life of pain and suffering. Most interestingly, the ecofeminist, particularly, the Gaia thesis churners would like to know that the whole of *"vasudhā"* or the earth is depicted in Hindu theology as "mother *Vasundharā"* and it has been said that the whole of earth community is one large entity, it is like a huge interrelated *family*. Hence, we speak of *"vasudhaivakutumbakam"*. If religious metaphors are no more nonsensical and not merely emotively meaningful to the post modern theistic ecoethics, the ethical course of action for the theist is to protect the earth community by sustained life, well-being and flourishing. It is interesting to mention that the theistic Hindu pantheon is so elaborate, deep and vivid (and yet so metaphorical), that the great saint Atharva had composed *Pçthvi Sukta,* hailing the mother earth and life on earth. It has been said in the *Sukta* time and again that all that exists are of equal value; there is no hierarchy as there is no intrinsic dualism in nature. So the ethical course in the earth community is clear enough. Welfarist duties and non-hatred are particularly mentioned for human beings.

Leaving aside the *śāstras*, something has to be said now about the more modern theistic ecological ethicists of India. We have chosen Gāndhi and Aurobindo to clarify the stance, particularly when both have powerful visions on *Gitā's* ecoethics, which has been touted as failing to have any such ethic, and we are trying to show that *Gitā* has one such ethic (the text notwithstanding). First to say is that Gāndhi was neither an ecologist nor official ecological ethicist. He was a thinker, a political activist, a social reformer, a law knowing and abiding person etc. but he did have an ecological vision and an ethic about ecological matters provided we read his *words* and *deeds* beyond their literalness. Though Gāndhi was a theist and drew much insight from the *Gitā*, he was profusely helped by the atheistic Buddhist and Jaina philosophies, particularly their social and ethical philosophies. The theistic justification of ecological ethics by Gāndhi stems from his basic principle of *"Satya"* or "Truth" about which he himself says: "But for me, truth is the sovereign principle which includes numerous other principles. This truth is not only truthfulness in word, but truthfulness in thought also, and not only, the relative truth of our conception but the Absolute Truth, the Eternal Principle, that is, God... I worship God as Truth only".[19] With the background of *è ta* and *Satya* at the outset of this part of Hindu ecoethics, it is clear that Gāndhi draws his inspiration from the theistic *Vedas* and *Upaniùads* to establish that God is the basis of existence. Gāndhi was aware of the truth that the term *"Satya"* is derived from the Sanskrit root, *"sat"* meaning "to be" or "to exist" and for that reason, he did not take it as spoken truth but conveying the sense of eternal Divine Self as the basis of all that exists. Gāndhi's *ethical* call for *"satyagraha"* is not a blunt political call rather, it is *"āgraha"* or earnestness towards the Truth or more clearly, an earnest (devoted) relentless pursuance of realization of the Supreme Being through thought and actions. The ethic of life is for a person—*"styāgraha"*. When this becomes a political weapon, it still does not lose the sense. Yet, Gāndhi is after the realization, the fullest realization of the Absolute, which is not possible but that does not prevent us from knowing it approximately in our earthy life. In this way we could combine the spiritual with ethical and make our abode firm. It is clear that the Absolute being there does not make much sense unless some *"āgraha"* or effort for realization is there. So the search for Truth is related a spiritual-practical life. As such, it may be realized in practical life when we practice non-violence or *ahimsā*, self-purification or *tapas* and self-rule or *svarāja*. In "My Experiments with Truth", Gāndhi says: *"ahimsā* is basis of the search for Truth".[20] It is thus a worthy means towards an ultimate end principle. Equally true is that *ahimsā* is embodied in the whole of existence because the existence as expression of *Satya*, cannot be anything less than Love and Compassion. In such loving and compassionate nature, the ethical guideline could be nothing less than that because we too care in and with nature. So *ahimsa* is the guideline of our practical life. It is an active force to live, love and realize. It is not mere negation of *himsā* or violence—it is selfless service of "others". But Gāndhi like many Gaia, feminist and ecocentric thinkers face so-called stiff challenge here. The problems are of violence in war and in case of killing a rabid dog. A learned reader and commentator of the *Gitā*, Gāndhi is unperturbed (as Leopold but in a different way), to say that the *Gitā* does not support war in *Mahābhārata*. The war is a *'dharmayuddha'*, which we face constantly in our practical life because we are at war always against evils or *asatya* and to tolerate it is not ethical. *Kçùõa* wages a war of righteousness because *Satya* has to prevail. We are by nature non-violent and when violence is needed for the nature to remain as it is, it is ethically just. Hence, the case of rabid dog or a snake has been carefully sorted out. Though they are not against us, they are against their own loving nature and for that reason, killed. Interesting is to notice that elsewhere Gāndhi remarked that he would not like to live where snakes are killed for they are dangerous for men, that they may bite or kill. He did not favour to go against nature, which is Truth itself.

From this it was natural for Gāndhi to deduce the truth that there is no inherent duality, discreteness and isolatedness in the nature as a whole. The universe is, therefore, a homogeneous and interdependent one where human beings have peacefully lived with animate and inanimate beings and things. In this universe Gāndhi does not find any point in seeing a cleaved human society on the basis of castes and religions. Once the basic ontology and ethic of life according to Gāndhi becomes clear, we may move to his conception of *"svarāj"* where his practical ecological ethic becomes more clear. Gāndhi says, a life of love and compassion as expressed in *ahimsā* is possible only if we abstain from such ways of living in cities and villages and abstain from such economic decisions and social policies which are averse to *Satya* or are violent. The ultramodern and the mercantile ways are thus not in accordance with love and compassion—they are against the peaceful environment and the earth community as a whole. Hence, some such self reliant, village based policies and economies are needed which sustain life on earth in most equitable way. It is due to this that self reliant village economy, *khādi* and non-heavy mechanizations were preferred by Gāndhi. To Gāndhi, *svarāj*, *svade-i* and *khādi* are pilgrim's progress to *Satya* because together they are some such ethical ways in life which cares for the sustenance of the whole existence.

Sri Aurobindo is his works like *Sāvitri* and *The Life Divine* has given a new dimension to Upaniùadic and the *Gitā's* metaphysics and epistemology. He also gives a new meaning to Advaita philosophy as propounded by Samkara. His views on ethics and his theism deserves special mention as well. For us, it is extremely heartening to find that his elaborations of existence, Absolute, conscious force, involution and evolution taken together projects an excellent account of ecological ethics based on theistic presuppositions. However, he was not consciously building one such ethics but our revaluation would reveal that he had one for us. To my mind, what is most interesting about Aurobindo is his deep analysis of the theism of *Gitā* which focuses its bearing on ecological ethics.

Aurobindo regards the Absolute or *Brahman* as the basis of reality, who is as such indeterminable and unknowable, that is, the Absolute is subject to our affirmations and/or negations. There is however no agnosticism here as the Absolute is known by our hidden infinite spiritual principle. Aurobindo thus makes it clear that his epistemology depends much on our spiritual cognition of reality—a unique experience or a unique realization. From his realization (and not theoretization) Aurobindo projects the Absolute or *Brahman* as *Sat, cit* and *ānanda* (*Saccidānanda*), that is, the triune principle of Existence, Consciousness and Bliss in our consciousness. Absolute *per se* or as self existence is manifested as *Atman*, *Puruṣa* and *Iśvara*, that is, Self conscious Being or Spirit and God or the Divine Being.

It is important to notice that *Brahman* or Supreme Reality is projected to our consciousness or Existence, and for that reason, many theistic thinkers like Aurobindo does not have any hesitation in calling the Existence as manifestation of Supreme Reality. One may argue that theism as seen in Aurobindo or *Upaniùadas* or the *Gitā* is only apparent, as there is no belief in God as the supreme reality rather, much is said about Absolute or *Brahman*, which is at best a kind of spiritualism or spiritual formalism. Aurobindo makes it clear at the outset that what we call "God", is not a dual reality rather, manifestation of God. Hence, in the theism of most of the *Brahmavādins*, God finds a place in the nature of the Absolute, though, to call it spiritualism is not a great mistake. Yet, it is not Platonic formalism of a contentless Absolute Reality. In fact, much confusion will be created owing to "spiritualism" in this case as atheists like Buddha and Jaina are also spiritual thinkers.

Aurobindo justifies that the Absolute is not pure Consciousness but *Cit-śakti* or Consciousness-Force which has the inherent disposition for the real dynamics of manifesting the Absolute spiritual form to existence itself and even to matter, mind, life and psyche. To be very simple, the Absolute is the basis of all possible and conceivable beings and things. Not only this, Aurobindo conceives of Absolute as Bliss and as Supermind. These notions are extremely important in his philosophy as conception of Bliss as inherent in all existence of every finite being and in the cosmos, make the so-called mute matter latent in spirituality. This latent spirituality of Absolute is explicit during the process of Supramental descent or *involution* of the Absolute in manifold things and beings. The crucial point of the initiation of the descent of the Pure Consciousness and Bliss to *ignorance* and the creative aspect of Absolute to take place, Aurobindo conceived of Supermind. It is through the Supermind that *Saccidānanda* manifests the universe out of its unity. It is the intermediary principle between *Saccidānanda* and the World, it reconciles the opposites of *Brahman* and the world. This is also the basis of Integral Philosophy of Aurobindo.

It is clear that *Sccidānanda* has to plunge into ignorance, a self-limitation for the manifestation of the world. The world of matter, life and mind though opposed in nature to the Absolute, is only apparent but not real because they are all self-limited and concentrated forms of *Brahman* failing to limit *Brahman* itself. However, these things are themselves imperfect, limited and divisible *now* but destined to achieve perfection when they evolve stage by stage to attain Brahmahood. It is evident that the evolution of matter in life, mind, spirit etc. is possible for they are inherent to the *Brahma*, the spirit, through involution. A latent force to evolve makes matter to evolve to higher forms. Aurobindo's conception of the spiritual nature of the whole of cosmic things and beings has been thus explained in the "The Divine Life": "The Divine descends from pure existence through the play of Consciousness-Force and Bliss and the creative medium of Supermind into cosmic being: we ascend from matter through a developing life, soul and mind and the illuminating medium of supermind towards the divine being".[21]

Aurobindo makes it clear that the truth of the *non-dual* and *non-hierarchical* philosophy as explained in the *Upaniṣadas*, the *Gitā* and the *Brahma Sūtra* (*Prasthānatraya*) should be rightly seen as an *integral* philosophy of the *Advaita*. Here the lower and the higher principles are all integrated in the Supreme Principle of the Absolute. There is no point in calling the World, an illusion. Aurobindo stresses on the *teleology* of evolution for it cannot be blind. Whatever is going on throughout the nature in terms of evolution is *liberation* of the spirit and matter as well—in fact, matter, mind, life all liberates in evolution for they are spiritual in nature and strive for their realization. The whole of nature is thus in an incessant moral journey, all performing what they should with a teleos of spirituality—the nature is not mute and lifeless, it is evolving, it is moving, it is tending to Infinite. This way Aurobindo puts the gist of the Upaniìadic theology and ethics in unambiguous way. It gives a new meaning to "liberation" as we conceive "wholesome liberation" of the existence (with existential), which is the ethical basis of cosmic reality. Aurobindo's theistic environmental ethics thus carries the simple truth that whole of existence being Spiritual and that all its components are striving for an ethical goal of liberating to the Infinite, all are interrelated, interdependent and all support each other for the ethical goal to be achieved. Hence, any human action that is averse to the spiritual evolution would be unethical for it will harm his or her own spiritual progress as will harm the ways in which natural entities help liberating themselves and "others". The holistic ecological ethics is at its pinnacle here with a theistic basis that clearly answers the problems of anthropocentrism and biocentrism.

If this idea of cosmic evolution is accepted, human beings do not remain ignorant and limited either, as their ethical goal is to realize the Infinite. Such evolving people reach a stage of evolved personality where their individuality is not at stake, which he calls the state of Gnostic being. It is here that we find something about the ethical behaviour of man towards nature. As Gnostic beings are guided to the Infinite realization by their inherent spirituality, there is no ethical error and evil. In fact, such people do not need any ethical principle to guide their actions. Such people are thus guided to an ethical behaviour towards other beings and things. "Marking and keeping all things pure... no separatist egoism inflicting by its ignorance and separate contrary will harm oneself or harm others, self driven to a wrong dealing with one's own soul, mind, life or body of others, which is the practical sense of all human evil".[22] Hence, for Aurobindo the essence of ecological ethics as based on theism is that the nature is left free to act according to its inherent spiritual *"svabhāva"* so that they all liberate together step by step, to a Gnostic status who like a *"jivanmukta"* would act for the cosmos to its spiritual upliftment through works and deeds. The deeds are ethical in so far as there is no egoism, no dualism, no exploitation and harm to any part of the nature.

The readers now awaiting a comparative account of the Christian and the Hindu theological environmental ethics would have to remain satisfied with mere exposition because the aim of the work is different. We are here more interested in knowing the different ways in which an environmental ethics is possible or may be justified but not to find out which justification is logically more strong. It may well be that none of the theistic and the non-theistic justifications considered so far are foolproof or are logically most strong or good enough taken isolatedly. A careful scrutiny by the readers of both the chapters (II and III) on "justification" would reveal that the anthropocentric, the biocentric and the holistic justifications have their own strong and weak points. It appears that holistic justifications of the non-religious/non-theistic type is logically the most strong one because the religious/theistic holism does not like the scientific approach argue that holism presupposes an ontology of unique type, the one immersed in theistic or non-theistic faiths. The inherent idea is to ask: how on earth do we come to realize that the cosmic existence is in entirety of worth safeguarding ethically? If we find an answer to this, we find worth of existential and the earth community as such because disregard to this truth grows some immorality, some sort of meanness towards nature and natural objects and we start to see them as of "use" and thus valuable, then the logic is not good at all for "using up" psyche has been proved beyond doubt fatal to ecology as such and thus unethical. To add "judicious" to it still raises some questions about the reasons for being judicious (moreover when is something judicious, is not a comfortable inquiry). The natural answer of the "using up" would be for the welfare of future generations, and then the question of "sustenance" would move circularly around "use" as the kernel, which may be acceptable by many non-religious/non-theists as well. If someone is more broader in outlook to say that sustenance of the whole of existence is ethically good in so far as the basic rights of nature are kept intact and doing so is just, equitable, good etc. one may still ask, why rights of nature is so important to keep intact and why at all doing so would be just etc? We cannot fall back on either rights as intrinsically good and thus desirable or keeping rights intact as "useful" for nature as whole because the first option leads to the question of knowledge of intrinsic goodness or worth of "rights" or "keeping rights intact" which might lead us to uneasy intuitionism or non-intuitive socio-economic, political, legal etc. explanations having their own compulsions to point ultimately towards "usefulness". The other option is evidently based on the "usefulness" thesis having no added greatness by adding "useful for all". May be owing to such compulsions of options and the problems in the concept of "use", the religious/theistic justifications tries to hit the central issue of why at all is existence so valuable? And why will it be ethical to strive for the sustenance of the whole nature? Nature as it has been looked upon as non-metaphysical in essence, has been thus glossed over by the religious/theists as thoroughly metaphysical in essence. With this the logical grounds to show why it is valuable and why we should care for it becomes evidently clear, thinks the religious/theists. The problem is, there are several other questions to be answered now, not so comfortable ones. The first among many is that when matters of ecology and ethics are concerned, why should metaphysics, religion and theism be the grounds of such things? Even if ethical concern about ecology be shown to be sensible, in what sense now metaphysics, which is beyond it, be dragged unnecessarily? Is it not true that without being religious and/or metaphysical some of us are deeply ethically concerned about ecological matter? What great addition religion and metaphysics does to ecological ethics when they are themselves a "lot of confusing and conflicting thesis"? Moreover, which noble intuition of whom would justify that nature is sacred or divine? Or, to take a more trifling one, how 'intuition' itself may be proved to show that whole of nature is sacrosanct in this or that way? If intuitions be the crux of the matter, why would atheistic Buddhist ecological ethics fail to prove its worth over the Christian or the Hindu? To give it a belligerent turn further we may ask whether or not feudalism and mercantilism themselves have given birth to the God of ecological ethics or the Godless Pure Being the same status?

I would not ask my already worried readers to worry more about such conflicting questions and replies. I would simply ask them to consider two things seriously. We are still in hand justification of ecological ethics as strongly based on both ecology and ethics about which we have discussed something and further we shall speak something more, but for that we have to wait till Chapter Five where we may find a new justification when the nature of people's participation in ethics becomes clear. When ecological ethics is not merely handmaid to theoretical passions, when it is on ground reality acting, as it should, we come to know why something called environmental ethics is possible, sensible and what it really amounts to. May be then we find replies to the problems we now faced.

NOTES

1. Santmire, Paul H., *The Travail of Nature: The Ambiguous Ecological Promise of Christian Theology*, Fortress Press, Minneapolis, 1985, p. 218.
2. Rasmussen, Larry H., *Earth Community, Earth Ethics*, Orbis Books, Maryknoll, NY, 1996, p. 184.
3. *Ibid.*, p. 185.
4. Paul (II), Pope John, *The Gospel of Life*, Random House, NY, 1995, pp. 60-61.
5. *Ibid*, p. 149.
6. Rasmussen, Larry H., *op. cit.*, p. 237.
7. McFague, Sallie, *The Body of God: An Ecological Theology*, Fortress Press, Minneapolis, 1993, p. 69.
8. Ruether, R. R., *Women Healing Earth*, Orbis Books, Maryknoll, 1996, p. 12.
9. Gutiérrez, Gustavo, *The God of Life*, Trans. J. O'Connell, Orbis Books, Maryknoll, 1991, p. 126.
10. Rasmussen, Larry H., *op. cit.*, p. 344.

11. *Ibid.*, p. 344.
12. Radhakrishnan, S., *Srimad Bhagvad Gitā,* Blackie & Son, Bombay, 1976, p. 213.
13. *Ibid.,* p. 215.
14. *Ibid.*, p. 220.
15. *Ibid.*, p. 215.
16. *Ibid.*, p. 215.
17. *Ibid.*, p. 203.
18. Nelson, Lance E. in C. E. Chappell and M. E. Tucker editors, *Hinduism and Ecology: The interaction of Earth, Sky and Water*, Oxford University Press, New Delhi, 2001, pp. 127-164.
19. Gāndhi, M. K., *The Selected Works of Mahātma Gāndhi*, Vols. 1 & 2 (eds), Shriman Narayan, Navjivan Publishing House, Ahmedabad, 1968, p. xxi.
20. *Ibid.*, p. 411.
21. Ghosh, Aurobindo, *"The Life Divine"*, Vol. I, Pondicherry, 1944, p. 333.
22. Ghosh, Aurobindo, *"The Life Divine"*, Vol. II, Pondicherry, 1947, pp. 857-58.

4

Some Other Problems in Environmental Ethics

Clarification of the "nature" of Environmental Ethics and its "justification" are to great extent available to us by this time. By this time, we have realized that environmental ethics is a branch of ethics, which deals with ethical problems pertaining to environmental or ecological matters. However, we have come to know that the range of "ethical problems" relating to environmental matters is not so narrow, it is rather, broad enough as we have said that environmental ethics being a branch of ethics, deals with important metaethical, normative and practical problems as well as problems regarding ethical values and virtues. We have obtained a fairly detailed idea of metaethical concerns in environmental ethics while we were discussing the problems of the "nature and justification" of environmental ethics. Though the double layered justification on the basis of ethics and religious-ethics have their own distinctions and problems. I have been non-committal as to which justification should hold well in logical terms. Or, does it amount to accept both justifications or reject them? I have said that without entering into weighing the logic of both justifications it would be better if we patiently wait till Chapter V, where we might be presented with a new set of arguments about justification of environmental ethics. However, it is clear beyond doubt that in environmental ethics, metaethical problems are of great importance as it is a matter of common sense that if we do not know about the nature and justifications of environmental ethics, we are at a position of basic ignorance about the subject. This is one way to justify how important 'metaethics' is within ethics because without it one is not able to justify how something is within ethics proper. Equally true however is to show that environmental ethical "problems" does not end up in discussing the above because if something is claimed to be within ethics, then it has to solve other "problems" such as: Whether or not environmental ethics involves *normative* discussion? And whether or not standard ethical norms need to be *applied* to resolve value-laden environmental problems? Let us now come to consider these, because we believe that we have still not clarified the ways in which environmental ethics can be said to fall within ethics proper. According to me when we have dealt with these "problems", we are also able to know something more about the "nature" of environmental ethics. Hence, to say now that environmental ethics is a 'branch of ethics' and it deals with 'ethical problems pertaining to environmental matters', is trite, more clarifications are needed.

Coming to the problems of "normativism" and "application", a careful scrutiny would reveal that these problems though different, are closely intertwined because when we ask in particular if environmental ethics is a branch of applied ethics, most of the time we have in mind normativism as the guiding model of applied ethics. We believe that if environmental ethics is a branch of applied ethics then, nothing but one or more normative theories like deontology or teleology are applied by us in order to solve the environmental problems raising ethical questions of desirability, good, right and just. Hence, we can take these "problems" together to find out what environmental ethics amounts to (or, what its nature is). An unequivocally clear answer covering both the "problems" have been provided by L. Westra like this: "Environmental Ethics appears at first to be species of applied ethics, like business ethics or bioethics, applying ethics to the problems of human interaction with environment. Unlike those disciplines, however, environmental ethics goes beyond the appropriate application of familiar doctrines to a certain species of practical problems: it requires that we extend or transcend our accepted moral doctrines because it forces us to rethink the boundaries of the morally considerable... Environmental Ethics is broader, more inclusive than other practical ethics; hence it is, in some sense, a new ethics, addressing as it does totally new problems in many areas".[1] What Westra tells us is extremely clear and simple terms need great care for understanding of the problems of including environmental ethics within applied ethics, where applying ethics is nothing but applying norms or ethics. It is also important to have a closer look at Westra's (or others) opinion that a species of applied ethics like business ethics does amount to an application of norms for its being a species of applied ethics. In this case, however, that is, in the case of environmental ethics, it goes beyond mere application of norms in business matters for we are not concerned with those problems of practical life of human beings in different societies.

An analysis of what Westra says and its evaluation will help in achieving our main objective as stated above but the evaluation of Westra's position and our conclusion demands some detailed reading of normativism in applying ethics and its related problems as discussed by me in Volume I of this Series called: the *Primer of Applied Ethics*. For the sake of evaluating Westra's position and to reach to our conclusion, an outline may be provided here, though its scrappiness may be excused. I have argued in the *Primer* and elsewhere that what applied ethics consists in, focuses what are its different species. I have also made clearer that applied ethics is a species of ethics because ethics consists in apart from performing the normative, the psychological and metaethical tasks, and the applied task as well. However, performing the "applied" task of ethics does not amount to applying norms of ethics by us because either ethicists or non-ethicists or both do not have any moral authority to act Chauvinists of morality and apply any one or more moral theory or theories to moral cases of practical urgency and revolve them or, to believe that in this way such cases may be resolved absolutely. I have argued at length to justify that in actual life when value-laden practical problems arise, deontology or teleology or whatever standard norms may be imagined, cannot and should not be artificially applied by anyone of us imagining that the theory is most suited or perfect in such and such cases and a resolution thus formed should be accepted by all of us in all times to come. I have argued that applying ethics is essentially free from "normativism" or "theory ladenness". What is crucial in this regard is to realize the truth that different people living in different socio-political-economic and geographical set-ups are also members of a vast *ecological community*, and they have both *moral experience* and *moral knowledge* by virtue of being members of this teeming community in which people act, react and intermingle in a web of dynamics to get influenced and to influence many people, life forms and even inanimate objects. The community life of people, thus gives them the essential moral experience as obtained within the mesh of actions and thus gives them moral knowledge of what is of worth or value, what is good, right, just, duty, obligation and desirable along with their corollaries. The 'moral knowledge' is, therefore, not necessarily those of 'moral theories' or 'norms' though any moral theory, whose abstraction by a trained ethicist or a wise person is a technical affair, nevertheless, demands the same 'moral experience' in this ecological community. Hence, I have argued at length to show that applying ethics is not application of ethical norms artificially but it may be done in another way, which I have called, "Intersubjective Corroboration". It consists of identification of value-laden practical problems by people of occupations and professions, but not only by ethicists. This is needed because as it has been said just now, all people, high or low, educated or illiterate have basic moral authority and moral knowledge to identify value-laden practical problems and participate in moral discourse to come to a closure through corroboration of views. This may be further given a legal shape or some concrete decision-making rules may be formulated for practical purposes. The process is a continuous one in which the moral dialogue may be initiated and moderated by ethicists on the basis of well-structured questionnaires. To avoid 'structuralism' here, the ethicists should always consult other knowledgeable persons to frame adequate questions to evaluate the moral views of people in discourse. Most important role for the ethicists is an *academic* one, which I have called, "post-corroboration analysis" in which the trained ethicists should analyse the entire moral discourse to find out the real dynamics of moral theories we already know or to find out some we are not acquainted with. This needs some teaching and propagation at the academic levels not for training, as did sophists of academic ethics, who like "engineers" solve problems, rather, to know the ways in which norms of morality function to be useful for social life.

Most important is to notice that applied ethics as it is or as it should be understood depends largely on the *model* of application and on *what* has to be applied if a model is accepted to be free from chauvinism and artificiality. It is clear that the "nature" of applied ethics, which counts for its "tasks" is fairly simple, it is *intersubjective corroboration* of moral views of the masses of various walks of life to resolve moral crises and formulate decision-making rules for people of occupations and professions. Clearly, 'what' ought to be applied, is our common moral knowledge of desirability, right and good but not moral norms as did the sophists or the casuists or the moral engineers deliberately. A moral philosophy, I said, should be morally based on some such models of application which has regard for people's ethical knowledge and corroboration of views provided democratic way of life prevails over those giving no liberties to common people in any serious social venture.

Another point is that applied ethics as we now understand it has close relation with metaethics and normative ethics. The first is true as we have to analyse the "nature" of applied ethics, the important concepts featuring as important tools for its understanding should be known for their meaning and significance and the logic inherent in justifying one model of application of ethics over the other. Applied ethics is also related to normative ethics but not "normativism" as we have to analyse the real dynamics and pragmatic value of standard norms in application of our moral knowledge. Not only this, applied ethics is closely related to "value" inquiry and to the issue of application of basic "virtues" of human beings. Keeping all these things in mind, we come to know that applied ethics is within ethics.

Now if we consider the problems of finding out the species of applied ethics, that is, what are the branches of applied ethics, we have to keep in mind the "nature" and "aim" of applied ethics as discussed above. Without having any disregard for what has been said and written (volumes after volumes) till now, I should stick to what I have found and what I have followed (to be discussed in Chapter V) about application of ethics. An important matter in this respect was discussed in the said book when I have considered the case of Professional Ethics and its relation with applied ethics. I have argued that professional ethics is regarded as species of applied ethics because apart from the resolution of professional value-laden problems (either in the way of engineering or in another way), it undertakes many things which applied ethics does not do.

In professional ethics one important task is resolution of ethical crises by applying ethics, but on the basis of this one does not conclude that therefore, it is a branch of ethics. We must understand that the metaethical task of professional ethics is quite different from practical ethics because the analysis of nature and kinds of professional ethics, analysis of key concepts therein and the justification of professional ethics is different from those of practical ethics. Most important is to notice that in applying ethics, unlike professional ethics, "group-affiliation" of a particular kind is not needed, that is, in applied ethics, we consider the mingling of trained philosophers and non-philosophers, the latter consisting of professionals, people having occupations and others for achieving the end of applying ethics but in professional or in occupational ethics, "group-affiliation" is distinctly marked. It is due to these reasons that dealing with biomedical ethics as a species of applied ethics may be closely related to professional ethics as the resolution of value-laden practical issues of the medical professionals. But applying ethics is one most important task of biomedical ethics though it is not everything as considering resolution of value-laden biomedical issues would not merely need application of ethics by a group of people only. Many issues of practical concern in this field cover people having different professions or having no professions' or having different occupations. Simple case of "euthanasia" would reveal that the problem is *not* of a medical professional rather involves, nurses, ward boys, attendants, the sick person, family of the person, acquaintances, lawyer and many others. In any case ethicists are also involved having some methods of resolution of the ethics. However, the same is not true when we consider educational ethics as a branch of professional ethics or engineering ethics as a branch of occupational ethics because apart from differences in other tasks, here the resolution of ethical problems by a "group" becomes important. Though it may be argued that in case of an issue like physical punishment in schools, different people other than the "group" of teachers and/or educationists are involved, and for that reason, like biomedical ethics, it is also within applying ethics, we may still show that that is not the case. For one thing we have just said that applying ethics is not taking up resolution of problems arising out of a particular field where a group of professionals are most concerned or are directly concerned, the converse is true in case of professional ethics. There is no doubt that though the physical punishment in schools needs for resolution of other parties, the problem is faced by a group in particular and the resolution and the decision-making rules are concerned with the group exclusively. Whereas in the case of euthanasia, it is not an issue only by the doctors because the resolution and decision-making rules cover many *people other than doctors*. Is there any doubt that the rules formed about euthanasia will cover the practicing nurses (sub-professionals), lawyers (professionals) and also common people who want such things but are not having any profession or even occupation? Hence, we come to realize why applying ethics is one important aspect of professional ethics but they are not synonymous. The question of the latter being the branch of the former, does not arise.

Moreover, environmental ethics is closely related to both applying ethics and professional ethics because there are emerging value-laden issues relating to environment which needs resolution by different parties and for different people, and also there are "group-related" unique problems for ecologists or environmentalists needing resolution for their group based work. It is also true that environmental ethics has close relation with normative and metaethical tasks. Shortly we will see that it has close relation with discussions about values and virtues. Yet, environmental ethics is neither a branch of applied ethics nor of professional ethics for apart from metaethical and other tasks, it does not merely aim at resolving value crises in environment by applying ethics and is only concerned with professional interests. Environmental ethics discusses nature-man relationship and ethicality in that relation as well as the ways in which though "nature" itself does *not* have a reciprocal or participative value in ethical resolution; it may be a "party" projecting the "influences" on it by human intentional actions. This is not anthropomorphism but scientific testimony to what nature reveals as such to us lest we have the acumen to understand it. Interesting is to notice that in the *Primer* my hypothesis of "intersubjective corroboration" as the proper model of applied ethics is largely based on the conception of human beings as participating ethical members of a teeming ecological community and influencing it in many ways. I believe that rather than arguing about environmental ethics as a branch of applied ethics, we should focus on the fact that applied ethics itself depends on conceiving an "ecological community" and "human beings having moral knowledge" in the community for further application. This may be a strong basis of professional ethical resolutions as well, particularly when environmentalists are concerned.

Concluding now, we must comment about Westra's passage about applying ethics, its species and that truly environmental ethics is not a branch of applied ethics, it transcends the tasks of applied ethics (also professional ethics) because apart from its call for non-normativism, it is broader in many ways than applying ethics. It adds a new dimension to ethics as well. Equally true is that environmental ethics having different morally considerable factors does not amount to say that it has nothing to do with a human venture resolving morally considerable issues by applying ethics because artificial normative obsession is *not* applying ethics. However, we must consider another problem: Is it not true that environmental ethics itself has assumed the normative status when it comes to resolution of value-laden environmental problems? Is it not true that though standard normative theory's preoccupation does not have much role in environmental ethics when it comes to value resolution of problems like large scale mining in tribal areas, environmental ethics, enriched with its different theories, some narrow, some broad and some elevated comes to the rescue? Most of these theories are based on human-nature relationship in narrow, broad and elevated terms. As such on the basis of anthropocentrism, the above problem is solved in one way (thought to be acceptable universally and absolutely) because the theory is the basis of justification of a viable ecological ethics itself and, therefore, acts as a valid maxim or principle as does universal theories of morality, each in their own ways to assume unilateral capacity to resolve *all* moral crises. But there are disagreements in moral theories as there are disagreements in environmental ethics (or we can say environmental ethical norms) resulting into proposals of biocentrism and holism as substitutes of normative principles to resolve cases like "mining". Hence, the decision-making rules or laws framed in legal terms, both owe their origin to some kind of a theory or norm in environmental ethics or the other. Clearly therefore, we should ask whether or not environmental ethics is free from normativism? It is interesting to notice that referring to the ethico-religious and metaphysical environmental theories or norms as the most "elevated" ones to resolve *all* environmental crises is nothing different from what has been said just now because they too assume the same universally valid and absolute norms which are believed to be sufficient aid to resolve ethical crises in environment. It is undesirable that the relatively young environmental ethics does have an inclination to build and apply its own "norms", which in reality are logical bases for justifying the ethic itself, and most quietly, the ethic, which had a hidden pride to shun normativism or at best to come out of its shadows, acted an incarnation of the fallen God. But what is the problem in doing so? The truth is that any decision emerging out of an ethical inquiry about environmental matters should be provided independently by the subject itself and not from outside. Moreover, the problem of "evaluation" of the good, just, right and duty of environmental decisions taken and actions done, will not be possible if the subject itself does not come to the fore. It is also true that the *sciences* of environment and ecology may provide us the *factual basis* of determining the locus of values and disvalues as well as the locus of problems—practical and value-laden. It may also be the basis of determining what repairs or make-ups *should* be done in what ways in the mucked nature. Hence, nature or natural facts themselves cannot be the bases of moral resolution and moral evaluation. We do need independent theories in environmental ethics. One might also argue that your (another's) justification in favour of having an applied ethics without norms is also based on the fact that man's life in nature is the basis of moral experience and moral knowledge which is needed for ethical application to resolve value crises; hence, environmental facts are truly the basis of our moral life.

Let us now come to the earnest efforts of some environmental ethicists to prove that environmental ethics should itself be the repository of norms for value-resolution. It is true that nature and natural facts are not the only grounds of moral resolution but they may be strong grounds to show the ways in which it is adversely or favourably affected by human international acts and affect their lives by reciprocating in different ways. Nature also stands as testimony to the intricate ways of life and web of natural dynamics which gives us the conceptions of the way life *should* be led and the natural states ought to be maintained. Truly, nature is the mother of all our knowledge of the "desirable" and the "good". I have deep faith in the truth that not a single moral theory or so-called environmental theory can ever emerge in a wise mind and take shape without the wise men being a member of the wonderful nature. It is good to follow the things that go around us and what should go around us. It has to be kept in mind that when I am speaking about nature as mother of all moral knowledge needed for moral evaluation and resolution, I am not justifying any environmental ethics. I am taking some cue from the basic environmental facts and phenomena, which have relevance to build norms of ethics, as much it is an aid to build an environmental ethics. I also accept that there is no meanness in coming out of the shadows of traditional normative ethics' theory-cooption, particularly when it comes to artificial application of theories by ethicists. What I want to suggest is that assuming an independent 'normative' status by the environmental ethical theories is always fraught with danger if and only if the said norms are built for the exclusive purpose of resolving value-laden environmental problem by means of ethical engineering by ethicists. We have to avoid normativism in environmental ethics of this kind. However, I am serious doubts about what is really going on around our esoteric ethical circles which is always obsessed with 'clan-normativism' and 'theory cooption in application of ethical knowledge'. 'Clan-normativism' cannot be ruled out as existing in environmental ethics. Take this from Westra: "environmental ethics is basic to social, political and economic policy-making and represents one of the major considerations required of business operations… when business practices are hazardous to human beings, through their environmental impact, corporations may simply respond by appealing to traditional moral theories to evaluate their activities".[2] What Westra asserts at the first part of the passage should be well accepted as it is true that ethical concerns in environment becomes extremely important when social, economic, business and political decision-making are concerned. However, it is also important to notice that applied ethics becomes important, too. It is also important to notice that ethical concerns in environment with regard to business etc. are not merely theory cooption or an artificial misadventure. Writing pages about environmental ethical concern in modern business would be futile intellectual gymnastics unless a proper model of resolving such concerns are not found out and made functional. Westra and other people having the conception of 'clan normativism' and 'engineering' in moral affairs thus asks the 'corporations' to refer to "traditional normative theories" for resolution and decision-making, that is, to find out what their actions should be like. There is no manipulation on my part here, as we proceed along the essay of Westra, normativism is imminent. It has been further emphasized that "utilitarianism" certainly acts in a limited way with regard to environmental ethical crises emerging from modern day business practices as communitarian maximization of good is desirable. Deontology too has great value as human rights and life value ought to be safeguarded for they are intrinsically good. Even Rawlsian "fairness" principle may be useful to stop environmental racism and to stop exploitation of the stakeholder. Here we are not at all interested in Westra's anthropocentrism or the like, we are pointing out how grossly the model of resolution and decision-making has been constructed and applied. Though it is comforting to know that many normative theories and not one might be useful in order to resolve environmental ethical crises, it is not clear how this simple truth has been discovered by Westra or others. We should accept the truth that any amount of intuition or wisdom or imagination cannot reveal this unless a member of this vast ecological community by virtue of his experiences regarding ethical matters and by virtue of his ethical knowledge does not come to resolve value-laden environmental problems not in isolation but in groups, small or large and thus contribute to a democratic life where decision-making is not "clannish" or "normativistic". *The dynamics*, I repeat, *reveals many theories* as involved in crises resolutions and it is not theories (one or many) which are applied by us reveals that they are important. Hence, traditional normative theories do not form the basis of ethical concern in environment. Then do we not say that the environmental ethical principles independently serve the same purpose?

We have seen that an ontological ecological ethics of Naess, the more deeply metaphysical ones, the naturalistic ethic of Leopold and even those of the feminists all draw some basic guiding norms in their ethic from the facts and phenomena of the ecological community as such, where all have to take major "shifts" from human orientedness to holism. Thus, none of the theories may be assumed to be absolutely true or useful such as anthropocentrism or biocentrism. However, the guiding principles may be constructed on the basis of holistic view which combines the theories of anthropocentrism, biocentrism etc. but the problem is, whether or not such principles might be really used as either thumb rules or impressive universal moral principles to resolve all sorts of environmental ethical crises? If yes, by whom and in what ways? Most of the time the principles constructed by the environmental ethicists are sweeping in nature, that is, they assume the status of universal moral theories find in the monolithic tradition of normativism. For instance, we find Rolston and Leopold constructing such principles. Leopold says, "A thing is right when it tends to reserve the integrity, stability and beauty of the biotic community. It is wrong when it tends otherwise".[3] The same is not true of more constricted theories such as of Naess and some others which enlist principles in the form of thumb-rules of morality to decide the do's and the don'ts. The greatest problem is with these rules and not any less is the problem with those of Rolston and Leopold if statement has been not made about the ways in which their principles are pressed into service. Otherwise, the academic value of these rules is of no use in ethical concern in environment. There must be no doubt in our mind that people showing ethical concerns in environmental matters are all supporters of activism, that is, they are not lovers of theoretical gymnastics but prefer to do some concrete things so that their real 'concern' comes to the fore. What if we are great thinkers, shallow or deep or deeper? We must know how to put our ideas into action. The question is that can our shallow or deep or deeper theory-cooption, that is, a kind of clan-normativism be of any use even if we are really proactive and not inert? The whole environmental scenario is almost like a house on fire, does composition of water and its properties help in extinguishing fire on the spot? Equally true is that we must know *how* to come out of this grave situation. What guidelines or models do we have for us? Thus far I am trying to press upon this crucial point that Naess' or Leopold's thumb-rule type or sweeping type normativism is of no use in the practical affairs or when the house is on fire if they mean to say that some wise people or many of them who knows or should know these environmental norms very well, nurture them as habitual states of mind, intend rightly, decide and act likewise, only then an active environmental ethical concern that comes to value-resolutions is possible, I am afraid to say that it is completely unfounded. In fact, ethical problems in environment are tricky matters; no clannish-theoretical approach can be of help. Number of countries across the globe, the third world nations, in particular, which have tolerated and is tolerating the brunt of the progress of other worlds, but infested with uneducated, illiterate, poor and helpless and yet not bereft of moral experience and moral knowledge, *should* be included in our practical efforts. Some intellectuals, some genius of high order who care so much for the principles in environmental ethics for resolving crises need to imagine that hundreds and thousands of morally knowledgeable people are always ready to fight out the crises but they do not have least idea about these normative theories. Yet, these people, a majority in fact, anywhere I dare say in the world, are theory non-co-opted but still have virtue enough to come together to resolve crises if we care for a model of 'corroboration of views'. When we do so, relatively intellectually well-informed people, the ethicists or others, when they analyse the moral agreements, disagreement and the closure, would be awestruck that Naess, Rolston, Leopold and many armed with their elevated theories have been nicely portrayed by these very ordinary ethical people. Where does this inherent power of the ordinary theory non-co-opted people come from? It must be from the ecological community itself, not from the heavens. A proactive environmental ethics is thus free from normativism, the one which we call, environmental ethical normativism. In any case, overdoing with such principles and theories by a party or some people is the same intellectual hypocrisy, which we find in applied ethical engineering. However, active people as Naess and Leopold and many others would, most likely agree to what I say, that is, they would accept the truth that theories are not merely handmaids of intellectual scrutiny but found from a "community" life where fire is fought by the "communitarian" public detached from chauvinism. Yes, there is no point in remaining intellectually bankrupt and illiterate. Still, one should remember that nations of high literacy and intellectualism, which scientists and Nobel laureates decorate with glamour, have been *actually* so foolish though they were no less voracious readers of high class theories of theoretical ecological ethics. Where did their wisdom go? In fact, knowing something good is not bad either, what is pointless is to think that 'first we know theories and then act', whereas theories themselves emerge in the milieu of dynamics—human and non-human. Let us accept the truth that any amount of theory cooption will not amount to ethically charged environmental thinker. Let us be wise enough to accept that the best trained mind cannot act as per one norm or many norms when it comes to resolution of crises. If someone does so it is artificial. The person is then a moral absolutist who does not want to learn from common people and must be a snob who does not face the non-philosophers or non-scientists who are believed to be 'without' moral knowledge for they do not know theories. An environmental ethics which is ethical itself should have an *ethical model* of resolution of moral crises in environment. This, I believe, is crucial to an understanding of the *nature* of environmental ethics, and it is here we also come to realize the close affinity among environmental ethics, practical ethics and professional ethics. Actual concerns in environmental ethics leads us to the acceptance of a model of practical application of human moral dialogue, where both the professionals and the non-professionals come together for resolution of several value-laden problems like building a mega-power project or mining in tribal land or many things like that. For the readers of this book, this conclusion is of special importance because the rest of the book relies on it.

Interestingly, we have slowly and surely discovered the "nature" of environmental ethics. We had discovered that being a species of ethics, environmental ethics has some metaethical functions to perform. It has to perform some normative functions as well, and most importantly, it has to apply a model of ethics in order to resolve value-crises in environment. It is time now to discuss whether in environmental ethics we have "value" consideration because that is one important consideration in ethics. Though we know many things about morally good, right, just and so on, it is believed that apart from other things, environmental ethics contributes significantly in understanding these value-notions in another way because we have seriously considered the opinion of environmental ethics that several values need a reunderstanding for ethicality is not merely a human affair though human beings have the ability to think morally and act, and they are also able to deliberate on the ethic of the welfare of non-human things and beings, about their rights and so on. To start with, I must clarify that what I have said before about the role of our life in nature and our conception of good, right, values and virtues as well as the valuable and virtuous ways of living life, which in turn are crystallized in the forms of some normative theories, clearly justifies that environmental ethics has to do with values and virtues of unique significance. There is perhaps no way to neglect these things in an environmental ethic.

Rolston is emphatic as is Fox about value to be found as objectively present in the nature in many forms—they are to be found in animate ecosystems and even inanimate objects. However, *knowledge* of 'values in nature' and 'value of nature' does not rule out the possibility that subjective determination of values are possible, that is, it is natural to impute some worth or value on natural objects or facts because that is the way human beings are and we cannot help doing that. It may well be that many animals like dogs and most of the higher ones, particularly domesticated ones have such subjective acumen to value human beings and their behaviour towards them or towards others but we cannot be sure about this or the way things happen at their level. What is of more value here is to notice that the values found in nature and as imputed by us are both possible, which carefully sidesteps the subjective-objective debate, which is rather outdated. We should believe that the nature and the natural objects can have both intrinsic and extrinsic values and the 'use-value' of such things does not make the user and the used higher or lower or ethical and unethical, provided natural objects are not *merely* conceived as those of human use, benefit and welfare. Rightly Rolston analyzes to find that the intrinsic value of nature as such depend on its being a 'self maintaining system', the several organisms are of the same nature, it does have a purpose which may not be felt and it has some goodness which needs to be protected for its worth. Equally true is that nature as such is good for community, that is, it is instrumentally valuable in a community for several reasons. Thus the parts of the whole of nature have the capacity to sustain the intricate system in their own ways, not in isolation, but in *relation*. Hence, each thing and being in relation to others becomes valuable not otherwise. This is true for the human beings as well. It is also of great importance to find Rolston urging the inadequacy of the traditional 'intrinsic-extrinsic value language' when we speak of the whole of existence, for the whole is the producer, the whole is the sustainer, protector and regenerator and reformer of all valuable entities imaginable. For Rolston, we may, therefore, use "systematic value" to speak about the value of the existence as such. It is in this sense that rocks, rivers and other non-living things though not having any capacity to defend their good, strive for welfare or ask for rights, have some value as being parts of the interrelated whole. Hence, we may speak of intrinsic values but importantly the interrelated whole urges for a systematic value. Values no doubt may be human-centred and non-human-centred, more important is that whether or not such things in nature urge for respect for their respective values, we have some *duty* to respect these values. With this we come to know that "duty" as a value concept need not be taken as something related to what we ought to do towards human beings but towards non-human entities and towards the nature understood as a whole. However, it may well be that duties of human beings may vary from time to time from locale to locale. Another point is that 'human values' and 'natural values' are vastly different and, therefore, humanistic approach to environmental issues are different because some issues involve humans, other involve non-humans and yet other issues involve both.

Most important thing to notice here is that a phenomenal change could be noticed in the ways in which environmental ethics has analyzed the notion of "value". It is due to this that human ways to understand the value notions like 'good', 'right' and 'just' have dramatically changed or is at least changing. Interesting is to notice that when we speak of things like "duty", it will be a kind of unethicality if we imagine that the different parts of the nature has no "duty" towards us or they perform no "duty". This is not anthropomorphizing nature or sheer imagination; it does not involve any poetic defect either, the interrelatedness of the system justifies that all species *does* something or is expected to do something for the sustenance of the system. Hence, there may be no point in championing the highness of human beings in terms of ability to conceive of duties. Truly, to ask whether rocks, trees and dogs have the sense of duty? Or, whether they perform duties or not? Are all matters of difficult inquiry but it is a matter of common sense experience that things and beings sustain human life, and this is important kind of "doing" as we believe. It is another matter that we feel equally responsible to "do" something for natural entities, but this does not prove that "duty" should be known for its meaning on the basis of what we do towards others.

Taking a cue, we can now realize that though rocks, rivers and dogs do not cry for their "rights" but we do so, does not prove that they have "no rights". Moreover, for the reason that we conceive of the "rights" of natural objects, these objects do not enjoy "rights", they enjoy rights by virtue of their inherent values in nature. We know that talk about duties involves talk about obligations and both are closely related to the talk of "rights" in the *moral sense*. The problem is whether or not *our* conceptions of duties, obligations and rights need some change for their understanding? The point is to conceive of duties, rights and obligations of the nature and the natural entities irrespective of human-oriented traditional understanding of these concepts. In environmental ethical parlance the morally good, right and just have acquired different meaning. It does not merely amount to human "language use" rather, phenomenal change in ethics-talk itself. This is to say that in ethics we do not now merely talk about human-human relationship based good, right, desirable and just, as also we do not merely conceive of duties, rights and obligation on the basis of human-human relationship. The simple reason is that human-human relationship in the ecological community is just one isolated relationship among many where human beings are seen to be closely related to animals, trees and rocks as well. In fact, relationship among more than two individuals apparently seems to be "human only" but that is not true as complex human relationship begs other non-humans relationships. We have argued that our lack of experience or understanding does not rule out the natural entities (living and non-living) "doing" something for us, which functions are generally for the sustenance of existence as it should be. But can we say that such "doing" and "acting" in any way are not subject to moral evaluation? The question is why should we support such parochialism? People reply that stubborn moral views, in the name of a new ethics-talk, have included some such things within moral periphery or moral discussion, which are completely out of place. Hence, there is no point in ethics to adopt a new talk like environmental ethical talk. I think we have justified environmental ethical talk beyond doubt but a final word is needed here. In the natural "system" the several non-human 'doings' or 'functions' or dynamics are not well intended or deliberate needing care for choice among the desirable and the undesirable, they are rather embedded in the existential as part of their nature, mostly, instantaneous, habitual and incited by extraneous conditions. Such actions are thus not like those calling for moral evaluation. Yet, this does not prevent us from conceiving non-humans as constituting a moral community for however instantaneous or incited their acts are, together and alone, they strive for a higher purpose as if, which is, "sustenance" of the system as such, on which depends our existence as well. Hence, their largely beneficial, desirable and valuable functions if overlooked by us or are destroyed by our actions, should certainly call for a moral-talk involving the non-human existence.

It is due to this that we speak of what is desirable and good by 'involving' non-humans. Apart from many evaluative assertions like "X is good" or "X is desirable" or "X is just" or "X is right", where "X" refers to human-human relation-based intentional actions, we now include some such instances of "X" where it refers to human-non-human relation-based actions where deliberations by humans are considered for evaluation not that of non-humans. Let me here repeat with some welcome risk that up to this date in the history of ethics, there is no foolishness in asking for a routine evaluation of deliberate actions and function of animals, trees and rocks, though we sometimes glorify what they 'do' to us. Till date, we evaluate human actions but there is a "shift" when we evaluate those human actions for morality and immorality which affect non-humans as well. It is in this sense (extremely qualified); we say that there are certain just and right actions done to non-humans. We do say that natural objects enjoy rights, *we* have certain rights to safeguard *their* rights and *we* have certain duties and obligations towards them as well. All in all, to be precise, our duties, rights and obligations assume significance when we talk about these things in case of non-human existence and not merely in case of human existence. There is, therefore, no respite from an ethic-talk in environment, as there is no respite from value and evaluative-talk in environment. A value-based environmental discussion is but natural. It is interesting to notice that if "duty", "obligation" and "rights" are taken in their strict *legal* significance or "official" significance, where there is institutional binding and sense of prize and punishment, even then the inclusion of non-humans is natural. People in civilized world having civil laws, imagine legal and official duties towards non-humans, they are believed to have some obligations towards them and no doubt certain rights or claim-rights (in the sense of paying what is due), are supported. We have seen these things clearly in the Hindu and the Christian environmental ethics. The non-religious environmental ethics need not be averse to such ideas. What I have said should be clear when we read this passage of R. L. Purtill (1976): "there are two classes of ecological obligations: The first, not to harm other human persons, including those yet unborn, by damaging the environment. The second, more controversial class consists of obligations not to harm animals or plant life, damage natural environment etc., where no harm to other human beings results".[4]

Going this way it is easy to conceive of double levels of understanding the entire major ethical notions and there should be no 'controversy' really speaking. But there are some controversies as many among us are far too worried about our own survival. These people are of these types—the omnivorous who, after mucking nature beyond recognition mug up what is suited for their good now but never care for environment. There are others who are environmental refugees and those exploited by the omnivores who do not prefer a tiger-project or a protected park as their own existence is at stake, and they believe that if people are not on earth, there is no ethical cry about environment. Yet, another group of people believe that rather than giving any importance to environmental values, including those having broad significance, certain human values like "knowledge" and "prudence" is of more importance. There is no need to continue with the arguments of the omnivores for any disregard for environment to care for them would be as Mill said, caring pigs than Socrates. The second argument holds a strong ground but the main question is, what would these people say when it comes to the stark-truth that they have been uprooted from what is their abode. Is it not *desirable* that the nature, which is mucked, gets back its lost treasures to some extent at least or most importantly, be now saved at any cost for there are no more exploitations and refugees? Even if it is true that when human life is at stake, they should be given much importance, but that does not rule out concern for environment. So we have duties towards nature, some responsibilities and obligations. Much need not be said about anthropocentric view supporting chauvinism. If knowledge is preferred over any other thing, the knowledge of environmental value may be needed to show that knowledge itself is incomplete without it. Hence, the above controversies do not make much impact on the 'value-talk' in environmental matters.

A final word about the concerns for "virtue" in environmental ethics but not before we have answered an imminent criticism that my support of environment as the basis of ethical value fall in the trap of "naturalism" or amounts to "naturalistic fallacy". Such charge may be levelled against Rolston and Leopold as well. However, we should first try to realize that though many of us support "naturalism", do not necessarily commit 'naturalistic fallacy' for we may not be either *logically deducing* "ought" from "is" or, *logically defining* "value notions" in natural terms. Leaving Rolston and Leopold, I must say that I have no way to be apologetic about the truth that devoid of natural existence, man has no knowledge of moral values. What does that prove? Am I trying to logically define any 'value term' like "good" in natural terms? Or, am I trying to logically deduce what ought to be done on the basis of what happens in nature? In fact, without taking any recourse to such technical things like logically defining and deducing, a naturalist might well hold his position by virtue of common sense, and some degree of heuristic understanding or informal ways of interpreting things that devoid of nature, there are no values though there may be numerous human beings, extremely wise by birth but live in a big void—in that case, there is no ethics, and as such, no environmental ethics.

Another charge may be levelled that my view amounts to 'anti-theory' stance much like those in 'virtue ethics'. There is no doubt that only 'virtuous' people care for the nature and themselves but vicious people do the opposite. However, for that, there is no call for virtue teaching, preaching and inculcation methods as advocated by me as there is no call for normative theories to evaluate our virtue of applying our ethical knowledge. What I have said is again a matter of simple understanding of the fact that virtuosity comes from our ways of living in this community life and need not be necessarily taught. Teaching certain virtues may not be liked by a group or groups of people as there are disagreements about racial and social virtues. Certain virtues like 'caring the environment', 'benevolence to natural entities', 'judicious in use of nature' does come from our community life, which should be maintained at any cost. Interesting is to notice that it is in this queerly complex life we live in nature that we came to know what are vices. The point is *should* we like to maintain a level of vice in our nature as we have experienced or should we shun it? It will be hopeless, therefore, to think that in nature we know about values and virtues but not vices and disvalues. It is so simple—if you look around, you will find people acting virtuously and viciously towards other and they are caring for both values and disvalues, but the problem is one of 'ought' or 'should', which is again a matter to be known from nature lest we care for the *consequences* of the virtuous and the vicious actions, and of disvalued decisions and actions. It is but natural for a naturalist to care for such consequences and build an empirically based moral philosophy, which is moral. Criticism against naturalists does not die as more charges like: "The knowledge of such and such things in nature around us are virtuous and others vicious depends on the proper criteria of evaluating virtues and vices, and when we say that the 'consequences' of human actions are sufficient empirical criterion, we speak of certain *quantitative and qualitative goods for man and existential in nature*, we are there taking the help of nothing but a normative theory called welfarism". Hence, we are either "consequentialist" or "welfarist" and in both the cases we are "normativists". The charge thus stated simply is: "naturalists forget that their obsession with empirically based knowledge in nature being the only basis of knowledge of virtues, values, vices and disvalues, accepts normative theory or theories". Almost on the same basis W. K. Frankena argued in his book "Ethics" (1988) that virtue theorists forget that normative theory or theories are needed to claim that something is a virtue or someone is virtuous. Frankena taught the obsessed virtue theorists (and us) an important thing, which we do not contest at all. Yes, there is truly no obsessed 'anti-theoretism' possible in virtue ethics because it amounts to beg one monolithic principle, not leaving much logic to support it. However, what lesson did naturalists learn from all this? We must remember that our general stand about ethical principles and ecological ethical principle does not change—they are based on "nature". We do not want to disagree that naturalism amounts to support consequentialism and welfarism for we have no obsession as monolithic ethical thinkers have. We will soon discover that *ethical pluralism* is natural to a naturalist for nature cannot be basis of discovering a true ethical knowledge. We will also discover at a proper place that ethical pluralism governs ethical application. An empirically based ethical theory while speaking about virtues and values would tend to agree that there may not be a need of any one but many bases for evaluating something to be virtuous and vicious but we must be careful here. In actual dynamics of such ethical evaluation, *we* (the best trained minds), do not and cannot consciously take recourse to a basis or some handpicked bases first and then declare virtuosity and viciousness. All these happens to us by virtue of being members of an ecological community though regional, racial, local and temporal contexts having their uniqueness does influence us because these are parts and parcel of both the thesis of naturalism and ethical pluralism we have been advocating. This is not to say that in changing space and time, region and locality there are no objectively valid or universally true theory and that there are no universal ethical principles. How can we forget that the universal value of "natural sustenance" is so important for our ethics? What we stress upon is that relative or universal ethical truths and theories all emerge in nature and that they emerge in nature is a matter of discovery. We simply do not disagree that there are not ethical theories but we disagree that there is not only one ethical theory. We do not disagree that theories are *needed* in our ethical life for several evaluations but they are so 'needed' is known to us when we find that we really needed them in such and such practical cases. We disagree that these principles are not 'needed' by us *artificially* either to evaluate something or to resolve value-crises as pliers are 'needed' to hold and cut electric wires, though we should never forget that people around the world have been using naturalism and welfarism for value-resolution and decision-making. Most likely, they will continue to do so but we do not argue that only this will be the course in future.

Coming to virtue ethics and anti-theoretism, I would like to say that it is old-fashioned to conceive of virtue ethics as necessarily 'anti-theoretic' or anti-normative because we know that to call some decision and actions having virtues we need some *basis* and the basis (one or many) are nothing but some normative standard or standards. Moreover, virtue-theory in any form, agent focused, agent based, agent centered or whatever is most concerned with an ethics based on human-human relationship and thus focuses the human virtues, the numerous traditional ones as important in our moral life and decision-making. If virtue ethics is just this, readers by now have realized that though in essence, virtue ethics need not be human centered, certain human virtues lie in the fact that people decide and act for the sustenance of the ecological community. Again, this is something we learn from our experiences in the community life we live. I do not reject the relevance of virtue ethics that speaks of three things: (a) certain virtues are basic for the existence of the nature as such, which needs prime care; (b) any human action that does not care for these virtues, is undesirable and (c) virtuous human is not virtuous by virtue of theory cooption or racial, social, legal etc. cooption rather, by virtue of *cooperating* with human beings and other natural things and beings for just ecological existence. Moreover, when it comes to decide what ought to be done in resolving value-crises, teaching values and inculcating values, it would be important to apply basic virtues in us in cooperation with people of various castes, classes, clans, status and earning. One important virtue needed therefore is to shun ethical chauvinism.

The last assertion takes us to the last problem to be considered here, that is, what should be the *model* of value-resolution and decision-making in environmental matters. We have just stated that over and above theories and 'officially' trained and propagated virtues, people's cooperation is needed. Here, the most crucial point is to consider whether 'rustics' can resolve value and become decision-makers? Let us face the problem so that when I move on in the next chapter to show what really has been *done* by me in this field there is no confusion about "models". The most trifling matter in the resolution of ethical problems in environment and to formulate decision-making rules is who we should entertain to do such things. In the Primer of Applied Ethics, I have said that to have an artificial 'division of labour' in applying ethics, that is, on the one hand giving some responsibilities to the trained ethicists and on the other hand to some non-philosophers, who are 'suppliers of problems' is a gross idea. Careful thinking reveals that the difficult task should be performed by both these parties. There is no doubt that in our societies some people (the non-philosophers included) are directly related to and concerned with value-laden practical problems and the problems being largely "their" or those of the professionals in different fields, these people cannot and should not be neglected. But there is a problem here. Let us consider value-resolution of environmental problems like systematic logging practice in a tribal land infested with rain forests. In this case the ethicists, the foresters, the workers in logging practice, the government and the private agencies doing business and most important is that the tribes or the village people are most concerned about such practice. They want to find a way out of this problem which is "morally charged". In fact, many among them believe that the problem is just not another practical problem; it is rather, an ethically charged problem needing some care. In such cases, the educated, trained and knowledgeable parties are the ethicists, the businessmen, the agents and so on but workers and tribesmen and villagers may be not so educated, knowledgeable and trained, though they are equally concerned and have some degree of moral knowledge. In fact, some so-called educated may not be ethically charged, they may be pragmatic people or prudent as we call them but others may be morally oriented and prudent in yet another sense. Naturally, there has to be disagreement in moral matters and in resolution but we have to face the situation *together*. Most of the time environmentalists face this problem and their intellectualism along with the same of the ethicists and the other parties, club to form an 'elite group' trying to avoid the imagined uneducated and non-intellectual "non-decision makers" in such cases. Any model of value-resolution and decision-making has to take up this dualism first. Who are the decision-makers? In democracy, who should be the decision-makers? Does less education or no education, a shame and failure painted dark on the government machinery, the policy-makers and the implementers, a great block in way of *people's participation* in value-resolution and decision-making? From where do people's representatives, the policy makers of nations derive power? If they are the same ill-fated uneducated as well as educated masses, can one be neglected for his opinion for no fault of his? It is not that unless we know what each one of the tribes say about the ethical aspects of environment matter, there is no resolution; rather, their representative views should be accommodated for the *moral debate* leading to *moral agreement*. This is not easy task to do. An ethicist has immense responsibility to initiate moral dialogues and to conduct them not in the name of post or position, priority or education, training or competence.

This issue is, therefore, most crucial when much has been said about the 'theoretical' aspects of environmental ethics. Holmes Rolston, Andrew Brennan, Robin Attfield and others have touched this issue which has taken a peculiar 'economical turn' than remaining "ethical". No doubt a lot of economical decision-making looking at the present and the future generations of human beings are involved but such human-centrism does not really help when it is an environmental issue. The non-human existential should be eared for their benefits, and well-being because, the future of their population are involved as well, whereas the tragedy is that they show no interest in such resolution and decision-making. It may well be that economically oriented resolutions and decision-making take serious notice of the poor, the exploited and the tribes men. Equally true may be that the political machinery is active enough to participate in the economic weighing mechanisms but their sympathies cannot neglect different moral opinions about the matter. Interesting is to notice that when the debate is morally oriented, it cannot be an exclusive economic-political affair. What "ought" to be done is always prior to the several benefit weighing, economic weighing and welfare computations. Though the technical weighing may be highly intricate to follow, the concrete 'steps to be taken' are unambiguous, about which we should debate. Our ordinary folks are not expected to talk only about the "ought" aspects, they do take serious interest in the outcomes of several weighing and decisions because they further raise questions of 'desirability', which do not escape the simple morally charged minds of people.

An environmental ethics to assume a full-fledged shape, therefore, should deliberate on the dynamics of value-resolution through people's participation, which is explained in the next chapter. Several queries about normativism we raised before find an answer here. It is here that considering non-intellectuals seriously in moral matters come to the fore.

NOTES

1. Westra, L., Environment and Environmental Ethics, in P.H. Werhane and R.E. Freeman (eds.), *The Blackwell Encyclopaedic Dictionary of Business Ethics*, Volume XI, Blackwell Cambridge, 1977, p. 205.
2. *Ibid.*, p. 207.
3. Leopold, Arthur, *A Sand Country Almanac And Sketches Here and There*, O.U.P., N.Y., pp. 224-225.
4. Purtill, Richard L., *Thinking about Ethics*, Prentice Hall, N. J., 1976, p. 109.

5

Some Environmental Ethical Issues and their Resolution through Intersubjective Corroboration

Our inquiry into the nature, justifications and types of Environmental Ethics gives us an idea that it is a branch of Ethics known for its specific interest in Environmental matters, that is to say, it has ethically relevant interests about environmental issues. In showing such interest, it takes up metaethical task, normative task and applied task. However, some interests in virtues and values may be noticed as well. We also found that environmental ethics is not monolithic as no one norm is supreme and sweeping rather, we conceive of narrow, broad, broader and extremely broad types within it. Though we did not rule out the possibility of justifying a most acceptable form of environmental ethics or what we may call most sensible way of expressing ethical interest about environmental matters, called "ecological holism" or simply "holism". Yet, we did not accept that "holism assumes the form of a universally valid ethical principle so that numerous ethical impasses might be resolved by virtue of applying this norm for all times to come. We have to understand that 'ecological holism' is one among many types of ecological ethical theories which in the main argues in favour of the possibility of or justification of ecological ethics itself, perhaps this is noticeably the most acceptable justification among all. However, that does not show that it is one among many norms that may be applied in cases of practical ethical urgency. We have said that when it comes to the applied ethical interest in environmental impasses, it should not be 'artificial' application of either a standard ethical norm to resolve the said imbroglios where parties involved are some chosen professionals and ethicists but not many others who come to a consensus through dialogue. I have argued in favour of "intersubjective corroboration" as a model of applying our common ethical knowledge about ethically right, good, desirable and just to resolve such crises. Let me deconstruct the model further for it is time now that we know the way in which environmental ethics becomes truly dynamic but not merely theoretical, and that its dynamicity is not concentric rather, "asterisk" as it spreads in all members of the society to reach out to more and more value-laden issues constantly. This chapter thus concentrates on the real dynamics of ecological ethics by considering some issues of practical concern for resolution. The chapter deconstructs the whole of the process of resolution therefore. Yet, this work cannot be encyclopaedic to cover as many issues for resolution or all issues. In this chapter I will take up three extremely important value-laden issues and explain the way "intersubjective corroboration" has been effective in resolving them. With this, we perhaps come to the fag end of discussing what environmental ethics should be like.

Intersubjective Corroboration should be deconstructed for its nature under the following sections:

A. Choosing a value-laden practical problem like value-laden ecological problem for resolution in a particular social scenario;
B. First hand or basic information of the problem and its importance;
C. Collection of theoretical information, that is, the moral opinions of different people;
D. Preparing questionnaire for moral dialogues;
E. Conducting dialogues and collection of views relevant to the analysis of dialogues;
F. Post-corroboration analysis;
G. Suggesting decision-making rules and working for their acceptance;
H. Theoretical information to the intelligentsia and other people interested in knowing the real dynamics in

ecological ethics and its several findings in different social institutions like schools and colleges; and

I. Always keeping an eye to disagreements in decision-making and consensus.

First point to notice is that all these nine steps taken together as constituting the model of value-resolution is important not for an ethicist or a trained academic philosopher, but for those who want to enter into this task. As such, there is no way to prevent non-philosophers such as the environmentalists to assume the role of a corroborator but the analysis of dialogues and explaining the ways in which several normative theories emerge as active is a technical affair needing sound knowledge of academic ethics, as also building up questionnaires for moral dialogue need some acumen for ethics. Hence, the model suggested here though not meant for a class or a group does have some role at least for ethicists. Yet, these things can be done by other interested parties, the non-philosophers if and only if they have basic knowledge of academic ethics. Similarly, an ethicist should have some basic knowledge of the 'problem in consideration', in this case, environmental problem, so that proper questions are framed, parties selected, and to follow properly the technicalities of arguments of different professionals and intellectuals. It is due to this that the model is wide open as an "interdisciplinary" effort. Here, one counterview charging some degree of "structuralism" might emerge. First is that this model does not leave room for people having no knowledge of ethics and/or fields like environmental studies, where it has to be applied and is, therefore, structuralism if not chauvinist. Second is that framing questions, selecting parties, conducting dialogues, analysis of views and suggesting ways are all left to ethicists and/or people like environmentalists, leaving no room for ordinary masses, which amounts to the same structuralism and not decentralism.

There is really no way to prevent untrained, non-intellectuals in the actual practice rather, we have claimed that we all have basic knowledge of morality by virtue of our experiences in the world. We should not forget in the mean time that "models" are meant for practical purposes but their construction, justification and application are largely "intellectual" matters to which non-technical persons do not and cannot participate. Though well meaning models have to be initiated by someone at some point of time at least and most often than not the common people, the toiling mass do not have the time or energy to do such things. Or else ordinary folks, the tribesmen for instance, might have their "unique ways" to resolve their crises, which are not official "models". However, these ways may be pragmatic and therefore, suffers from some defects but not widely applicable. Nevertheless, the ordinary people might also actively participate in an ethical venture if called upon and most likely they turn out to be the most important parties in moral dialogues that pave way to resolution and decision-making. Hence, the deliberate liberties given to the ethicists and the environmentalists in this case is to at least start from somewhere than going nowhere. Interestingly, "intellectuals" and "intellectualism" need not necessarily amount to "structuralism", it all depends on the approach and aim of intellectuals and their toils in the society. An approach and aim that is for the people and by the people so that a qualitative community life is sustained, a life that cares for need of the ordinary and the poor as also urge for growth, is not structured. Let us not forget that theoretical knowledge of the intellectuals is structured if it is class-constructed or individual constructed but does not care for both the empirical truths in nature of which people's knowledge is a part or the opinions of the technical and the non-technical people. Why should we always imagine that "social scientific" methods like this one, which owes it origin and initiation to some intellectuals be individual oriented? There is a way when choosing a problem, constructing a questionnaire and coming to consensus in a dialogue is taken up by the intellectuals who care for the basic empirical truths around him or her than getting stuck to the theories known to them. It should be remarked here that the best known iconoclasts like Gandhi, Marx, Engels, Lenin and Mao who always fought against structuralism and anti-people intellectualism were themselves great intellectuals and did not least but did lots of things for the people and with the people with the help of their model and their theory. There is little doubt that Dialectical Materialism and Communist activism are brainchild of intellectuals and not that of any ordinary folk but that is not structuralism as there is not a shade of classism and individualism as well as anti-people attitude. Most importantly, their constructions did not lose sight of the empirical truths around them in which we find what ordinary folks think and do, lest we are informed and take care of them and not care only for some great people who have their intuitions.

To conclude, we may say that to save "intersubjective corroboration" from "structuralism" the initiator of the model for application needs to be one of declassed mentality, that is, free from the obsessions of his or her so-called true knowledge or valid knowledge and the idea that common men have no authority to apply any part of the model conceived by them. In fact, a careful scrutiny will reveal that the parts constituting the whole of the "model" owe greatly to the ways of life and ways of thinking and acting by different people.

This compels one to write that the first step, that is, "choosing" a value-laden practical problem in the field of ecology is itself a matter of keeping one's eyes open. If we are blind to what goes in and around us in this vast ecological community and we are not conscious about which problems are not stark trouble-making events rather raise serious questions of "desirability", "ought" and "should", we are not able to "choose" one such problem. In social isolation and intellectual hegemonies one cannot choose some problems in social life needing urgent call for value resolution. Most important is that our societies are complex and huge, if one is restricted to one's own clan, class, caste and creed, if one is stuck to fixed strata in society, then he does not know properly which problems need immediate care for value-resolution. Hence, "choosing" a problem is tricky affair. However, the efficacy of this model depends on the concerns of the specific people of a nation and a state. We cannot escape this because resolving the value-crises of creating plastic gardens in India may not be as urgent a value-laden environmental problem as is the problem of building a high dam across a river for a multipurpose project. Needless to say, there are some problems which have become globally urgent. Further, there may be problems of different states and different places within a nation, each demanding care for resolution; "choosing" a problem has to depend on such urgency as well though its resolution might have wide national impact. There is no doubt that the resolution of the issue of execting high dams may be a general problem in India or elsewhere, still, choosing the case of Narmada Project would be different from choosing the case of

Koel-Karo. Though I have used this "model" as a starter and cared much for some problems of national interest. Let us remember that choice of a problem for application of intersubjective corroboration to resolve it takes care for views of mass, not class. Environmental ethics in this sense is not 'class ethics' but 'mass ethics'.

Further we ought to have some basic information about the "problem" chosen for ethical consideration because we have to know the nuances of the theoretical, the practical, the technical and the ethical aspects of the "problems" clearly. One who does not know properly the cases of multipurpose projects, their objectives, about the dams, the factors involved in its height and extent, and the socio-economic, geographical, political and cultural issues involved with the "problem" to some extent at least, there is hardly any understanding of the ethical aspect involved in it, though this aspect is clearly different from others. This is needed not so much for an "ethics-oriented" questionnaire preparation rather for understanding and evaluating the opinions of people active in moral dialogue as well as for the stages of disagreement, agreement and closure of the dialogue to construct consensual decision-making rules. It is very important to notice here that the so-called non-intellectuals, including the illiterate mass, have low or high degree of awareness of the said technical matters, particularly pertaining to their locale and in some cases to the nation as a whole. There is no chance for snobbery for us to believe that "we" best understand them or are the best parties.

The next step is still more important because now we assume the important task of collection of as many moral opinions as possible by either talking to people directly related to the resolution of the problem or reading literature available about such things. To make it more clear, an ethicist or a non-ethicist interested in environmental ethical dynamics should first perform some "theoretical" task, that is, gather some first hand information about what some concerned people think about moral matters. This helps us in two ways: first, we have an idea about whom to approach for a moral dialogue and second, what types of questions should we put in the dialogue so that best comes out and yet participants do not feel uneasy. In fact, I have seen that many highly educated people have some hesitation in speaking directly about ethics because that is too technical for them but they would fluently go on unfolding their moral opinions when "ethically oriented" questions are put to them. We are thus free from "structuralism" when in the construction of questionnaire some homework has been done, otherwise "our" intelligent and interesting questions may not be liked by ordinary people, and they would quit the dialogue. It is due to this that best minds in philosophy and ethics are not best people to participate in a moral dialogue about Narmada or Koel-Karo because they are highly "theory co-opted" and they appreciate intelligently founded questions with much hegemony. This is true of those ethicists who do not go for the "homework" and frame intelligent questions for intelligent people and talk at length believing a great duty has been performed. Volumes of books of applying ethics are a testimony to such "intellectualism". Interesting point is that "reading" certain materials, mainly those published in newspapers and elsewhere gives us idea about what many well read people believe about the subject and how to accommodate questions which would be relevant for all participating parties. It is easy to understand the limits of the work undertaken by us as there is no complete information and there is no most satisfactory questionnaire available to us.

In the preparation of the questionnaire, we should be very careful. It should be first of all fairly straightforward and simple, clearly understandable to the illiterate participant if needed. It should be framed in a language understandable to all parties, and if there is a need, it should be translated in the local dialect. The questions in this case should be 'non-technical', that is, not much of theoretical jargons should be dumped for the participants. Questions on multipurpose projects or whatsoever should be also "ethically" relevant for here we are concerned with the problem of whether or not certain things "should" be done and not what technicalities are involved in it.

Next, we arrange for one or many dialogues with one party or many parties (sometimes called "panellists" in seminars). For this we should depend on a person or a group of persons (ethicists and others) who have been responsible for undertaking this task, done the homework, prepared questions and selected the participants. These people should be better called "corroborators" or "moral corroborators" who put questions to the parties, carry on the dialogues in steady direction leading to a consensual opinion.

During the actual discourse there are heated debates; questions are put by one party to the other and also by several others who listen to it. This makes the "question asking" task of the corroborator easier, more people-based and thus sidesteps "structuralism". However, the participating party cannot be cacophonous; it needs a direction which the corroborator needs to control by applying his or her wits. There are delicacies here but you have to be honest to the aim of the dialogue and you should rule out the unnecessary talks of any party. The dialogues may be gruelling, and my experience is that it can go for 3 to 4 hours without even a notice of the period spent.

The discourses should be recorded mostly in handy tape-recorders and also in writing for the "analysis" of the views and for informing about the several suggestions and decision-making rules. The "analysis" of the moral dialogue does not help much the people who are really concerned about a problem in consideration if they are not also well trained in ethics or at least have some interest in its theoretical matters. Yet, "post corroboration analysis", as I call it, is useful for all *academic* purposes, which is certainly *not* anti-people by any imagination of the term. If there is something like philosophy and ethics therein, and if they work sensibly in their field, where is the harm in knowing that the theories of ethics are not mere handmaids of the fertile thinking rather useful as well? There is no problem whatsoever for the dialogues reveal most interestingly a range of normative theories known and unknown to us in action. This also gives us an idea about the ways in which the theories may be beneficial in constructing suggestions or decision-making rules. The dialogue might also reveal the relative advantages and disadvantages of these theories in actual ethical contexts. This may be further informed to the intelligentsia and other interested people by means of teaching in schools and colleges and in other social institutions. In fact, a spreading people's education might gradually adopt such things though teaching such things does not guarantee a nation full of ethical people. Ethicality of a nation depends on a lot of 'doing' than piggy riding on the heaps of information supplied.

All these do not guarantee that after corroboration and closure the matter ends and we do not have any need for further dialogues because ethical disagreements are truisms in social life. We have to be serious about disagreements, which are of great value. They are needed to revise the decisions taken before. The process of ethical dynamics is continuous if we have urge to remain ethical and not traditional, stereotyped and archaic.

Let me now clarify about the reasons for considering some value-laden problems I have considered for ethical resolution. In the year 1997, I was awarded Major Research Project of the University Grants Commission, New Delhi to work out a Project entitled: "Applied Ethics in Indian Context", the first part of the project was to consider some environmental ethical issues for resolution, for which, the model explained by me was used extensively. In the course of the Project work eight important environmental ethical issues of national interest were chosen and the said model was applied successfully at a preliminary level. As the funding was not generous and work power limited, I have to restrict it to field work in and around Ranchi district. At times, when I went to other places I talked to some environmentalists to gather their opinions and about what they felt about the consensual view emerged and decision-making rules formulated. The findings are now presented in this book in three sections only. I would thus consider here the following value-laden issues:

Section I	:	Urbanization and moral crises in India
Section II	:	Ethical issues in Harnessing Rivers and Building Dams for Power generation and other purposes.
Section III	:	Moral Problems in India's Nuclear Option

The reasons for considering these problems are fairly simple. For an Indian it would be easy to realize why in our times these problems need urgent attention. First, we have to accept the truth that the environmental catastrophe towards which India is slowly and surely heading is a practical problem in modern times which needs some care. Second, there are some practical problems as those mentioned above which raise serious moral or ethical questions as well for a careful look at these problems reveal that we often stand in moral dilemma as common people, ethicists, other professionals and ecologists when we ask: "what should be done?" Or, "what is the most desirable decision to be taken?" In fact, all these issues are so complex in terms of value-based decision-making that a casual one-sided reply as: "only this needs to be done" is not proper. There are certain decisions taken where "this is done" or "this is practical to do" or "this is useful to do" are too simple, at times foolhardy, obstinate and immoral because in society we do certain things, certain things continue as usual, certain states of affairs are taken as if they should be like that but that does not justify that such things should be like that. Isn't it simple to realize that in so-called civil societies we do muck nature in many ways? For instance, we dump filthy and poisonous things in waterbodies, which are sources of drinking water in many cases. The question is "should" we continue doing that? Is it "desirable" to do so? Yet, one will hesitate a little before answering it emphatically in the negative because the person faces a "moral dilemma" or a "dilemma about desirability" because he or she is immediately reminded of many associated facts like, booming population, lack of power generation, public conveyance and so on, mostly reminding the several human needs and wants. We are thus apprehensive about what really "should" be done. This is the crux of the problem for us here.

It is extremely difficult to differentiate "what is" from "what practically needs to be or should be done" or "what ethically needs to be done", but to differentiate their meaning is crucial in understanding the difference between "practical problem" and "value-laden practical problem" and also "resolution of practical problem" and "resolution of value-laden practical problem". Hence, to say that all practical problems like practical environmental problems are hardly any different from value-laden practical environmental problem is a great mistake. It is important to find out whether or not apart from being practical, useful, pragmatic, sensible etc. something is "most desirable" as per people's decision of "ought". If the "ought" or "should" is largely concentric, at times obstinate and derogatory because larger opinion for value has been neglected, then the decisions are unethical. This is extremely important for us to understand because we often fail to realize that concentric resolutions for a "practical should" may not be desirable at all or may not be "ethical", as most of the time, "other human interests" are completely neglected narrowing down to a group, a class, power holders and is thus restricted to the "now". It may be thus "practically desirable" to dump filth and poisonous effluents in the Ganges or the Yamuna for "now" we (you know who!), need power plants most or it may be at least "practically desirable" to dump these things in a pool of bad water or marshy areas, but there may be some hitches in acceptance when a large group is considered for what is "ethically desirable". It is not a matter of great argument but a matter of looking at the history of Indian policy-making through the ages that an eye on "practically desirable" has neglected the "ethically desirable" because "ethical dilemmas" have been misunderstood merely as "practical" ones and peoples' decision-making followed by a discourse about "should-value" of the "practical-is" remained an alien idea though in democracy, powers of decision-making blossom not in 'clannish prudence' but in 'association'.

The reasons for considering exactly these problems are also fairly simple. All of these problems are of great concern for a huge chunk of the concerned Indians. Even the ordinary people, the illiterate mass though unaware of our technical terms like "urbanization" are concerned about exactly what we mean by them for they raise moral dilemma of acute type, which they aspire to resolve. This is not prophecy or assumption; any one of us whose eyes and ears is sensitive to information, knows about such things. Here it is interesting to notice that the common mass in India is not necessarily talking in ethical parlance. They do not use moral vocabulary or are not deliberate in using 'practical should' and 'moral should'. Then does it not follow that "value-laden practical problems" are "intellectual discoveries"? Is it not that what is value-laden is so by virtue of intellectuals finding it like that? Isn't this "structuralism"? In fact, no intellectual ethicist or others can put a value (ethical) to a problem and make it 'value-laden problem' though our moral experience in society enables us to "discover" that certain problems as those mentioned above are not merely about thumb-rules for decision-making rather, calls for "moral debate" in the public. Sometimes we feel that people are worried about not leaving certain powers of decision-making to some professionals only because it is not as simple as doing something as we are used to do. This is not a blank-talk, if one comes to the public life, ever eager to "know", one "discovers" the "moral concerns" of public about environmental practical problems mentioned by me (there are many more nevertheless). If we feel that it is too much to believe that ordinary people are "morally concerned" about the "nuclear option" of India, I would ask them to seriously consider their ways of life; most likely they are people bound to the cosiness of their isolation. Let us also accept that even if people do not use exact moral vocabulary, they may be still concerned about moral issues. Perhaps we do not care about the concern of the common people of a nation.

If one glances through the problems selected for moral resolution and their arrangement, one finds that most of them involves humans, animate and inanimate existential at the helm of moral affairs. These problems, therefore, make good cases for environmental ethical discussion. The arrangement of the problems here is nothing special, but I have considered the problems related to human-other relationships as most important. The problems selected here are few in number. I shall give outlines of some more when I come to the prospect of people's ethics in the last chapter of the book. I will first take up the case of urbanization and moral questions related to it for resolution through intersubjective corroboration. In each section to follow, I will first introduce the problem briefly, the parties selected for moral dialogue, state and analyze the questions framed, responses to these questions, the consensus emerged, the findings of the post-corroboration analysis of decisions-taken and conclusion.

SECTION I—URBANIZATION AND MORAL CRISES IN INDIA

Introduction

It is a truism that the Indian society is predominantly rural, though in the recent times there has been phenomenal spurt in the urban population. Quantum shifts of people from the rural to the small and big towns and cities is a prominent feature of the post-independent India, particularly after heavy industrialization across the country owing to five-year plans and of late owing to the opening out of the Indian economy, most fondly known as globalization of economy and free market economy. Urbanization is a complex social phenomenon mostly consisting of growth of human population in a relatively dense form in a limited area owing to remarkable changes in the social, political and economic activities within that area. The concentration of population in urban areas is owing to human migration from rural areas and also due to influx of foreign nationals for several social and economic reasons. It is a phenomenon of changes in demographic patterns, changes in racial, cultural and class stratifications, and noticeable changes in economic activities and forms of life. India has witnessed heavy urbanization through the ages gradually due to which all the changes mentioned above are clearly noticed. Urbanization thus calls for many things together, one most important thing is change in moral perception and value systems of the public, to which we shall turn our attention shortly.

It is a matter of common sense that sudden rise in population, changes in social structure, economic activities, political forces and cultural diversifications along with multifarious forms of life have deep impacts on the entire urban environment. As none of the urban areas we now find in India were in that shape perennially, each had a changed environmental face due to gradual urban developments, therefore, each of them has to lose something to gain something else. The loss is prominent enough—the urban areas lost their serene environment or, if not so serene previously, at least lost its original form. Most of the time we find that the loss is phenomenal in environmental terms and it is worse with the speeding up of the process of urbanization. With this, there is a parallel idea that though some perennial patterns and serenity have been lost and the environmental changes are prominent nevertheless, gain is more than loss because many underdeveloped or neglected areas have gained unimagined status through quantum development.

Let us first take a look at what environmental changes we find in India with the spurt of urbanization and then evaluate whether or not they are really worth for humans and other beings and things which together constitute the environment. One important aspect of moderate to aggressive urbanization are marked by change and "development", where "development" is argued to be of great worth for a nation but it is also taken as completely using up already developed and yet undeveloped spaces by building several things. A developing country like India takes this sense of development as most important. The result is that a prolific, building activity and using up spaces marks Indian urbanization. We have thus consumed agricultural land at the rate of 3 to 5 per cent per year, resulting into a huge loss of our cultivable land and with that the people and their habitat that were dependent on that land for primary economic activity has been snatched off. Hence, "building" activities, infrastructural developments put severe pressure on air, water and soil, those perennially there in the areas which have been defaced (or, have been face lifted!). Understandably, urbanization in India caused loss of lots of open spaces, inequitable distribution of economic resources, risks of public health and loss of a sense of "community". Urbanization has caused limited regeneration and renewal potential of the natural resources over a long time.

Urbanization in India has moved along the global urbanization in almost equal footing. Reportedly, in 1980 London was the only city with one million inhabitants, whereas by 1990 we had one hundred huge cities all over the globe accommodating five hundred and forty million people. The urban growth was mostly in the northern hemisphere. Today, the world's largest cities are growing in the Southern hemisphere because of population boom of unimaginable magnitude. Migration from rural areas and unprecedented industrial development add to it. Modern cities are defined by high demands for energy owing to concentration of intense economic processes and growing personal consumptions. The result is that city folks in majority have become monopolistic users of fossil fuels and metals. We now consume nearly half of the global photosynthetic cover-plants and trees. Our cities have thus affected the ecosystems as a result of such huge demands for costly timber and rare plant species for domestic decoration. We have passion for caged birds and pets removed from their perennial abode.

Understandably, urbanization has affected traditional rural economies and their unique ability to adopt a life among biological diversity. Rural population thus abandon their own cultures; forcibly adopt urban culture and ways of life. They are forced to adopt urban standards of living and a new mind set. Urbanization has taken a good toll on the vast waterbodies. Indian cities consume huge catch crop and they are responsible in passing filth and dirt into waterbodies. Urban expansion in Indian has also devastated mangrove forests, coastal reeds, marshy areas and vast swamps, abode of several species. As we know that urbanization worldwide is marked for its huge population growth, it has tremendous effect on the environment. Reportedly, world gets one billion people every twelve years, of which a large chunk comes from India and other Asian countries. Population Action International argues that current growth rates in India is posing serious risk to human welfare. The dreaded population growth in India has encouraged several nuisance, among them release of greenhouse gases is one which contributes to global climate changes and can alter habits and loss to plants and animals on whom human life depends. Urbanization in India has devastated the weather of the country, almost uniformly. An analysis of temperature data from one hundred and sixty-six stations show that the maximum temperature has raised in several areas in recent decades. India has seen sizzling summers, one in 1998 and the other in 2005. Talcher in Orissa recorded 55°C as maximum temperature and Purulia recorded 50°C and above for quite some time in 2005. Such drying effects are worse felt in the concrete jungle of the city for added to unbearable heats are air and water pollution. Such pollution in our cities is not due to alien reasons. Take the case of urban buildings in India; they reportedly contribute 49 per cent of sulphur dioxide, 25 per cent nitrous oxide and 35 per cent carbon dioxide in urban air to pollute it.

More frightening reports follow now about the most polluted Indian cities, which were erstwhile believed to be paradigm of urban growth and modern civilization. World Health Organization Report of 1995 states that air pollution in India kills one lakh people every year, out of which ten thousand people die in Delhi alone. Heavily industrialized city Kanpur shows that its SPM limit is the highest (470.9 mcg), whereas in Delhi it is 410.5 mcg which is heavily urbanized. World Bank Report states that number of deaths in urban India during 1991-92 increased by phenomenal 28 per cent and the rate is steady. Slow Murder Report of the Centre for Science and Environment of 1995 states that fifty two thousand people died due to pollution related complications in 36 Indian cities as against forty thousand deaths in 1991-92.

Let us take a look at the phenomenal urbanization results in Delhi, the capital of India. Reportedly Delhi is the fourth among forty-one worst polluted cities in the world. 65 per cent of its air pollution is due to vehicles and thermal plants. The territory bleaches 2,200 metric tonnes of pollutants every day. Yamuna, the life-line of Delhi, which is almost like a dark muddy nullah receives

18,000 million litres of domestic and industrial wastes with toxic chemicals though Supreme Court has asked for removal of several industrial units. The result is that Delhi tops in lung diseases. Every tenth child is asthmatic. Lead pollution surpassed the WHO limit in Delhi which damages brain, liver, kidney and blood. The cases of Kolkata, Mumbai and Ahmedabad are not good either. In Mumbai due to heavy automobile flow 50 per cent population have 30 mcg of lead concentration in 100 ml of blood. The Institute for Research in Reproduction in Mumbai states that sperm count in Indian males fall from 60 million per millilitre to 20 million per millilitre causing abnormalities and abnormal babies. Nothing is very bright in Kashmir valley after the Kargil war for there is heavy pollution in the air. Its growing urban areas are becoming more and more polluted. The Dal Lake in Srinagar is fast turning to a swampy area full of weeds. And our Taj in Agra has been damaged considerably due to the Mathura refinery and coal based brick kilns. To add to it, its approach areas are unmanageably filthy.

All this has been written not to paint all black on the glimmering urban life in India. A big fool will not realize the truth that in India, we have already irreversibly moved to urbanization and we are continuously doing so. Here is the question: what can we do now? Better asked, should we go for too unharnessed urbanization in India? Is it still desirable given the dark state of environment? The point is not what has happened or what certainly is happening or most likely to happen, the point is what should happen now or henceforth? We ask this for the urban and the rural Indians are witness to massive environmental disaster which they know is not only impractical and imprudent rather, unethical. Or, is it the other way? Let us now come to this "moral dilemma".

Selecting Parties of Moral Dialogue, Construction of Questionnaire and its Analysis, Brief Statement of the Participants' Response to Moral Questions and Consensus

Any moral talk like this one might start at any part of the country. The most desirable thing is to spread its network rapidly and widely in India, which needs a massive and aggressive ethical programme nationwide. Hence, intersubjective corroboration is not one-man effort. Still, someone has to start somewhere before it becomes a national creed. I started it in my city and covered a population in and around it. Environmental ethical talk was extremely easy to start here for most of the value-laden problems considered, are those, which trouble the population of entire Jharkhand state (earlier in Bihar). Ranchi being the second capital of the then Bihar state had good number of environmentalists and other interested parties. However, I did not have enough financial back up to cover a huge population. Yet, I thought, I should start at least with a low-key dialogue, which was practically possible because wider dialogues needed a lot of fieldwork as well.

The value-laden problem of urbanization is particularly important for Ranehi is no more a small city, framed as picture on wall. The most important parties to contribute to resolve this problem are no doubt environmentalists, environmental activists and professionals in the government organizations who are directly responsible for taking decisions related to urban development and planning. Of course, a number of common people who are simple citizens of this state also make an important party. Keeping this outline for selecting parties for the dialogue in mind, I visited some people, talked about the project, its aim, its value and tried to gauge their opinions, reactions, emotions and rational acumen by talking plainly about some moral matters about urbanization. It is important here to learn not to give your lengthy moral opinions but to listen to them so that you come to know whom to select and what exactly should be your questions to bring out the best out of the parties. A golden rule is not to become a moral preacher or aggressive propagandist because at times many intellectuals and ordinary people are either hesitant about participation or they simply try to ditto your views thinking that the gathering would like it. It is, therefore, extremely important to talk to a number of people before selecting them for a dialogue and prepare your questionnaire (which does not suffer from structuralism).

After a lot of survey, I selected Dr. D. K. Tiwari (IAS), Project Director of Bihar Plateau Development Project; Mr. Rajiv Kumar (IAS), Managing Director, Ranchi Industrial Area Development Authority; Dr. N. Dayal, Environmentalist and Head of the Botany Department of Ranchi University and Dr. S. D. Singh, Political Scientist and former Head of the Political Science Department of Ranchi University as main participants. They were joined in the dialogue by a number of students, teachers and common men of different walks of life who asked many searching questions, talked about the matter at length and helped in the consensus to emerge. Interestingly, the Vice Chancellor of Ranchi University, Dr. C. R. Laha himself was present along with the head and teachers of the philosophy department. The debate went for some three and half hours before moral consensus emerged. I myself participated as corroborator and analyst.

On the basis of first hand information gathered about the subject from books, articles, notes, pamphlets and by talking to different people, I framed only six questions (in English as well as in simple colloquial Hindi) for the discourse, which runs as follows:

(1) Which alternative should be chosen as goal of human life and why: The development and welfare of the human race or sustainable development of the whole of existence around us?
(2) Do you think that rapidly growing modern and ultra-modern areas have any moral worth in the face of widespread environmental disturbance?
(3) Why does urbanization appear more desirable than the state perennially existing? Is it good for the attainment of some other thing or is it itself good?
(4) Do you think that urban growth is oppressive? If it is so, what does it oppress and why? In what sense will it be called bad? Or, if it is not oppressive, what does it contribute to and why will it be called good?
(5) Many people are against an urban life form because that is not ethically good, particularly when it comes to India, do you think that an ideal urban life is still conceivable and is most desirable in India?
(6) On the basis of your replies, what conclusion would you draw about urbanization in India and what things would you suggest?

A close look at the questions gives us an idea that they are 'ethically charged' for only then does it induce in us an urge to express our moral views about urbanization. If you analyze them, you will find that answer to these questions cannot be about what is urbanization, what led to urbanization in India, its history, its forms and various ecological effects owing to it. Here, our focus is to know whether or not urbanization in India is morally good. It is undeniable that answer to these questions often compel specialists in that field to discuss elaborately many technical matters, however, the task of a corroborator is to frame questions and control the dialogue in a way that laymen are not given good doses of technical things. Nevertheless, there are some technical ethical questions disguised here but I found them answerable freely by most ordinary men participating in the discourse. The first question, which is closely related to the third one, tries to unravel the normative position of the participants. It is interesting to notice that wise Indians do not get bluffed by given options in the first question for many of us are sure about welfare of man and of whole of existence being morally good. The second question on the other hand tries to bring out more clearly the views of the participants about the "value" of urbanization itself, that is, whether or not it is valuable as means or an end. Any answer in terms of means and end drags them naturally to a "why" question, which in turn confirm their normative position or exposes contradiction in their views, which needs to be pointed out for they may defend or change it. Putting the second question, the corroborator wants to know whether or not the speaker follows his or her basic moral position while replying to the "value" of "ethical worth" of urbanization. Question number three in fact completes the answer while contrasting the non-urban state in India. The fourth question is more searching; it is deeper as it initiates more intense debate about affixing "oppressive" label to "urbanization". It gives the speakers, protagonists and antagonists to show full range of their moral arguments to not to call it "oppressive" or to call it "oppressive". This should again be according to what they said in first three answers or else self-contradiction should be pointed out. The fifth question makes the moral debate more specific by citing some views because others opposed to these views have equal chance to speak about fittingness of urbanization in India. It also brings in the "urban life style" as most intensely debating as that adds another dimension to the earlier answers because by now it is expected that speaker would talk about effects on environment and urban way of life. Noticeably, there is always a strong "ethical should" colouring the dialogue or giving it a steady direction. The sixth answer brings the speakers to conclude their moral opinions and to suggest measures.

Now let us come to brief statement of the replies of the participants and some others and the consensus emerged thereby. To the first question, Dr. D. K. Tiwari, a trained doctor and now an administrator related to urban projects in the serene plateau region of Chotanagpur said that he perceives "two worlds", one the natural world created by God and the "other world" built by us. We cannot escape the truth that in "building" the "other world" we have to seriously intervene in what has been "given" to us by God but human life is one of urge for development and well-being in the nature by not disturbing its basic harmony to an irreparable limit. Hence, the goal of moral life is to grow, flourish and care for the whole of existence judiciously. We do not have any moral right to disparage God's world for our benefits alone. Dr. N. Dayal did not agree with the fact that the ethical aim of human life ought to sustain the biotic community as such including human beings and to see that human beings grow to optimize their welfare because God has created everything. He believes that man is a spider in the cobwebs of existence, whose interest fulfilment is extremely important. Yet, if he forgets sustenance of the ecosystems, many damages are caused to living and non-living as he cites an example of a bird who has to consume certain seed until it drops with faeces on soil to germinate. The tree that flourishes depend on a simple bird, and they depend on our wits to grow with them. Dr. S. D. Singh's opinion in this matter was not much different but he had two important points to make. One is that morality with regard to intentional human acts like constructing an urban area does not overlook the truth that whatever he does is moral only in the perspective of human beings because what he does to the natural existence as such, including the animate and the inanimate existence, is extremely important to evaluate the morality of his actions. Another point is that though human morality is evaluated with respect to the changing locales when urban development is concerned, there is a "universal morality" which transcends the locales and is true for all places, times and people. When we keep this "universal" element in mind, the moral end of human life is to grow judiciously in nature, to develop, and to take care for the biotic community as such. Human development need not and should not harm nature in any way. Mr. Rajiv Kumar put the question asked to the participants in another way. He asked whether or not we can at all conceive of human existence without eco-sustenance first? For Mr. Kumar the questions of development and morality are only of secondary nature because if there are no human beings existing on earth, no one develops and no one bothers about morality. When it comes to the existence of human beings in nature, we find that it is impossible without conceiving the existence of the sustaining the whole of existence, the existence of human beings becomes impossible. Possibly due to this, our ancient verse in Sanskrit prays for the good of water, air, fire, soil, vegetation, humans, atmosphere and the vast expanse beyond us. So "good" is for all as no one's good alone guarantees that good in the real sense. Truly, development is one concept which comes here as good for man but one cannot forget the greatest good, that is, the sustenance of entire existence. Morality only secondarily comes here, as it is conceivable that the sustenance of nature as a whole is alone desirable. Naturally, the point of the morality of sustainable eco-development would be secondary to the basic point that the 'greatest good' is 'the unharmed existence of the entire nature'.

The second and the third questions pertaining to urban modernity's moral worth and about the means and/or the end-value of urbanization, the responses of the participants were as expected, not cleaved but correlated. Let us have a look at the replies. Dr. Tiwari was extremely conscious about his basic moral opinion as expressed early on and said that though the western moral conceptions is largely about 'end justifies means', we should keep in mind that the means need not be evil as well. In so far as the whole of nature is God's handiwork, the world we build does have a definite goal or end, which is good as such. This end is 'the realization of the full potentials of the natural entities'. If this be true then urbanization as our handiwork is 'good as a means' towards the higher good of the flourishing of the whole of natural entities. It is, therefore, important to notice that urbanization is a good means through which not only do we serve our purpose, our good, a wholesome good of our environment is made possible. In this course the modern aspects of urbanization vis-à-vis more simpler way of life in rural areas is not evil itself because one parameter of urban development is that by virtue of the fruits of the scientific discoveries, it has to be modern, which may be a boon not adversely affected by the urge for our maximization of good. Hence, urbanization as means should be controlled and systematic with a moral face, that is, should care for the wholesome good of an environment.

It is at this juncture that the Professor of Botany and environmental activist Dr. S. K. Sinha asked all the participants whether or not it is possible to imagine scientific ways to urban development and its modernity to care for the nature as a whole and the environment where we exist? To this, and to clarify his position about the two questions asked, Dr. S. D. Singh said that scientific development of any kind, be it urban development essentially leads to some kind of modernity, but that cannot be taken as something against the non-modern, serene and perennial states of nature because modernity itself becomes a means towards the higher good of the welfare of human beings. Here we have to remember that when we use modern means for a modern development resulting into modernity, our basic *humane mindset* is not changed. If modernity of the urban life is humane, it does not disparage the perennial life forms, the rural life forms and the agrarian set-up to a great extent, it is good itself and not evil. The world is shrinking owing to such rapid development, it is good for all of us, but that does not mean that our set of mind becomes dominantly aggressive towards maximizing in any way our good and not that of any other.

Mr. Rajiv Kumar retorted along his original line that morality, modernity and other issues become 'secondary' in the face of the main problem of urban development as a means to the higher goal of laring the basic existence of whole of biotic community. Keeping this in mind we can say that urban development, its correlative modernity and the so-called change in the quality of life can never be means to an end called, 'the pleasure maximization' of man or benefits of man or even welfare of human race as a whole. In fact, much of damage has been done both to urbanization process and development owing to our aggressive hedonic mind-set. In the name of development and that too scientific as well as modern, this end of life has ruined everything. So, morality of human welfare would be insufficient to evaluate the good of the means called urban development if we are not keeping in view the ideal of the welfare of the entire ecological community through its basic subsistence.

Not surprisingly that Dr. N. Dayal followed by his earlier line of thinking hold that the good end of the means called urban development is the optimum realization of the urge of human beings to grow keeping in mind the need of the poor and sustenance of the ecological community. There is no doubt then that man being the centre of the cobweb of existence, has to grow, of which one way is to expand its habitat in the urban areas. There is no harm in living a good life in the urban culture, where the good life is free from suppression and exploitation of the poor and the needy.

The participants were further asked whether or not urbanization amounts to some kind of oppression and what it may contribute to. They were also asked to compare a Gandhian rural life *vis-a-vis* the urban life form. Mr. Rajiv Kumar was emphatic that when we are speaking about issues like urbanization and its moral implications, we are not much involved in caring for the poor and so on. It is a matter of fundamental concern that first the whole of existence, including ourselves have to exist nicely, for that we need to develop, where 'we' does not refer to the human beings, the whole of existence needs some care. Hence, human beings have to first fix their "value system". It is not fixing the "value system" of Gandhi or Pol Pot or Mao, who spoke of "rural values"; we rather speak of the "value of the existence as such and to care for it". Keeping this in mind, both the rural and the urban developments are good enough provided they are not for some values like pleasure or welfare of man or poor man. It may well be that some rural ideals and values are extremely good but that is not against urban development if the basic idea is not lost sight of. Hence, urbanization itself cannot be oppressive. It cannot be anti-poor or anti-people if it is not anti-existence.

Dr. S. D. Singh argued that an ethic that speaks of a rural life form as essentially better than an urban life form does not have much basis if the rural form of life is itself not caring for the human welfare. Truly, the restriction of the entire ecological balance is important but a humanitarian ethic cannot be overlooked. Mao, Pol Pot, Gandhi, any thinker supporting rural economy and rural form of life does not amount to neglect urban development either if it truly has a human face. A judicious development of urban economy and urban ways need not be thus seen as immoral in itself. Gandhi and the other thinkers would never afford to say that both the needs of the locale and that of the greater territory are not equally important. The development of the locale cannot be isolated and predominant; it is related to the development of the greater needs of the nation. So, urban development may be equally ethically desirable, it cannot be stated as oppressive. It contributes to human welfare.

Dr. N. Dayal added that though ecological sustenance is important along with development of the rural and the urban areas, what is supreme is the existence of man in these areas in an exploitation free state. Dr. Dayal has no doubt about the truth that both the glossed rural life by great thinkers mentioned above and the relatively looked down upon urban life are "oppressive" when we do not cater for the ideal of need of the poor, urge for development and eco-sustenance together. If we now take one aspect, that is, "need of the poor to be fulfilled", we do find that since time immemorial till this date, there has been widespread oppression of the poor by the bourgeois be it in villages or in cities and towns. This is immoral. Hence, urbanization as a process is not immoral; it turns out to be so when in that process there is loot and plunder of the migrated poor masses. It is immoral when the helpless are exploited for welfare of a class. At this state, certainly eco-sustainability is a glossed idea. Hence, an exploitation free urbanization is morally desirable not that a pure rural economy and rural life is desirable.

Though Dr. Tiwari agreed to this point, he stressed that a rural economy and a rural form of life is much desirable in India. It is important though to make a balance but an exploitation free rural life would be much better. The great ideals notwithstanding, human focused growth cannot be most important as we have to take care of the entire community in which humans subsist. It is a matter of exploiting the whole of existence and not merely a section or a class of people, which is supreme. An urban development which oppresses the community as such is immoral.

On the basis of the moral discussion, the participants (including those who attended it) were asked to come to a consensus about the exact message they would like to convey the policy makers and implementers in India regarding this matter. The general consensus was that morally speaking: "a balanced rural and urban development in India is needed wherein primarily the ideals of non-oppression of the poor, the needy and the common people along with the beings and things of nature are kept in mind". Some specific suggestions came from other participants like Dr. Ahmed Sajjad of the Urdu Department that "an urban development in India should also restore the vastly extinct valuable cultural and metaphysical values because these things cannot be scapegoats of development". Speakers agreed upon the "renaissance of rural values and forms of life for which most needed are village oriented aggressive development, phenomenal management of the population, spreading education, basic infrastructure and health care so that massive migration could be stopped". It is obvious that "comprehensive urban plan should take care of the exploited migrated poor man".

We may now come to the ethicists' task of *post-corroboration analysis* of views and the consensus to explore the dynamics of "normativism" in ethical discourse. It is undeniable that if our questions are "ethically oriented" in the broad context of a practical issue, only then the moral opinions flow more easily, but this is not an indirect tactics on the part of the ethicists to compel people to orient their views in one direction. What he does is to channel a practical problem raising moral questions in a manner so that in the midst of all technicalities and practicability, morality does not remain implicit but comes out prominently, the result of which being a panorama of ethical theories transversing in a gamut of disagreement and agreement to make us believe that (a) all of us apply our ethical knowledge involving the known and unknown theories not deliberately but spontaneously; and (b) irrespective of disagreement among known and unknown theories, value-laden issues are easily resolved through wide agreement among non-ethicists or non-philosophers. This point needs some more time to be clarified.

As we focus to explore the moral thesis inherent in the dialogue we find that the prominent ethical thesis maintained by the participants is:

(i) Welfare or well-being of the whole of the existing reality is the ethical basis for evaluating some means such as urbanization to attain the same (Singh),

(ii) The ethical basis for evaluating certain human decisions like urban development is that whether or not it disparages the harmony and intricate design of the nature created by God (Tiwari and Sajjad),

(iii) The ethical basis for evaluating urbanization of some moral worth is the welfare of human beings primarily though in doing so biotic community's well-being should be guaranteed (Dayal and Sinha),

(iv) For evaluating the moral worth of urbanization, sustaining all animate and inanimate beings and things is important (Kumar).

It appears apparently that these are four different norms to be found in environmental ethics and not in ethics as we traditionally speak about because all norms enjoy a different status, they are regarding the evaluation of unique environmental decision-making of man. However, a close look at the four so-called 'norms' reveals the truth that there are two broad normative theories in interplay—one being *theological-ethics* and the other being *non-theological holism or holistic welfarism*. In fact, what Tiwari-Sajjad thesis upholds is essentially not different from Dayal-Singh-Kumar thesis. To realize this, first we have to see that reducing Dayal and Singh as one view is not at all difficult because there is hardly any ethical difference in maintaining a holistic welfarism where human beings are most important in ecological cobweb (Dayal) and to maintain that well-being of the whole of existence is important (Singh). Had Dayal maintained strictly that the well-being of human beings is only important, then it would be really different from what Singh and also Kumar says. In fact, Dayal starts with the welfare of human beings as supreme but does not rule out the well-being of the entire "cobweb" including the "spider". Though at a later stage, he presses on "welfare of the poor" as supreme, he still does not lose ground to quickly mention that for this the whole of the "cobweb" needs great care. Coming to Singh, his view is extremely straightforward holistic welfarism, which is essentially different form of utilitarianism because he never speaks of "pleasure or happiness maximization of human being (or other beings)". This "welfare" though covered in a thick gel of ambiguity as is of many "welfare" proponents nevertheless is not averse to pleasures, benefits, flourishing, may be even happiness of human beings and others. This nicely includes the 'rights-talk' and 'duty-talk' about other than human existential. What Kumar says first about relegating "moral-talk" as secondary is gradually promoted to most important thing when he continuously painstakingly repeats as "welfare of existence", which is no different from uttering the same holism of Singh. His examples of Sanskrit verses are clear in this matter. Hence, to act for the basic existence of all of us would be basic holistic welfare of all of us, which includes Dayal's "poor man" as well. It may well be that Dayal has another point in mind which is often maintained by humanists but he does not toe that line particularly as he is exclusively worried about the good for man. Does it amount to say that he would not mind some degree of harm done or disturbances in the ecological community? Is it true that his view amounts to the same as saying: 'saving nature is all good but only if human good is done, otherwise not'? Dayal's apparent anthropocentrism sometimes found in many radical ethical thoughts (with his addition of the good of the poor) never gets a proper exposure most expectedly because as a trained ecologist and an environmental activist, he always reminds that even the welfare of man depends on or is not to be cut off from the whole good of the nature as such. Hence, Dayal-Singh and Kumar forms one normative thesis, we may call it "holistic welfarism". It is at this point that the basic difference of Tiwari's and Sajjad's view comes to the fore because they think that our "moral-talk" is not disjoint from our "theistic-talk". According to them human beings are obliged to act morally to take care of the fine tuned nature given by God. If we lose sight of this basic thing, there is no point in talking about morality. It is not clear with Singh and Kumar (also Sinha and many others) whether or not they take the 'whole of existence' as 'given to us by God' because "God" never figured in their nearly scientific-secular views or we can say that they uphold a predominantly "moral view" rather than "theistic-moral view". Yet, Tiwari and Sajjad were at serious dilemma to conclude about the value of the "use" of the 'given nature' by man. Interesting is that Dayal would face the same dilemma relatively less intensely than his supporters of holistic welfarism (all not so clearly non-theists) for Dayal would have least hesitation in "using natural entities" for man. For him, therefore, using trees, stones, rivers, soil and many other natural resources including the human resource for larger benefit of man or their welfare is but natural. Nevertheless, for him the point of "judiciousness" in "use" is extremely important as without that there is in the long run no welfare of man. So it is not a view of ruthless or optimum use of nature 'only' for man.

Even Singh and Kumar did not differ here because for them too, the natural things and beings need to be "judiciously" used for our "existence" and/or "welfare" when, for instance, we constitute a new urban settlement or developing existing ones. How can a holist speak of only human good and for that urbanization as an important means? Truly, on the other hand, how can he speak of urbanization in any way for the used-up nature? So all participants face this dilemma about the good of the means, that is, good of urbanization causing some harm to the nature. Interestingly, none of the above dare speak that "urbanization" is evil as a means or should be shunned completely if holistic welfare makes a sense because this is the most extreme view, which was most clearly ruled out by Dayal. But why do we forget that fewer radicals that the theists are did not hesitate in supporting the using up thesis? As we stand then, urbanization is a must and there is a 'should' of it in so far as the 'means' thesis holds well. It is good as a means of human welfare. Here is a gappiness about whether or not the good means is good for all, when that is the moral basis of our deliberate choice of means like urbanization? To put it simply, how do we solve the moral dilemma of good of using-up inherent in the above thesis of urbanization and the holistic well-being of the nature as a whole? Does "least" intervention make any sense as does "judicious" in the Leopoldian answer to the above dilemma? Our proponents all appear to be Leopoldian naturalists in the final run as theism is also not a big hurdle to maintain that urbanization is morally desirable when using-up natural things and beings are least and/or judicious for the welfare of all. But how is there welfare of all? Did we get that answer? Perhaps Dayal is best facing the Leopoldian ethical dilemma as even strong intervention of man in nature is morally viable as the good for man has been primary, and when this is possible by not satisfying an effluent class of people but the poor and the needy and others in a balanced way only then do we say that we take care of welfare of all. This was coming out in a hazy fashion in Kumar's position as well, though it was covered with Singh-type "welfare of all".

Still, in the midst of all these controversies and gappiness, came out truly in flying colours without any doubt two normative theories or perhaps only one if Leopoldian naturalism is not anti-theistic. Though in the name of Leoploldian holism or naturalism, we forget that nothing but the normative thesis of "welfarism" (a non-utilitarian type) ruled supreme about which our mostly well-read speakers and contributors did not have any formal training. Did that prevent them any way to come closer to such ethical resolution? No, perhaps.

What our participants suggested, though covered in an ambiguity of "judiciousness" and "least" I mentioned just now, have been concretely shaped after they conceded to their common point of agreement about "welfare-by-using":

(1) Planning in India should consider the moral aspects of urbanization, that is, why such and such things should be done or should not be done.
(2) A balance in rural and urban planning and development is needed—the "balance" is to be quantitatively as well as qualitatively weighed.
(3) Urbanization should be non-oppressive, that is, it does not oppress people or a class most poor and needy.
(4) The urge to develop and the need of the poor should be well adjusted with ecological balance while urbanization is going on.
(5) Urbanization should restore the values of the locale; the cultural, the spiritual, and ways of life need not be oppressed with development process.
(6) Rural economy, rural values and rural ways of life should be heavily backed up to stop migration.
(7) The infrastructural and other physical benefits of the villages should be phenomenally increased to discourage shifts to town and cities.
(8) Aggressive management of population growth and management of human resource are needed.

(9) Comprehensive urban plan should take care of migrated rural population for their several physical and mental demands.

(10) Most importantly, urbanization should be backed up by aggressive ecological restorations and the bulk of development be checked—it should be a green urbanization.

(11) Eco-unfriendly means of living should be turned into a "green mind-set", a rural mind-set essentially where we use certain things for our good, which do not harm nature.

(12) Using alternative sources of energy is another green-mind-set to be adopted in urban areas. It is also important to put a check on our ruthless use of electrical energy.

(13) Urban architecture should be of "green character" and "gentrification" causing massive migration of low, middle and lowest income group people in cities to sell of their plots and old houses should be prevented adequately.

SECTION II—ETHICAL ISSUES IN HARNESSING RIVERS AND BUILDING DAMS FOR POWER GENERATION AND OTHER PURPOSES

Let us now face the second value-laden practical problem of building high dams, allowing multipurpose projects and generate hydro and thermal energy in India. First, we should get a first hand idea of the problem. In this regard, I have to clarify my choice of the issue at first. I am here more concerned with the Koel-Karo Project along with reference to the Damodar Valley Project for some reasons. First, all these projects fall in the state where I grew up and now work, so they concern me most. As such, I shall not draw much from Narmada or other projects to clarify the problem and its resolution. The second reason for considering this issue is that it is closely related to the issue of urbanization of river-valley areas infested with deep woods, rich culture, and tribal life and above all, perfect biodiversity. The third reason is that the issue has in the recent times tormented the nation as such including its top judiciary, legislature and administrators. Finally, Koel-Karo is a typical dream dead before it was seen in colours, which is a good prolegomena to the fight going on in Narmada.

Introduction

The problem under consideration is one that falls under the broad issues of water management and power management where crucial thing is "hydro-human relationship". The issue also involves the rightness of large-scale capital investments as large dams of great height have to be built along with a number of power production units and so many ancillary projects. People are concerned about these things as they are environmentally destructive, cause social injustice, economically unfair and destructive to perennial cultures and life forms. There are also serious problems involving human-human relationships as well because a tension among planners, administrators, migrants, the refugees and the original people of a place persists for a long period in an intense fashion. All this started in British ruled India with the urge of urban demands and expansion of irrigation and saw a tremendous burst in post-independent five-year plans under the leadership of Nehru who envisioned a transformed industrialized India. The immediate result was that our perception of hydro-human relationship changed vastly. We started dominating natural, freely flowing waterbodies and looked upon them as "commodities" evaluated for their worth to human happiness and affluence. The state machinery was mainly responsible for such "commoditisation" and maximization of use of water, which was deemed necessary.

Much tension is due to the fact that post-independent hydro-based planning was guided by an economic philosophy or a philosophy of development which may be termed as "city oriented welfare" causing large scale migration of rural people owing to neglect of the value of rural economy. When such philosophy guides us and money floats in air in bulk, multinational companies come in to cash, particularly when we were not technologically updated and tragically now, when we claim to be updated. The national power grids has been thus systematically hijacked by urban mind-set people, the affluent and the greedy though 'power' has been touted for the need of the rural people. There is no doubt that the urban people use electrical energy the most than the rural people whereas rural India is historically prior, factually larger and strategically most important. Most crucial is that the urban centred growth and need of power went right against environmental sustainability, the needy and the poor. The social consequence of water resource management has been disastrous where environmental refugees and their plundered cultures tell the story. The forced resettlement causes great problems for the local people and most benefits are taken up by the outsiders not the local people.

The story is however not over as one sided statement of the issue would be an intellectual bluff. To judge the graveness of the problem we have to consider the counter positions regarding this matter and then start our moral discussion. It has been urged by a large section of people worldwide and in India, who are directly involved in large scale water and energy management that eco-sustainable development of a nation need not be overplayed when greatest problem is to manage the welfare of a huge population, particularly that of the poor. If an eco-sustainable project does not cater to this basic need, it is of no use. In the recent times in India and elsewhere, environmentalists have in fact hijacked the opportunities given to our planners and developers when it comes particularly to water and power management.

It was Nehru's foresight after independence that created many mega dams and multipurpose projects to help millions of poor Indians and make India self sufficient. The result has been unbelievably good as these projects were responsible in helping millions of Indians to fulfil their needs. There are arguments about the huge success of mega-dams worldwide where larger human displacements were caused without much suspected fears coming true. The US and Japan built huge dams in great numbers but now nobody cares about who were displaced and what were negative environmental impacts. In fact, in India, a large number of multinational activists and NGOs have infested to make heat out of nothing. Whereas in not so far poor and underdeveloped places like Laos and Vietnam a changed perception has helped them avoid some of their acute problems by building dams. Greatest thing to keep in mind is that the World Commission on Environment and Development in 1987 states clearly that the survival of human on earth ought to be ensured first and foremost for flora and fauna were made for humans and humans were not made for them. With the survival of man on earth, many things survive. Much misconception about dams create advanced psychopaths as once lot of them feared environmental disaster when Aswan Dam in Egypt was erected but now it is believed to be the most environment friendly. If we need to feed billions of Indians today and want to cater their growing needs, we do need such dams. Though population control would be most important thing in this regard.

Coming to the Damodar Valley Project taken up by the Damodar Valley Corporation (DVC) in 1948 towards the conceived development of large area in the state of Bihar (now some parts in Jharkhand) and Bengal. It was taken up particularly in view of the devastating flood of 1943. In the year 1948 by an act of constituent assembly greatly influenced by the Tennesse Valley Authority in USA the project saw light. Its main objectives were flood control, providing irrigation facilities to a number of people depending on agriculture, generation of power, enough drinking water for domestic use and water for industrial use, forestation and large scale social and economic well-being of tribal population who were affected by the stations that mushroomed along the Damodar river in some years and banked on it grew network of coal mines, washeries and iron and steel factories. A new steel city in Bokaro came up. The project started producing around 2,535 megawatt thermal power by 1998.

Let me now speak of what I saw while I grew along one of multipurpose projects in then Bihar state. It is called the Bokaro Thermal Project which came up at the first stage of the DVC's planning on the river Konar. I came in with my parents as a small child in one of the coalfields of Bermo around 1960, moved to two more coalfields very close to Bokaro to stay till 1980. So there is perhaps no need to learn from sources to write what I personally saw here. By 1975, Damodar, which I visited from West Bengal to Bihar, almost its full stretch, was a dead river. Yes, there were no regular floods in Damodar but flash floods did occur owing to heavy accumulation of rainwater in barrages both in Bihar and West Bengal. I saw its fury and had some idea what Damodar was like. Very near to my first high school in Bermo, the stretch of Damodar was huge but at places, it was like a brook, and now water is in and around the barrages not where it flows. Water is black, muddy and too shallow at many places. It is full of coal dust; fly ash of thermal stations, effluences from washeries make it a dirty nullah. Around 1970-75 when we regularly went for picnics beside the banks of Damodar, we used to walk through the breadth to find a sand island in the middle where some clean water could be found for our cooking and drinking for we never cared to have ready-made picnics and water bottles from homes. Yet, once we were forced to use the semi-dirty water and quit the spot forever. While we walked or cycled to our high school at Bokaro Thermal, we had to make a shortcut through fly-ash area, which was like a huge grey pond. The fly-ash made the colour of some dwarf trees and plants and rocks unbelievable. We had regular eye problems due to fly-ash and our homes had to be cleaned four times a day. Many among us had breathing problems and dysentery and diarrhoea were common. We are witness to what happened to the villages and their population. They were marginalized, in fact, non-entities at the pressure of the power station colonies and coal colonies, which were small towns in black and white. The tribal could not be seen in their perennial jobs, most of them were induced to coal extracting and other manual jobs most alien to them. They had to live in some 400-500 square feet homes, which together made a kind of town slums. The outsiders were settled in good numbers which could be seen easily in any open space or work place or institutions. The local culture and ways of living were mostly missing; we had either a peculiar admixture of outsider-culture or pseudo-culture. Most interestingly by 1975 onward there were frequent power cuts even in and around Bokaro Thermal area; coal washeries and mines were losing business and the power plant used to shut down any one unit frequently. The whole water body in and around this area is by now unimaginably unusable but people use it. Damodar is toxic basin, it is dead. I do not know whether people still find some fish in it which some of us enjoyed way back.

Well, what are the arguments to glorify it? Where did the well-being of the population (first) and others (next) go? Did we not see who made money out of what? Did we not see what we got and the nation got out of erecting the huge dams and a number of huge dams along the serene valley? Did we not see a complete biotic plunder? Yes, we did. We do not need a testimony to progress, a long talk of success and national pride. We better now talk about the ethicality of all that has been done and that is done till date. The unfailing spirit of the protagonists would still say would you then love to live in dark serene valley? Don't you need power? The moral dilemma is there but there are many things to face it. To make a passing remark here: "is power production related only to huge dams?"

Koel Karo is not like Damodar Valley Project because the latter is an old horse which flogged itself and the river beyond recognition to wait its death, whereas the other survived the initial flogging which was made ready to continue but saw an immature death. Interestingly, I have some experience of this as well as we started living in Ranchi, the summer capital of then Bihar and now the capital of Jharkhand. The above project was vigorously opposed since their inception in 1960s after the Second Five Year Plan. The more talked about Koel-Karo Project is spread along the rivers of Koel and Karo in tribal dominated districts of Ranchi, Gumla and the West Singhbhum which planned to erect the biggest hydro-electric producing plant with huge dams. An estimated 732 megawatt electricity was to be produced to contribute to the national grid. The first barrage was planned at Basia in Gumla and the second in Lohajimi in Ranchi district. The first on south Koel River at estimated height of 44 metres and the second on north Karo River at 55 metres height. The projects involve complete evacuation of 121 tribal villages where thousands of mundas, hoes and santhals live. Within twenty years of its inception the estimated cost grew up to a whooping 1338.80 crores. By an order of the Supreme Court in 1989 and the approval of the Department of Environment the work was expected to start at full flourish but then the people of Jharkhand who were witness to the Damodar Valley and many others encouraged by the Narmada movement came together to stall the proceedings, which succeeded. The movement was so proactive that the then prime minister was forced to skip the initiation of the work. It did not work even after another prime minister vowed to create a separate state for the tribal dominated areas. The furious prevention of making Koel and Karo other Damodar was mainly based on social, cultural, religious and economic grounds. It also was based on the heavy displacement of tribal and the greedy exploitation of their state by several people and agencies. Till date, the movement has been successful but that does not mean that many people including commoners as well as tribals do not have another view. We now turn to the dialogue initiated by me with a number of people to resolve the moral issue. It was one of the most gruelling, educative and successful dialogues ever arranged. Let us first come to selection of participants and the construction of questionnaire.

Selecting Parties, Constructing Questionnaire, Moral Discourse and Analysis of Discourse

My experience in an ecologically devastated area owing to the spurt of multipurpose river valley projects and widespread mining and the supporting reading materials enabled me to understand that I should first of all talk about this matter with the common people, the tribal and others who have been turned to ecological refugees and withstood the brunt of development. Apart from this are the serious environmentalists or others having remote self-interests, the political leaders, administrators having direct experience in this field, engineers and planners. The moral issues were emerging spontaneously as I talked to people, convincing them successfully that they should help resolve this issue, though many were suspicious about its consensus part as several protagonists of dams would never accept truth of devastation. Even some people suspected that I am representative of the official exploiters and asked for concrete proof that the UGC really did fund such a project. This happened particularly when I met many common people, the activists and the radicals. They would tell me that the UGC may not have realized what you wanted to do and, therefore, they awarded you at least a little amount. Any way, while I talked to them at places where you need some risk to be taken, they agreed. A group of activists, including some radicals participated finally. They listened to everything nicely, contributed positively, did not create any mess, came to a consensus and helped the nation immensely as I believe. Though as soon as I have introduced the problem to the house, one young man asked about ethics of such theoretical manoeuvre by calling people who are not well read and feel shy about these things. Yet, before I answered, he said that at least after long twenty years of ground level activism regarding Koel-Karo they have been called at the premier institution of knowledge to speak freely among academicians and students, which is of great value to the them. In fact, I had not to take much pain to answer any more, but one point I now keep safe for a final discussion which is about the ethical justification of people's participation in academic ethics when something is on fire. Do long talks help any way when the house is burning?

The discussion was arranged on the Fourth of April, 1998 in which a large number of teachers, students and common men participated. The panel of contributors in this case was not small as it demanded views of many parties to be incorporated. Among those who participated were Mr. S. K. Chand, IAS, who had vast experience in river valley project administration and a devout Gandhian thinker; Ms Dayamani Barla, a tribal environmental activist who is also associated to news magazine "Janahak"; Professor Indrajit Dey, a renowned ecologist in the Society for Rural Industrialization in Ranchi; Sri Meghnad Bhattacharjee, a social activist of much repute; Dr. Ramesh Sharan, economist and social activist; Dr. S. K. Sinha, botanist at the J. L. N. College, Ranchi who has research experience in the field in India and Japan and noted environmental activist and awarded documentary film maker Shri Sriprakash of Ranchi. The discourse was led to a consensus by Dr. N. Dayal, noted botanist and ecologist of Ranchi University. As I said, many activists of the Koel-Karo project contributed in the dialogue immensely.

Let us now find out the nature of the questions raised to evaluate the moral position of the participants in this regard. The questions should be in this case simple, straightforward and searching for moral opinions regarding the matter rather than becoming technical. The framed *questionnaire* is provided below:

(1) What kind of a relation do you conceive among man, nature and development? What according to you is the morally acceptable relation?

(2) Do you think that harnessing freely flowing rivers by constructing big dams to produce electricity pose certain problems? What these problems are? Is there a problem of moral desirability as well? If yes, why and if not, why not?

(3) Do you agree that man as a being of higher and wider abilities have every moral right to grow and develop? Do other beings and things like rivers and trees have the same rights? In this context how do you look at the moral aspect of the need of power for man and the rights of rivers and other things in nature?

(4) Kindly compare the moral value of the modern ways of producing bulk power at huge investment and the scientific ways of producing power in alternative ways at lower investment.
(5) Kindly shape the points of agreement and the decision-making rules you desire to see effective in power and water management in India.

Let me now explain briefly the moral position of the participants one by one in the context of the replies given by them to these questions. Mr. S. K. Chand (who died recently) is of the opinion that harnessing rivers is a problem closely related to the displacement of the natives living along a river and the irreparable loss of the natives of the areas alongside the rivers. Mr. Chand holds that tactful and at times forcible land acquisition causing displacement to initiate a mega power project is a great vice because it is against the broad social interest. We can call it immoral on the basis of the Gandhian ethical thought that both means and end of human decisions and actions should be valuable or good. We, human beings are in inseparable relationship with every part of nature, animate and inanimate, and our end in this relationship is to restore it to its full and not break it but equally true is that we need to develop and flourish, which however can never be at the cost of our end of restoring the relationship said above. Hence, any means of human development that goes against the realization of the good end is evil itself and, therefore, immoral or vice. For example, one aspect of human development is to generate power then such thing should not be done in a manner adverse to the natural relationship as a whole, it ought to be immoral. Another point is that if power generation is one social goal of development its means like harnessing a serene river by building mega-dams, inundating large areas and displacing people from their perennial abode is essentially evil and thus immoral, though the goal of development by generating some power in itself is not evil but good. So the point is the way power producing should be taken up in our societies.

Mr. Chand thinks that there is a lot of debate by the government machinery to justify the value of the means of displacement of people in greater economic benefits backed up by "proper rehabilitation" of the affected people. This is according to him a complete misconception on the part of the planners because neither displacement should be evaluated for good on economic basis nor should complete rehabilitation be taken as building some homes for living as among many things lost, the lost cultural values and ways of life are never to be regained by the affected people. This is not a blank talk because one, who also has worked in this field, knows the truth. In fact, the immorality of the 'means' as he calls it, is justified in another way, that is, the exploitation of a class of people by others or giving benefits to a class at the cost of others, which is against the moral end of a 'just natural relationship'. Sri Chand presses the point that if we want to understand the morality of developmental projects, like building a dam, we should keep human-nature relationship, the fundamental of ecological morality and the end-means value in mind. Over and above all these things Mr. Chand holds that the basic philosophy behind all mega projects like mega-dams and multipurpose projects is "violence towards the entire ecological relationship" because "harnessing" or "taming" is nothing but "violence" against nature, which as per his Gandhian morality stands out to be immoral because it is non-violence which is good itself not violence. Gandhian morality, therefore, requires first to put a moral check on ourselves to curb maximization of needs and demands and on optimization of ecological exploitation for unharnessed human urge to develop and to get pleasures that encourages violence in us.

Ms Dayamani Barla emphasizes an intricate relationship of man and nature but says that the need for human development, particularly those of the poor cannot be neglected and as such there should be every opportunity for them to enjoy their basic right to grow and develop in nature. Nature in turn takes care of human beings if they do not make nature a scapegoat to their benefits alone. Hence, morality in human nature relationship is to maintain a perfect ecological balance by taking notice to rights of natural things and also the rights of human beings, particularly the needy ones. Ms Barla said that while we speak of a widespread development as claimed by power projects which needs harnessing rivers at large scale, we should keep the "cultural" factor in mind because a disregard for the cultural values of those people who are ousted from their peaceful abode for the so-called development, is immoral. A moral talk without the restoration of the cultural values of people is hopeless in this context. For Ms Barla the greatest damage multipurpose projects do by building big barrages is systematic destruction of the cultural identity of the original inhabitants of a river-valley civilization. An intentional attempt to destroy such a thing, which is irreversible and never to be paid back, is unethical if benefits of a few people are kept in mind. Not only this, it destructs the human nature relation, which is so old and original among the inhabitants of these areas. This destruction of relation is also immoral. When Ms Barla was informed about provocative article in a leading Indian newspaper entitled: "Is it proper that greens should always go red?" She retorted that these big house newspapers publish what is for the interest of a class who need most of the produced electrical energy at their offices and homes but do not know or do not care about cultural values and cultural identities in relation to ecological destruction by undertaking large power projects. A class that prefer maximum exploitation of the natural resources as well as human resources at any cost does not have basic knowledge of morality. They do not know that the maximization of use also ends up in exploiting the poor, needy and down to earth people for whom not power but nature's abode is most important. The creation of a class-based tension and providing benefits to one class is immoral when developmental programmes amounts to just this. Ms Barla said that many intellectuals and technical persons try to estimate the losses of people in multipurpose power projects by scientific method, mostly economic cost-benefits analysis, but these people are apart from river-valley culture and not a part of it, who thus fail to recognize the non-scientific ways of human loss calculations and, therefore, acting immorally. The ground truth of displacement of aboriginals owing to construction of big dams etc. is usually "forced" where a lot of evil facts are under play. When humans are forced to come out of their natural settings, with which they have a blood relation as if, they cannot emotionally withstand it. Lack to understand such human element is also immoral because we are not inert identities. Ms Barla was critical about a lot of hue and cry of the modern ecologists including some foreigners who would not raise their voice of protest in the ecological disaster of the deserts of the Rajasthan but take much interest in places where river flows and life is teeming with greenery and beauty because these people are veiled exploiters not eco-lovers and protectors. They think that they can live peacefully in such places by raising their voice and yet, coolly exploit people in their own systematic ways. These people and others do not get a basic ethical teaching that those who are parts of a culture do not need an ethical-ecological teaching from those who are apart from that culture. Simple people have simple ethic of life—not to harm nature for maximizing their benefits.

Mr. Sriprakash started with a flurry by asking whether or not harnessing those people who uphold the benefits of large dam and multipurpose projects by bringing people to their social, cultural and economic devastation was not the most moral thing to do? It would be a kind of middle-class intellectualism to ask whether or not harnessing a river is morally good than not asking whether or not harnessing people who harness rivers is morally more desirable. Any way he continued to say that in one sense the answer to the first question would amount to answer the second one. Mr. Sriprakash thinks that the real problem of harnessing rivers is with the small, water scanty rivers flowing form rocks and hills like those from Chotanagpur plateau and not from icy mountains. Such water scanty rivers when harnessed raises basic moral questions related to ecology. Coming to the relation of the people who live beside these rivers with nature as whole, it is one of parent-child relationship, one of close dependence where development is not anything adverse to the other for people know how to care for mother nature. But the tragedy is that people outside this relationship think that developmental projects do no harm to nature for there are greater benefits of people of a nation. These people do not know that in the course of constructing the Kutku dam while we were celebrating fifty years of our independence, fifteen labourers uprooted from their own soil died being drowned in fury of gushing water. Still great economists would ask: why should greens always see red? Why should they not? Is the moral question. In fact, conscious people should harness marauders because they are doing everything just for the fun of maximizing their pleasures, and thus immoral. Mr. Prakash was also critical as was Ms Barla about pseudo-environmentalists who for their self-satisfaction and gains come to the simple men ousted in the river-valleys assuming status of leader and make these people means to their ends. Their activism should be thus treated immoral. For Mr. Prakash a good human life is one of non-exploitation of the nature and men for ego-satisfaction. He said that he is most concerned about the loss of cultural identity of a class of people most oppressed and yet possess original rights to live peacefully in a place where they belong to since a long-long time, and to deprive them of such things is immoral. To bring in outsiders in large numbers in their place with all sorts of diseases, nastiness, politics and pseudo-culture in the name of a multipurpose project is immoral. However, Mr. Prakash reiterates that being a filmmaker he cannot say that we are in no need of electricity. Yet, he reminds us that for generating power, we do need eco-friendly, small-scale and alternate ways. Eco-sustainability is crucial to such development. In fact, this point of alternate ways of power generation has been stressed both by Mr. Chand and Ms Barla. What is most important according to them in this regard is the bulk of power needed for reasons not beyond the basic philosophy of sustainable development.

Coming to Sri Meghnad Bhattacharya, the noted social activist, we find more reflections on moral matters and alternative power management. At the outset, he made clear that any degree of rehabilitation of dam ousted in quantitative terms may not be ethically desirable as it fails by the specific standard of ecological ethics. For Mr. Bhattacharya, any utilitarian and cost-benefit analysis of the good for man involved in mega power projects cannot be morally justified because it is averse to the basic morality of eco-sustainable development. He said that rivers follow the basic principles of water, namely dynamicity and static. When we build a dam to harness rivers flowing for millions of years for the sake of human benefit at a very less extent for fifty years or so and that too for a paltry population in towns, cities and some stray villages, we kill an entire civilization, perennial cultures and harm the nature of river. It is due to this that such projects are immoral. In fact, eco-sustainable power management apart from several things demands restoration of life in waterbodies and alongside its course as also the original culture of a river-valley and its civilization. Mr. Bhattacharya holds that harnessing rivers for power generation is part of a greater problem of development. He thinks that there are two views of development, the *patriarchal* and the matriarchic or *feminine*. The former being a model dominated by scientific chauvinism propagated in the West depending much on mercantile mind-set, optimum utilization of goods, to benefit a group or some groups or a class of people who have the capability to purchase the produced commodities or who can pay the costs of development. The feminine model is one of care, nourishment and preservation of nature on the whole while the judicious growth and development are carried out for basic subsistence and welfare of people irrespective of class, caste, creed and ability to buy what they desire most at any cost. Mr. Bhattacharya says that ethics of development anywhere and in India, be it in the name of power management or whatever, should be feminine. One way of following the feminine model of power management is to change or divert the course of rivers in natural ways to store some of its water where it is needed most for small-scale power generation. Similar thing can be done by converting the brooks and tributaries of a river into a small water deposits to generate power. Alternate ways of power production follows the same model, which is now better known as non-conventional ways of producing energy. What is most undesirable according to Mr. Bhattacharya about the multipurpose projects is that they are commissioned often on the basis of big 'scientific bluffs' to invest huge money that goes on increasing in recurring fashion so that some people digest a bulk of public money to give something at last to some people by killing others, which becomes a complete flop-show by fifty years. The purpose projected of a multipurpose projects is mainly bluffs for intensive budget provisions. The whole exercise is self-defeating. The unethicality lies in such untruth. Reality is that if we harness a river it cannot supply adequate unpolluted water for a long time for regular use, irrigation and aquatic life to sustain. Interestingly, quite often than not when projects are completed either people have fled villages giving up farming or the land has been barren leaving no scope for cultivation or water released is too scanty and polluted to be used for farming. Dams thus systematically kill rivers by thirty years or more. Reality is that big dams also perpetuate flood and does not control it for its storage capacity is less and sudden release of water when necessary to save the dam causes huge damages. Even the river that has been harnessed becomes shallow due to silting and effluences to cause inundation along its path. As all these things do great harm to eco-sustainable development and are immoral.

Professor Indrajit Dey said that there can neither be an absolute definition of morality nor can there be an absolute standard of morality because morality depends on existential/situational factors. Time, place and persons are most important factors for changing moral perceptions and moral evaluations. For Mr. Dey morality is a state of 'dynamics' for it gets involved in dynamic affairs of human decision-making as well as in the human knowledge of nature and urge for development. These things are continuously controlled by the awareness of need generation, utilization of natural resources and rehabilitation of nature on the whole for they are basic to the harmony as inherent in ecosystems. Morality thus creeps in exactly when human beings are involved in maintaining the basic harmony through their decision-making and acting. Hence, morality should be seen as power of persons to discern the good, the bad, the right and the wrong to take proper decisions about development helped by knowledge of the ecosystems. Such subjective moral discernment tells us that the whole of eco-systems and its harmonious relationship with human beings and their development is essential. An action done in accordance to it should be termed "moral". However, there is a continuous tension among man, development and ecosystems' harmony because human knowledge of reality is growing and his desires to grow and develop is also growing, amounting to exploitation of the natural resources to use them. The result is that there is a "tension triangle" which pulls the ecosystem more and more until matter-energy kernel within the triangle collapses in a big crunch. Any human contribution to encourage the collapse is against the whole of nature, of which we are part, and thus immoral. In fact, the more is the probability of crunch, more intense are the social, economic, political and ethical crises. We can control this by ecological awareness, ecological rehabilitation and ecological harmony. It is human beings who realize the truth and yet they are responsible for environmental depletion leading necessarily for a moral discernment. Morality is thus inseparably related to our urge to address to ecological harmony. Agreeing to what has other speakers said about rivers, their harnessing and power generation, Mr. Dey reiterated that we are heading for an ecological crunch in India which is mainly due to the plundering of river-valleys in the name of multipurpose projects, which is immoral. He added that India has enough alternative power generation capacity and we do not need any big dam for power projects. What are important are generation, control and conservation of natural resources. It is interesting here to notice that Mr. Chand's 'Violence' thesis and morality as based on non-violence was an opposition to Mr. Dey's proposed model of human, development and ecosystem's relation, to which Mr. Dey did not reply when he talked at length about these things. Though Dr. Ramesh Sharan reacted sharply to Mr. Chand's proposals on Gandhian lines. The disagreements were largely on their theoretical basis of calling something unethical, which had little or no bearing on the general agreement that after all harnessing rivers is an unethical act by any standards.

Dr. S. K. Sinha thinks that morality is concerned with human actions concerning the entire ecological community and not human beings. Ecological community is a self-growing, self-sustaining and self-dependent system of which we are members. We thus establish an inseparable relationship with this community. Water systems are also closely related to us and they sustain life systems on the whole. In so far as human beings are concerned, moral actions are preceded by *conscience*, which is "God-gifted" and intentions which are either good or bad. At times our conscience, intentions and actions are in conflict situations. Often lack of moral conscience leads us to wrong intentions resulting into wrong doings. Such lack of conscience is mainly owing to an old dogmatic prevalence or pressure of cost-benefit economics in modern societies. Hence, any human action that is against the Divine spark within him, which calls for intentional eco-sustainable actions are "immoral". In this background, then we can say that all types of harnessing rivers are bad because an excess of water needs to be judiciously harnessed at times. When our intentional decisions and their ends are unconscientious or dominated by the cost-benefit mind-set only then it is unethical. He further tells us that though the problem of human displacement and cultural devastations are important, they need not be isolated from the broader issue of ecological disaster. It would be a little narrow to perceive immorality of harnessing rivers in the context of human losses; much greater loss may be perceived when humans are seen in the broader ecological relation. Prof. Sinha argues that the ecosystem is an incessant flow of energy conservation and use of energy. When we intervene in this 'flow' negatively, the entire system falls in crises. One such intervention is harnessing rivers by big dams, and is thus immoral. It is immoral for there is a deliberate creation of disharmony in the entire system. The green activists have moral authority to see red because a class of people cannot destroy the whole system. In fact, it is this class which is now barking at top about "social forestry" which is big joke to restore the damage done to river-valleys by plundering the green areas. Ecological harmony does neither depend on such jokes nor does it need mega projects.

Dr. Ramesh Sharan said that "man centred morality" is a gift of Christianity and Western modernity, which is undesirable. We face ecological crises because man can neither be taken as the rule of nature nor can he be taken as one who is in isolation self-sufficient in ecological community. If you believe that this vast ecological community is God's handiwork, it is immoral to act against the very harmonious creation of God for dams are sure to disturb the harmony. On the other hand, if you believe in evolutionism, acting against the natural ecosystem that has evolved through so many ages, is immoral as we find in case of building large dams. If you harness rivers for power generation for the sake of human welfare, it is immoral because for the evaluation of welfare-value of projects not merely present time-frame but the future time-frame and benefits of generations to come are most important. In fact, when we harness rivers, the present cost-benefit evaluation is grossly immoral when larger benefits of classes in the society in future are kept in mind. Further in case of displacement and compensations, morality depends on situational morality because the situational demands in Indian society are such that they rule out such projects to be morally desirable. When it comes to the greens looking red, Prof. Sharan says it is natural to do so because public money is plundered in bluffs. Multipurpose projects like Tenughat and Hirakund have proved to be bluffs for which people and money were massacred. This is immoral by any ethical aspect of economical decision. Coming to Gandhiji (in response to Mr. Chand, which followed a heated debate for some time), Mr. Sharan said that history tells us that even he prevented the green activists by saying that Hirakund should be made by sacrificing some of our freedom, whereas what Hirakund has done to us by now is clear. Hence, time has come that green activists should be more aggressive to prevent any further economic and ecological plunder. All these activism are moral for they fight the negative challenges thrown by modernity and by the exploiting classes. It is grossly immoral to view everything as this class views as useful.

At the end of the discussion, it is time for ethicists to go for a brief *post-corroboration analysis* and then to the *suggestions* emerged after consensus. A close look at the nature of the questions framed for the discourse reveals that they are mainly aimed at exploring the moral opinions about the relationship between man and nature on the whole. Interesting aspect in this relationship is "development" of human beings in complex societies and "power production" as one means of development. Answer to this question was expected to bring out clearly the moral basis, which people think should be needed for evaluating the morality and immorality of choosing the said means of development. The other questions were also put to the participants in view of knowing about what they think about the moral ought and moral good of protecting the rights of people living along river-valleys, their basic needs and urge to develop and grow as also the rights of nature on the whole to maintain its balance and inherent ability to sustain living and non-living beings and things. This enables us to know whether or not it is sensible to speak about the morality of ecological sustenance. It also enables us to know how people face the tension between themselves and the environment when particularly tension presses on our ethicality. Another question was about the ethic of environmental activism, in this case, those related to water and power management as also management of the greenery along rivers are pressed to show why their activism is morally grounded. Questions also press participants to look deeper into the ethically oriented decisions to be taken by policy makers, planners and administrators when we can neither forget that power is needed for us nor can we forget that parts of nature is certain to face pressure if we do so.

When we come to the analysis of the moral views of the participants we find that they may be divided into three: (a) Gandhian moral theory, (b) Theological ethics, and (c) holistic welfarism of the non-utilitarian type. It is interesting to find that the moral views of the participants were extremely clear and at times appeared to be text bookish. It is not unlikely that some of them are well acquainted with one moral theory and tried some kind of engineering in this regard being careful about not to contradict themselves when they do so. However, that hardly makes a difference as their presupposition in morality was thoroughly challenged by others. The first two theories were particularly under scrutiny by the holists, but the Gandhian theorist was equally strong in disagreement about a particular type of explanation within holism (Dey), which was no doubt extremely scientific, academic and technical. There is no doubt however that such theoretical moral differences in the justification of the value of "large scale" or "mega power projects" were not of much worth because what unanimously evolved is that deciding to build mega power projects are unethical. This was particularly clear in Dr. Sharan's analysis of the different bases of evaluating the worth of multipurpose projects. He made it extremely clear that without going into much fuss about the value of evolutionary ethics or theistic ethics or utilitarian ethics, what is unique is that there is *no moral disagreement* about the immorality of multipurpose projects to be undertaken at such large scales in India. This is one place where I would remind readers that I have earlier on raised a question about the viability of applying any one moral theory to resolve a moral crisis artificially. The dialogue here reveals beyond doubt that any obsession with one moral theory and its unilateral application is neither acceptable nor desirable. Though several moral theories are in fact dynamic in a moral resolution, the resolution itself, which is a moral consensus, transcends the differences in justification of why such and such resolution is most desirable. It also shows that rather than theory cooption in dialogue, moral consensus is most important in our *practical life*. Yet, one comes to know that some decisions of "moral worth" are made possible by almost equally forceful but essentially logically different theories of ethics. Hence, each might have some value in moral resolution but moral consensus is not reached only by virtue of what theories hold but what the proponent of the theory finds to be most ethical when his or her "moral knowledge" is placed with many others to transcend the varieties of logic of theories and intellectual scrutiny to come to something that is morally most viable. A staunch theorist or an obsessed intellectual or a snob ethicist cannot do it alone or with others. Only social people who have ability to declass themselves in the milieu of common men transcend logic of theories to the end result of discourse—the consensus. There is no doubt that in vast nations like India, group or groups of people might come up with other sets of justifications based on different theories to come to some radically different "consensus", but that does not prove that the logic of theories count as most important. I would like to stress that in different sets of moral "consensus", transcendence of theoretic obsessions shall rule supreme and a different consensus needs to be further evaluated along with earlier ones not by engineers but by people coming together in yet another "consensus" through the transcendence of logic of moral theories. A divided house in moral dialogue, which has no consensus because logic conflicted for supremacy, proves that theories are under play and that they remain isolated like that by whims of people and not by themselves. Several moral theories have inherent disposition to evaluate the moral worth of certain things *uniformly*. Strangely in some dialogue between gang of looters there may be moral consensus and a Nazi moralist might find something to agree with democrats (!) Does it not show that "consensus" itself is not sufficient because the logic of theories that made it possible, counts for separate worth? Theories apart from consensus would count most for an independent scrutiny. No doubt we can do this as we can now search for the strengths and the weaknesses of Gandhian, theistic and other theories here but for us most important is the consensus. The disagreement about theories was never so aggressive to reject the basic claims that emerged uniformly in actual dialogue situation leading to consensus (stated by Sharan). It has to be understood that consensus among a gang of robbers and Nazis is not moral as other non-robbers and non-Nazis were absent in that. Yet, when some such people are included in the dialogue such as the proponents of "greens should not see red", who hold gross hedonic utilitarianism as the best theory, they are hooted out for their glossed logical theory or simply for their anti-environmental conclusions. More often than not they do not become a party to consensus for they are unable to transcend several intellectual and mental barriers, as Nazis and robbers cannot. A democratic consensus is thus not by all people including fool heads, gross egoistic hedonists and intellectual snobs. I would like to conclude that several moral theories in actual dialogue situations may lead to a uniform consensus if and only if theories are not obstinately pressed on for unilateral resolution. Consensus about important value-laden problems does not beg one theory therefore. However, different theories are opposed in moral debates for yet another important reason, though the debaters do not doubt much about the efficacy of the theories to reach an ethical consensus. The reason is that many theories have inherent flaws in so far as sticking to them might lead to some such suggestions or prescription or rules for decision-making, which we may conveniently call, "fruits of a theory", which are averse to the basic consensual position the theory helped to reach in a moral debate. It is due to this 'logical flaw' that certain theories are opposed by others. Perhaps much heat in an environmental ethical debate is generated regarding such conflicts among what would the fruit of a theory lead to if only that is accepted for formulating decision-making rules.

Some Suggestions for Decision-Making

In so far as consensus reached, in this case of harnessing rivers has been unanimous. On several grounds harnessing rivers ruthlessly for large scale production of power and for providing other benefits has been declared as immoralp. However, there is little doubt that we do need some power for even the poor people and the needy along the river valleys or elsewhere but they may be obtained in many other ways which do not involve much capital, exploitation of nature and population and their cultures. Power and water management should be checked for ecosustainability and ecofriendliness. Mr. Asoke Mukherjee in *The Statesman* (2001) says:

> "The application of technology has always faced resistance from risk perceptions, often combined with objections on humanitarian grounds. It is nobody's case that the threats are all unreal or that objections on humanitarian grounds are all emotional and irrational. Far from it."[1]
>
> "Mr. Mookerjee continues: "Humanitarian considerations in each specific case have to be weighed against the larger

> social cost-benefit factors guided by the principle of greatest well for the greatest number".[2]

Some specific suggestions that came up in consensus are worth considering in the light of what Mr. Mookerjee says:

(1) A careful survey of the areas for which power is needed as well as for which water is needed is most important.

(2) A careful survey like this should take into account the population for which power and water are needed and the exact targets in providing generated power to the population.

(3) A careful survey as above should also take into account the nature of the waterbody to be used for power generation and its entire course along with its tributaries.

(4) Such survey should accompany study for the cultures, needs and development possibilities of the population along side the waterbodies.

(5) If using waterbody is unsuitable, other means like alternative methods of power generation should be taken into account.

(6) In case of using waterbodies for power generation, other alternative ways best suited in that area should be equally stressed.

(7) Use of rivers and lakes for thermal and hydro-power generation should be "low-key" or "low budget". Small captive generation houses avoiding displacement of native people, dumping effluents and proper fly ash management are most essential.

(8) An extensive and exhaustive blue-print for "green power generation" should be morally the top priority.

(9) Influx of maximum outsiders in an area populated by tribes is undesirable if for that force to shift local people is needed. A compulsion for adopting industrial culture or changed ways of life should be avoided.

(10) There is an urgent need to care for population control. Rural economy should be encouraged with supportive infrastructural development and development in providing

goods and services to prevent migration to urban areas causing more crunch in power management.

This may not be the end of controversy however. We may go on debating about the suggestions and the theories to be considered for ethical justification of harnessing rivers. Morality is not a static affair; it grows with dialogue which are non-chauvinist. People's participation in moral dialogue will be hardly chauvinistic as I believe, what we have understood from the debate is that there is no need to rule out using natural things and beings. Consensus arrived at does not oppose limited harnessing of waterbodies for the production of power because in our times, even the tribes should have electricity to fulfil some of their basic needs like education, information, agriculture and some domestic comforts. I do not find any aversion to these things among the tribes because these things are not against their ways of life. We may now consider the last value-laden practical problem for resolution.

SECTION III—MORAL PROBLEMS IN INDIA'S NUCLEAR OPTIONS

Introduction

Both the value-laden problems considered so far for resolution involves widespread massacre of the vegetable kingdom in India. Trees have been ruthlessly cut to settle urban people and thermal and hydro-electric projects also did the same. In fact, there are a number of projects taken up in India which involve ecological disaster particularly owing so cutting of trees in dense forests and elsewhere. One activity called "mining" and the other called "logging" is involved in plundering trees including those growing abundantly at their perennial places of growth as sal, teak and gamhar in Jharkhand. We are all witness to it. Mining activity in Jharkhand has been vigorous for it is home to plenty of coal, red hematite and mica apart from many precious and scanty minerals like uranium. The uranium deposits in the Jaduguda hills have been systematically mined in post-independent India, Indian Uranium Corporation's Jaduguda Mines, in the Singhbhum district, quite close to Tatanagar and Ghatshila, known for steel industry and copper mines respectively. Some quantities of uranium for different purposes are still mined in once lively green and dense forest area of Jadugoda. It is a place which if people are visiting after fifty or sixty years, would hardly recognize. Even we will wonder at the rocky and barren looks while travelling by train. Reportedly, the place is widely polluted, extremely hot, dangerously inviting nuclear radiation and thus unliveable by some standards, but people do live there. The state of the tribes who once infested the area could be well imagined. The state of coalfields is however worse. Most pitiable is that in Jaduguda tribals and others reportedly have deformed babies or have lost reproductive capacities.

Monazite mineral, which is so needed for nuclear power production is abundantly found in India, particularly in the South-West of India, which gives out thorium as a rich alternative for uranium ore for the breeder reactor. Rare Earth Ltd. in India has set up plants to do so. However, researches are still going on. In fact, plutonium works as the main bridge of atomic energy and atomic bomb, uranium being the building block. Nevertheless, separation of plutonium from uranium ash is extremely different. It has to be done in plutonium plant, which involves great human risk owing to strong alpha rays emitted from plutonium. Moreover, an accumulation of quantity greater than the critical mass causes spontaneous fission ending up in explosion in the plant resulting into huge losses to human being and property. India has set up such plants in Tarapur, Kalpakkam, Rana Pratap Sagar, Narora, Kakrapara and Kaiga which are sleeping Vesuvius. It is important to know that in India, uranium is still the chief source of nuclear reactors, which is enough but not unlimited. Any uranium reactor is designed to serve for 25 to 30 years, some 20/30 years less than hydro or thermal projects. It has to be charged thoroughly for more activity involving huge cost, the burden of which would certainly go to the people of the country who are not so affluent. Though in India, Thorium is more abundant and suitable as well, reactor designing and manufacturing is more intricate and problematic. For atom bombs we need fast breeder reactors capable of converting either U-238 to plutonium 239 or Thorium to U-233 because plutonium 239 and Uranium 233 are highly fissionable substances needed for atom bombs. In Kalpakkam we have one FBTR or fast breeder test reactor, involving high cost of production. Moreover, heavy water is needed in this process which works as a moderator and coolant because it is difficult to enrich uranium type reactors rather, use U-239 reactors in which heavy water is a must. Yet, heavy water is extremely harmful for living organisms like fishes, microbes etc. and its disposal after use may cause several practical problems in biotic world.

In the face of all this let us see India's march to nuclear option. Sri Jawaharlal Nehru, India's first Prime Minister said during the first conference of the Atomic Energy Department that so far as *peaceful purposes* are concerned, we want atomic energy for the generation of power. Power is the most important thing in a developing country's resources. In 1971, the Atomic Energy Commission director Dr. Sethna disclosed in the "Atom for Peace Conference" in Europe that India was consolidating three types of peaceful experiments involving the use of atomic explosive to obtain copper from deep underground mines, to construct canal linking Ganga and Kaveri and to construct big water storage at Kaveri as well as at the upper end of the North-West part of the country. Perhaps the "peaceful use map" was laid down perfectly. Yet, on November 15, 1973 Smt. Indira Gandhi replying to a question in Rajya Sabha (Upper House) said: 'nuclear explosion's possible effect on the environment and ecological conditions are studied'. And on May 18, 1974 at 8:05 am in Pokhran in Rajasthan the Atomic Energy Commission of India detonated a nuclear device equivalent to 15,000 tonnes of TNT. Smt. Indira Gandhi said 'this is the part of our research and study and we remain deeply committed to the peaceful use of atomic energy'. Interestingly, the strategic Pokhran is only 150 kms from Indo-Pakistan border.

Now the "nuclear option" of India is surely not "bomb option". The option covers knowledge, expertise and application regarding theoretical nuclear physics, nuclear isotopes, nuclear warheads and nuclear engineering expertise. Hence, to consider the moral aspect of nuclear option is not to talk merely about India's bomb option. There is no doubt however that the research on atomic energy have far outwitted the research for universal unity of mankind in a multipolar world. In fact, the former research has been officially taken up for our development and to counter threat perception for territorial integrity, which outwits the glossed utopias of peace and universal brotherhood. Nuclear option thus gains both political and economic boosters and both the agencies are controlled by handful of people, most probably like us but unlike our decisions taken. Well, there may be no harm in some knowledgeable people (!) bossing us in this regard but what if common people are misguided and bluffed for ulterior gains? Should we not care for such things? If we give some liberty to common people in such matter in a democracy, our lack of moral scrutiny comes to the fore for we are often not aware of this aspect. But if we do not have any regard for this option, we have no moral responsibility in a civil state to safeguard many among us from exploitations and bluffs. Man has opened the Pandora's Box, there is no going back now and he feels that he needs to survive some lurking dangers he has himself invited. He has to survive the threat with full social, economic, political and moral might. The present work is to concentrate on our "moral" might needing no training of moral philosophy. We have to ask that when we, the moral agents, deliberately choose to go nuclear, what should be (not what is) our moral goal knowing fully well that 'going nuclear' includes having more knowledge of theoretical nuclear physics, applied nuclear physics, nuclear isotopes and their application and expertise of nuclear explosions for human need satisfaction?

Greater trifle is that why did man start putting theoretical nuclear physics to serious lab tests? Is it due to the necessity of proving certain hypothesis to be valid? If that be so, why did they not stop after proving? Why nature as a lab was used further? The trifle is acute when we ask other questions like, what do we do then? Should we stop inventing things useful for mankind? Should we not take some steps of deterrence to prevent nations from bondages? Let us face these questions, which take us to the heart of what we call "moral questions" in India's nuclear option.

Framing Questions, Selecting Participants, Discourse and Suggestions

The problem in consideration calls for two groups of people for immediate participation in moral dialogue. One group should consist of military personnel who know ground realities in war situations, it is also important to include psychologists and military doctors in this group. The other group should be a mixed one consisting of nuclear physicists, social scientists, Gandhian thinker, ecologists and also educationists. Though it is important to include some scientists and planners who have work experience in atomic research centres, I could not manage one in Ranchi. Though such views are found in several articles, which will be used by me as supportive to the views where it suits most. Keeping these things in mind I selected Father Johnkheere, Rector of the St. Xavier's School, Ranchi; Sri Anil Karna, ecologist and Gandhian thinker of the Society of Rural Industrialization; Dr. N. R. Roy, nuclear physicist of the Ranchi University; Dr. Amol Ranjan, psychiatrist of the Ranchi Institute of Neuro-Psychiatry and Allied Medicines; Dr. K. R. Banerjee, formerly doctor with the Indian Army and later on director, RINPAS; Mr. P. R. Bose, Retired Major of the Indian Army and Director of the Heavy Engineering Corporation, Ranchi; and Mr. Ravi Kumar, Retired Captain, Indian Army. Other participants included a number of teachers, students and ecological activists.

Let us now talk about framing of questions. I have raised some basic questions which give us some idea about moral dilemma regarding this issue. As I talked to some people about this matter and read some opinions of people, I had an idea that the moral dilemma emerges for the obvious misunderstanding about nuclear option being a bomb option. Many people, who are not in the military service and are not politicians, believe that even if nuclear option is a bomb option in modern wars, there are certain strong moral arguments in favour of that option which we cannot pooh-pooh by being sheer peace campaigners and environmental activists. On the other hand there are many people who not only find these justifications essentially immoral, they disregard value of a "peaceful use" of nuclear power as well. Much moral controversy is about the notions of the 'territorial integrity', 'freedom of the nation', 'threat perceptions', 'basic deterrence related to self defence', 'modern wars' and 'strategic international policies in balance of power'. The moral controversy is also about the politics and economics in interplay of arms race, arms deals and the survival and growth of war industries in nations much depending on it. There is thus a lot of hullabaloo about nuclear option being humane or rather, extremely beneficial to human race as such and on the other hand much heat is generated in its anti-human and anti-environmental stance. However, nothing much can be gained in the play of words unless something concrete has been derived through a dialogue among divergent views or else most of the time a nation's decision-makers in such matter are swayed by several non-moral and anti-moral factors.

Keeping these things in mind, I framed the following *questions* for discussion:

(1) Kindly evaluate the moral value or moral worth of the nuclear option in its broad sense keeping in mind the following questions: Is it good in itself? Or, good only as a means? Or, is it evil in itself? To answer this clearly, the basis for calling it either good or evil is needed.
(2) Is the first order theoretical nuclear physics prior to second order applications of it? Does the second order application, including those in explosions, warheads etc. have some worth? What are they?
(3) Is "peaceful nuclear use" a misnomer? If yes, why? And if not, why? Kindly refer to the effect of such use on environment.
(4) Kindly evaluate the moral aspects of war in view of human and environmental disasters. How would you now fit in the morality or immorality of nuclear war in this context?
(5) Keeping in view the consensus emerged regarding this matter, kindly enlist some decision-making rules.

As usual the first question tries to bring out the normative basis of the nuclear option as well as the proponents' main arguments about its moral value, which should go along with the normative general structure that the proponents have built. The second question incites a dilemma in proponents' mind by inviting them to accept the historical truth of the priority of theoretical nuclear science and its laboratory proving and then wants to know what instigated the world and the environment at large as a laboratory for its proving, particularly through explosion-experiments and as warheads. Expectedly, in answering the later half one should recall what has been said in the first answer. Careful scrutiny of any contradiction therein is necessary. The third question tries to bring out clearly the sense in which nuclear use may be called "peaceful" which is one strong basis for calling it morally worthwhile. It invites the antagonists to attack the protagonists' position or else come to a handshaking position. The fourth question is more searching about the ethic of war and peace, and the role of nuclear warheads in war if war itself is desirable. It also invites protagonists to shelve peace for war or admit the moral value of peace and curb war as well as nuclear war (also warheads). The final question tries to formulate some suggestions for our future decision-makers regarding this matter.

Let us now come to some details of the moral dialogue. Father Johnkheere was of the opinion that the nuclear option in any sense, narrow or broad, is not an end-in-itself rather, a means towards an end, the end of nuclear option should be the "general good of the people of the country" which is according to him the most valuable end and as such good itself. However, Father Johnkheere was quick to remark that though nuclear option may be seen as a means, it is good morally speaking if an only if it is used in peaceful ways because the general good of the people of a nation cannot be achieved by a violent use of nuclear power. For him, in no case, 'war' can be justified as a good itself or morally good because it does not aim at the general good of the people, which is the highest good. It is unacceptable for him to say that for general good we need war because general good depends on our freedom and territorial integrity. A nation's strength depends on peace, non-violence, non-interference, non-expansion, non-aggression and universal brotherhood and not on its fighting strength. Hence, nuclear option as one option in "war" cannot be ethically good because it overlooks the above moral values as essentially for the conception of "good of a nation". Father Johnkheere said that historically nuclear physics is prior to nuclear tests, and as it grew, nothing in it could be called "vice" or "evil". In fact, it is hopeless thinking that we can stop any scientific advance as this one by just projecting their adverse uses by mankind. All scientific inventions, theories and justifications are good theoretical-mechanical efforts mostly aimed at the "general good" of man but it is we who have to control our sensibilities and our moral sense to make better uses of scientific advances for our good, that is, our peaceful, non-interfering flourishing of character as well as for our material and spiritual progress. In this sense there will be no reason to use nuclear power for our good, the sense of which is clear by now but it never includes the option of a bomb in a broader sense. We have to learn to control the use of nuclear power barring its use in wars. Bombs can never be glorified and justified, not even in the name of saving a nation because a nation with a strong morality and ideology faces no threats. Our violent perceptions cannot lead us to use science to satisfy our whims. Father Jonhkheere recalled in agony as a Belgian that he knew what war means, and yet, he knows why after the second Great War entire Europe was able to avoid another such war. It was due to growing moral power of nations, the urge for good of people, removing false perceptions and with it using nuclear deterrence to a minimum extent to ward of threats on nations. Though still tinged in some kind of fear perception owing to two wars, which was but natural to them, a bare minimum deterrence cannot be declared immoral bluntly. What is most important now after so many years is that we better shed residual fear of mind, be morally strong, and flourish our characters than more deterrent use of nuclear power. Father said that there is a greater state of environmental urgency in the war torn and the war-prone east to prevent nuclear tests for deterrence and so-called peaceful purposes every now and then because we are more and more causing damage to our natural environment. It also hampers our nation's air, water and biotic life. The environmental fear is now more than fear of a bomb. It is thus of great urgency that all concerned professionals come together to corroborate views as we do today to find out the ways in which we should use nuclear power so that our environmental serenity is protected and yet our general good is realized. Father reiterated that the domination of aggressive capitalist war industries over the world community should be stopped somehow by more purposeful dialogues and moral teachings.

Mr. Anil Karna was of the opinion that we should not perceive human life in an isolated way rather, it should be perceived in the context of a teeming "collective life" or a life lived in huge ecological community. However, if we look human life in isolation, there is no denying of a tremendous energy inherent in them which is mainly directed to know reality more and more. Next, human energy should not be directed to an end which disregards the value of collective life in world because among other things human pursuit to 'know' is to know the value of collective living. Now keeping these basic things in mind, we can say that human endeavour to make use of the nuclear energy is to put it as plainly as good as means because it helps us in knowing how human good could be enhanced without much environmental disturbances. This urge to "know the ways to enhance human good in the world by environmental safeguards" is morally valuable end. In another sense, when we look into nuclear science as one human venture to delve deep to unfold the basic mystery of sub-atomic particles, it is also good as means. Clearly, the good of nuclear science and nuclear option or use of nuclear power depends respectively on contributing to "growth of knowledge" and to the "good of human beings" who have urge to live a collective life. This shows that if the nuclear science as a means to know some fundamental truth does not do so rather tends to know how to kill or destroy, only then it is evil as a means, and if the nuclear option is against the whole of existence, it is certainly evil as means. Such evils should be warded off anyway, but neither the science itself nor the option itself should be discarded. This proves that a blanket ban on the nuclear science and nuclear option when it is not against the collective good is immoral. There is a point in this regard. When we speak of 'nuclear option' as inculcating the option of 'nuclear explosions' or 'nuclear explosives' (not in the sense of a bomb explosion), it is good as a means *i.e.*, needed for human flourishing or collective human good because if we did not know the inherent power of nuclear fission and fusion and the ability to blast of stiff resistances, several human good will not be made possible. Many nuclear explosions at a controlled level is thus good as means because the basis of evaluation as has been already stated, is collective good, which includes human good. Mr. Karna disagrees with people who think that morality and moral education has nothing to do with people suffering from glorification and use of nuclear warfare. Though he thinks that ethic of nuclear deterrence depends on much moral churnings than its bare acceptance. The process is however a continuous one because ethical churnings about such things continuously control our egoistic desire fulfilments and to know about the value of a flourishing biotic life. The ethic of non-violence and self control is most needed to evaluate the value of even minimum nuclear deterrence. Mr. Karna thinks that the ethic of war has transformed after the great wars. In the Great War eras the ethic was 'peace through destruction', even nuclear destruction if needed. This has changed dramatically now because we have more of wars in our mind or psychological wars because by it we fight for subjugation of a so-called free nation having territorial integrity. Hence, freedom and territory does not mean anything if a nation is psychologically at war with others and then loses it. It is a global war in another way violently. It is here that nuclear deterrence and nuclear blackmail comes out with their negative values. There is a wide economic war in the name of war industry and nuclear warhead production and sales, all based on the same psychic warfare. The fear and the gamble for war at the mental level cannot be ethical either for it is against "collective human good". What destroys man from within is thus immoral. The ethic of war should also be valued in the context of world citizenship and ecological communities' welfare. The ideals of community life, world citizenship and ecological community's larger good are ways to flourish us horizontally and vertically because it has a strong ethic that covers all and moves up. The neo-American ideal of a vertical civilization through imposition of power including nuclear power to blackmail is, therefore, not ethical. Time will reject this unethical growth-war-based-civilization. A global civilization will thus have a nuclear strategy which is morally valuable and ecologically unharming through meaningful dialogues.

Let us now come to what Dr. K. R. Banerjee says. According to Dr. Banerjee moral value of a thing depends on how it is used for the sake of humanity at large, that is, for the "welfare of human beings" on the whole. Another thing is that the intention with which action is done is the measure of its moral worth. If the intention of man is to contribute to the "well-being" of man, the intention is morally good. Keeping this in mind we can say that an intentional human action which does not care for environmental protection or protection of the value of an unblemished nature, does not contribute to the well-being of man and, therefore, such actions are not morally good. Similarly, if our intention is to do well for a section of human population or only a section of nature by harming other section of population and nature, it is morally bad because in the final run, the end of human actions should be, "well-being of humanity at large". In fact, environmental well-being is good only in this broader sense. If a nuclear choice or option is intentional, it should be good and not bad as means to attain the said end or else it will be also morally bad. However, the intentional nuclear option today is either destructive of well-being of human race or a part of it and against the environment or a part of it and, therefore, unethical. Nevertheless, if that intentional choice also means "self defence", it is good as means because well-being of humans is not possible without protecting their self against harm. If one party wrongfully wants to harm another party, the other party should protect from harm. In this sense, if nuclear attacks are intended to uproot people from world, people have moral right to retort adequately or else they will be doing wrong to themselves. It will be foolish to wait for an environmental protection than protecting own life and nation in the face of a nuclear threat. Though there is little doubt that pre-bomb research and development had nothing to do with deterrence and ethics, the mass production of bombs during war as a measure of not self defence and deterrence but of domination, was morally wrong. Though the position of those under threats of being dominated during wars used bombs as deterrence for their well-being and was thus ethical. As the bomb episode is now over, we raise the ethic of nuclear bombs afresh because in view of global welfare which has to regard environmental warfare, we speak of ethic of further bombs as deterrence. Bombs as deterrent value has really no value now because we now think of wider ethics of protecting human life, human welfare and ecological well-being. In the view of our renewed understanding of nature's care and not to kill, we evaluate its well-being with that of human beings and thus disvalue even nuclear deterrence. Dr. Banerjee asserts that to imegine he is a supporter of the philosophy of "world citizenship" and "universal brotherhood" and thus disvalues deterrence will be a wrong estimate. If a poor nation like India glosses these high ideals for being anti-nuclear and against deterrent use of nuclear power, it will be a wrong conception. The world is changing; we live in a world where morality is highly relative to the well-being of our people. Truly then, for our nation, our people and myself, first we must acquire enough power to develop, grow and flourish. The good of world community and the ecological community will figure surely but next to it. In view of this, if we go for basic nuclear scientific research, development as well as nuclear non-violent uses including preparation for self-defence they are all morally good for they all keep in mind "our good" first.

Dr. Amol Ranjan agrees with those who think that "the survival and welfare of the human race" is what is morally desirable end of life; he adds that the means towards the attainment of it should be good as well, the means cannot be evil. He believes in the capacity of a doctor that nuclear option is a valuable or good means to the end said above when it is used exclusively in medicine. He comes to it while formulating suggestions. Dr. Ranjan thinks that even research and development of nuclear bombs may be seen as good means if that is for the larger good of people of a nation, its environment, where the good is guaranteed by 'no use'. Hence, deterrent value of nuclear bombs is undeniable. In fact, most of the morally conscious people of civil societies and even those of the nuclear club are using bombs as deterrence. They are all conscious about the environmental good as well but for that nation's welfare cannot be a scapegoat. Basic deterrent use of bombs is not against our environmental good. There is no doubt here that a loss of balance of deterrence leads to anxiety reaction and combat realities. We have to learn about this balance of deterrence. Dr. Ranjan dealt at some length the Freudian analysis of deterrence and peaceful use of nuclear power. At the initial stages of our moral development, we are at the stage of 'Id' where we act for our desire fulfilments. While we were at the levels of nuclear research and development and lab explosions, we were at this level of moral maturity, which did no great harm. At later stages of continuous research and maturity of the theory and its various capacities, we were at the stage of ego with regard to our moral development when we tried to master nature and others on the basis of fruits of research to boost our 'ego'. It is at this stage that we became immoral about using nuclear knowledge to plunder nature and dominate man. At our ego level of moral maturity we went unethical therefore. Dr. Ranjan thinks that mankind is swiftly passing this stage to reach the stage of superego when moral development of man will be at the height, his personality will grow, his thought level uplifts and gets refined. He will then surpass all to think of the welfare of the whole of nature to scrutinize more the value of nuclear option as a deterrence. Dr. Ranjan says that nuclear science will reach the climax of its growth at this final stage of moral development, when no agency can boss on our free will to go for even a deterrent use of nuclear power. Our environmental concern in nuclear option is one instance of such higher moral development.

Dr. N. R. Roy thinks that nuclear option is an intentional option of man. We, the rational beings, are thus most important parties with regard to this option and for that reason; it is for us to decide what should be done and what should not be done. The intentional human choice for opening up the vistas of the nuclear world opens in turn (for the rational beings), to choose between the constructive and the destructive aspects. In fact, the constructive and the destructive aspects both have their origin in the fission of heavy nucleus. Any act such as fission and fusion involves the nature on the whole, of which we are a parts and, therefore, in case of the nuclear option man and nature should be considered carefully to evaluate the constructive and the destructive facets of the nuclear option. If everything depends on our intention, its control and purity must be important, and that is possible if an only if we remember the basic guiding principle of human intentionality which is the principle of "peace and love in the nature on the whole". When we keep this in mind, we can easily see that our destructive intention, which is an evil itself, is so because it is against the said principle. Hence, we can conclude that what is in itself evil is against nature and man, should be stopped forever. In so far as nuclear option is for the maintenance of peace and love on the whole, it should be encouraged forever. Here, Dr. Roy dealt at some length which neuclear choice is in accordance with this principle and listed them clearly. What Dr. Roy suggested in the main is incessant harnessing of the unlimited energy for human race to live and grow in the world peacefully and in a most ecofriendly manner. However, he believes on several scientific grounds that the non-explosive use of nuclear fusion is more desirable than nuclear fission. Dr. Roy says that it is more emotional than rational to say that the nuclear option has been deliberately kept open in the name of peaceful use by people of vested interest. Even the scientists do a lot of business and blackmail by keeping this option open. In fact, the value of nuclear research does not lie in its economic use and benefits but in the furtherance of peace and love in world. A true love for unblemished nature leads us to such options though for basic minimum of mankind we do use nature; manipulate it even at times by the nuclear technology available to us. Dr. Roy argued that it is a historical truth that the first post laboratory, large scale industrial manufacturing and use of nuclear power was destined to make and use a devastating bomb because some people thought that only this way peace would prevail. However, this is a half-truth, the research was not only to make a bomb and use it. It is only an unethical use by scientists from 1925 to 1950 which we project too much. Truly, the famous "The Physical Review" journal published from the USA devoted 150 pages out of 200 pages in one issue during this period on the devastating aspect of nuclear research but that is not everything. Most of the later American scientists after that retorted against it which was further published in the Frank Report on moral grounds. With this the ethic of nuclear research came to the fore in the light of a lot of cry of peace and love. To cry for peace and love is not enough, ethics needs action, and for that we need to use our knowledge of nuclear physics constructively. Dr. Roy thinks that unless our politicians and decision-makers do not have an ethical vision of research and development, speaking about the ethic of nuclear research by us will be of little use.

We finally come to the moral opinions of both the heroes of our national defence who, like us, do not talk about these things on paper for they have seen ground realities of war and peace. To start with Mr. Bose, he holds that any value judgment of nuclear option should be made on the basis of the end principle of "knowledge of nuclear dynamics". Clearly, a nuclear option which is in accordance with this principle is good as means. In case of nuclear choice only as weapons for mass destruction and environmental devastation, it is obviously morally bad and not good as it is in complete abeyance of the basic principle. It is not that nuclear choice for mass destruction etc. is not backed up by knowledge of nuclear dynamics but knowledge properly so-called has been lost in the midst of the destructive intentions which is most predominant. Yet, we can by virtue of our intention-transformation completely transform our knowledge of nuclear dynamics per se and not for any other reason. A complete knowledge of nuclear dynamics is thus appreciative of the means of nuclear use in nature most judiciously and without much intervention. There are many instances, where nuclear option is good as means particularly caring for man and nature on the whole, such as, knowledge of nuclear medicine, nuclear isotopes, and nuclear reactors for power generation and so on. According to Mr. Bose there should be no moral cry over the nuclear science and the basic structure and purpose of nuclear research, technology and development. Ethics-talk is viable when in the course of nuclear related studies and production; negative elements like "mass destruction" and "environmental devastations" are intentionally encouraged. But this is in our control for we are responsible in giving our intentions a negative swing and it is we who has to control it. Moreover, any ethics-talk about nuclear science or research as well as production of nuclear warheads for deterrence is meaningless when our existence is at stake for first we have to exist and our nations should survive at the face of cruelty imposed on us by others. It is at this point that we understand why an ethical truism that 'to stop a war, blast a bomb' is good enough to uphold. In fact, when we are free, we know, do research, develop and talk about ethics. A talk about environmental ethics also depends on our freedom first and then deterrent value of nuclear warhead production could be realized. It is due to this that Mr. Bose disagrees with some views expressed here regarding the value of nuclear deterrence. He says that nuclear option is also an option of deterrence and in that way good as means for apart from putting into action what we know about nuclear dynamics, we answer fittingly to those who want to disparage our peace and identity. It is obviously moral on our part to do so. This is understood when we refer to one example cited by Sri Rāmakçùõa Parmahamsa in *"Kathāmçta"*. It is urged that a cobra should at least hiss and show its hide if not bite when people thrash it. This makes clear that rather than much hue and cry about environmental disaster, nuclear warhead production for basic deterrence is most judicious and moral because deterrence prevents greater evil. Yet, if deterrence crosses threshold, it passes the moral checks to become an evil. In so far as the concepts of "world citizenship", "ecological community" and the ethic of war are concerned, Mr. Bose thinks they are good slogans and utopias but not realizable because the state of affairs today are such that ethics is mostly ethics of survival first with all honesty and dignity. For dignified survival, therefore, nuclear deterrence as an option is ethically desirable. Summing the entire position Mr. Bose said that the nuclear option has three aspects: purely intellectual, peaceful technological uses and judicious deterrence for survival with dignity. The first two have tremendous moral value, ecological moral value creeps in both. The third one may be largely against the ideal of an utopian world community but it is moral when national dignity is viewed as important as utopias. Truly, he says that we have developed from a tribal to a national culture and may be we grow to a global culture but as we stand now, nuclear deterrence is good as means, as good is the peaceful uses of knowledge of nuclear dynamics.

We finally come to the views of Mr. Ravi Kumar. He said that there is no doubt about the nuclear option as a valuable means but the means might take two forms: constructive and destructive. In the common parlance in the fields of power generation, medicine and in cases of high speed ships and submarines the nuclear choice is outstanding also in view of our environment and for human beings. However, we should also realize that the destructive aspect of the nuclear option such as it shows in test explosions is valuable to a country's defence but we need intensive research to harness this aspect. There is no doubt then that the overall good neuclear choice with some checks cannot be labelled as immoral on several grounds given to us and on environmental reasons as well. In this regard then nuclear test explosions and warhead production to a limited extent is morally valuable because here the basis of morality is national identity and safety. Yet, equally true is that in the name of this we cannot morally allow use of nuclear warheads at wars anymore. Deterrence in that sense must have stopped greater evil like another Hiroshima and, therefore, moral. In fact, the judiciousness of nuclear research, development and use needs great moral sense and moral checks worldwide. It is not unlikely that such checks include caring for the environment as any amount or degree of test explosions may be as devastating as a war. Mr. Kumar is not prepared to generalize that we are global citizens and members of ecological community needing no interference. He rather prefers to say that he belongs to one community called India, whose environment is very valuable to him primarily, and as such he is concerned about the well-being of the people of India including himself. If some nations do not like our perception of freedom, peace and flourishing, I cannot idealize myself as a global citizen and for that reason, despite some risks in my environment, I would retort most ethically for the sake of my country by using nuclear knowledge as minimum deterrence as well as warheads though people like me elsewhere may be affected. One can thus, in the name of a global pseudo-morality, impose a test ban on us as is USA doing in the name of NPT and CTBT by providing war-psyche worldwide. Similarly nations like China and Pakistan are continuously putting threats on us. What is immoral if we retort to them? In dire situation as those, we cannot perceive global morality and ecological morality though there is no doubt that peaceful and deterrent uses of nuclear knowledge is morally just.

At the end of the dialogue and a long session of reactions of other participants in the house, a consensus could be reached. It was beyond doubt that morally speaking the following are right and just: (a) taking nuclear science and knowledge of nuclear dynamics as good itself; (b) using such knowledge for greater human well-being; (c) nuclear power generation as an ecofriendly step; (d) judicious nuclear deterrence including test explosion and warheads to an extremely limited degree; (e) growth of moral perceptions among nations regarding peace, non-violence, non-interference and no nuclear use in wars. It was thus largely agreed that "nuclear option" in India is morally desirable as a good means to national welfare which is not too exploitative of the national environment and national health. A limited exploitation cannot be ruled out for national interests, particularly strategic defence interest.

We should now analyse the moral dialogue to find out the normative element inherent in it. The dialogue reveals an interplay of the following normative standards: (i) human welfare at large; (ii) human well-being of population of a nation; (iii) welfare of a collective or the whole of ecological community; (iv) maintenance of love and peace prevalent in nature on the whole; and (v) knowledge and use of nuclear dynamics for national identity and esteem. Though they may appear as so many to us, we may conveniently club some of them to conceive the norms of *anthropocentric welfarism* and *holistic welfarism*. There are reasons to do so. The said standards as expressed in (i), (ii) and (v) above, may be clubbed without much logical flaw to "anthropocentric welfarism" as speaking about human welfare either of nation or of nations taken together, is essentially "human centred" and the human centred morality hinges on one important aspect called "welfare" or "well-being". This concept as we have seen before, have some inherent vagueness for its meaning. Its meaning is nevertheless clear to us when we gradually unfold the sense in which we use it in the moral contexts. In this case our participants have used it variously as 'beneficial', 'useful' and 'flourishing materially, spiritually and morally'. It hardly makes any difference to mention about a nation in this regard for any nation taken separately would go for the same welfare as stated here. There is also nothing different when we speak about knowledge of nuclear dynamics for the last part of its qualification is most important. It clearly states the usefulness or benefit of such knowledge is evaluated in terms of people of a nation. Even though all these anthropocentrists are not blind about ecological welfare, like naturalists, they would speak of nature's judicious use as well. However, they would further qualify it by saying that doing such things is for the good of human beings. Really speaking, the debate reveals (iii) and (iv) as clubbed together for in these cases we have ecological or holistic welfarism where not merely benefits and flourishing of people are important but equally important is the flourishing of the whole of nature where we live and grow. It is undeniable that elements of love, peace, brotherhood, unity, communality etc. add different academic flavours to such welfarism; what is most important is that the inherent love and peace as found in nature is much *metaphorical*. A plain thinking reveals that nature on the whole is viewed as sustainer of living and non-living in a fine-tuned natural manner, which is its basic structure and to disparage it is morally evil; whereas, any act that helps in or contributes to maintain this structure or at least backs it up considerably should be evaluated as morally good. The difference between the two standards lies thus not in the concept of welfare or flourishing rather in who needs to flourish. The latter is broader than the former, which however, rules out any constricted national welfarism. If we look at anthropocentric welfarism, it does not comfortably agree with nationalistic welfarism, which might soon turn to be regional, overplaying with the factors of existence and esteem. Though our participants did not have such ideas because their nationalistic welfare outwits all kinds of regional welfare for they are ready to sacrifice regional benefits for the good of the nation. Still they would not in any case entertain ecological existence as a greater problem when nuclear option is in question. Truly, if humans do not exist, nations do not, not even their environment but equally true is that some civilized nation as India is, should at least strive for a global moral journey. In cases of global turmoil, when nationalism is supreme, total denial of world's existence and its welfare is not possible. Moreover, though nation's environment may be so dear a matter, some nuclear tests are needed nevertheless. But this does not cover the interests of the dear nature completely because the "judiciousness" which is much touted today in nuclear tests and use of warheads is always governed by a now-growing and now-shrinking arms race, threat perceptions and blackmails encouraged by different uncivilized nations. What if the global citizens all raise up against this for yet another global fight for the good of environment as a whole? Our participants did not completely whisk off such inquiry as a 'mere utopia'.

We may go on for number of debates again. Let us leave something for days to come but here are suggestions found morally most desirable about the nuclear option in India:

(1) Intensive work in the field of Radioisotopes are needed for quality improvements in radiopharmaceuticals, fertilizers, leakage locations, casting, welding and forging, crack detection, silt movement study on seabed and polymers.
(2) Intensive research work to be done in the field of radioisotopes for quality improvements in plant nourishment and hybrid, pest control and storage of food stuff.
(3) Intensive research work in the same field for quality improvement in locating several disorders in human body as well as in various medical treatment.
(4) Careful, ecofriendly and limited uses of nuclear technology in civil engineering works. This should be done when it is most needed for the development of a backward or remote area whereby most needy and poor may be served by providing infrastructural and agricultural facilities.
(5) Intensive research should be conducted on planning design and eco-conscious commissioning of nuclear power houses. Serious thinking about worker's safety as well as the safety of local people is needed in this regard.
(6) Cost effective designs of nuclear powered aircrafts, rockets and submarines, particularly for basic defence needs should be evaluated for environmental backlashes.

(7) Limited production of nuclear warheads following no first use in wars but effectively projected for deterrence in so far as present global political situation continues.
(8) Blanket nuclear test ban on India should not be accepted in so far as threats on nation's safety and security do not completely wither away.

However, equally true is to remember what Mr. Mahesh Daga writes in *The Times of India* (2002): "The inherent logic of 'coercive diplomacy' ... an all or nothing affair. If you fail, there is nowhere to go but war, tainted or otherwise. Second, it suffers, like all good things, from a fatal flaw. In theoretical economics they call it the law of marginal returns. In life, it is common enough experience for men and mechanics. Try turning a screw beyond its last thread, and it turns into a free, rogue screw".[3]

With this we come to the end of our discussion on environmental ethics. Before I conclude the work in the next chapter, I must inform the readers of this book that I have presented only a few problems resolved following this methodology. Many other problems did not find a place here for all practical purposes. I have tried to present the readers a view of environmental ethics which is not merely theoretical and also not merely pure activism rather, including both. I have tried to show that though ethicists form an important party in environmental ethical venture, they are not all because environmental ethics is closely connected to application of ethics where people's participation is most vital. We found that the entire effort was closely associated to professional ethic as several professionals ought to have some basic environmental ethical concern.

NOTES

1. Mookerjee, Asoke, "Development Plans: Some unreal fears about new technology", *The Statesman*, Kolkata edition, 3.1.2001.
2. *Ibid.*
3. Daga, Mahesh, "Limits of Coercion: Tighten the screw till it is loose", The *Times of India,* New Delhi, 7.6.2002.

6

Postscript and Prospect

We have reached the last chapter of the book to start considering fresh set of issues which have been still not covered in the discussion so far. Most expectedly after some brainstorming some of these issues will not be left unresolved. May be in the final run we shall end up saying what is glorious about people's participation in environmental ethics. More hopefully, the end of all this will generate fresh set of issues for consideration, perhaps some other person or persons might take them up. Hence, in this chapter we have two broad things to be considered: Some problems awaiting explanation and the possibility of people's participation in environmental ethics. I have discussed these things under the heading, "Postscript and Prospect".

Postscript might start with an exclamation: Haven't we finished as yet! The answer might be another exclamation: Well, don't you feel its incompleteness! Both expressions have their own grounds though, for this work on environmental ethics on the one hand still does not do the following: (a) weigh "metaphysical justifications" in such ethical contexts; (b) weigh "technological progress" in such ethical contexts; and (c) weigh "human interventions" in a maligned nature for some patch up work. On the other hand, others might feel that after a thorough discussion of "nature and tasks" of environmental ethics which includes "some religious justifications" and even models of applying ethics in environmental matters for resolution, much has been said about it. The other part, that is, Prospect of people's participation in environmental ethical matters does evoke some concern for its incompleteness and not completion as we have to do the following: (a) to weigh intersubjective corroboration itself with regard to application of ethics; and (b) to answer if such things are at all desirable when some people feel that they are in socio-economic and political bondage.

Taking the first thing, that is, weighing "metaphysical justifications" for further consideration will be good enough even after we have seen in a separate chapter what some religious or to be more specific, some theological-religious justifications are. We know that not all religions are theistic as are the cases of Jainism, Buddhism and some Tribal religions, though they do have deep ethical considerations in environmental matters which are nevertheless, "metaphysically" based. Yet, what these so-called "metaphysical justifications" have in store have to be scrutinized. What I mean here by "metaphysical justifications" covers the theistic-religious and the non-theistic-religious justifications which have in common, the idea that human beings should act ethically in their natural and artificial environments because in doing so human *transcends* all *physicality* or moves beyond worldly relationships. This 'beyond worldly' end of life is eternal, whose value is restored in the recognition of the greater value of "another world". Clearly therefore, what marks almost all "metaphysical justifications" of environmental ethics is that our ethical concern about environment truly speaking, is "trans-worldly" though by that, neither worldly life is negated nor worldly life (with world itself) is abnegated. Not all 'metaphysical justifications' are of this nature for we are here cautious about some possible views which though metaphysical, are not 'trans-worldly-metaphysical', they have another interesting point to make. Some unique "metaphysical-ethical justifications" are concerned about sustaining nature in its totality not for the sake of its several material values but that it is truly "spiritual" and for the reason that humans are also in essence spiritual beings, who need to act spiritually. In the first case, therefore, "metaphysical" has a different import from the latter as in the former, we conceive of a 'metaphysical end of ethical life, which is trans-worldly' whereas in the latter, we are dealing with environment as essentially "spiritual", and for that reason what comes out to be ethical end of our life in the final run is to "act as a spiritual entity". We cannot trash this option any more because a lot have been said in our advanced cultures and religions about the spiritual nature and goal of human beings. It is most likely, therefore, that some 'other-worldly-metaphysical end of life' thesis happily embraces the thesis of 'spiritualism'. It is due to this that though the nirvāõic thesis is essentially different from the non-nirvāõic spiritualism, both come together to "metaphysically justify" our ethical concerns about environmental matters. It would not be altogether out of way to remark here that the views caring for God, divinity, deity and nirvāõa might more or less care for an ethical end in life that preaches "not to be related with this physical world any more, though coming here is not a curse". Whereas the other view we have just seen does not preach such things as in the 'spirit's entity' it is all a matter of eternal relation where no one cares about coming and going because "here we live, and here do we go for what is, lives here with us". It may well be that we do not "come and go" at all, "we live". No doubt we "should" continue doing that or else there is a deliberate "immorality". Then do we say that the "trans-worldly" nirvāõic ethic approaching environmental issues is "immoral" in a sense? Certainly not, for it holds that we transcend the physical to converge with the spiritual, be it Brahma or Allāh or God or whatever Supreme Reality, or else, attain Buddhahood. The ethical goal is thus no less "ideal"—no less "spiritual". Yet, the difference lies in the non-nirvāõic thesis not supporting the ideal of "attainment", which is not "here" but somewhere else. The former view would not agree as the nirvāõic ideal does not speak of the said "attainment", "here" or "there". It urges that in this life we realize the ethical goal but a return to the material is not possible, it is eternally transcended. The crux of the issue is that any "metaphysical justification" of environmental ethics necessarily and sufficiently refers to a "metaphysical" goal of human life, which is for some reasons called "ethical" goal of life—the cases of remaining here or not to return here though important, may be kept in abeyance. Let us take this crucial point made now by the metaphysical ethicists with regard to environmental matters for further analysis without further weighing of relative strengths and weaknesses of the two theses.

What we obtain from the mind boggling arguments (only a glimpse of these) is that the judgement or evaluation of a human act as ethical, particularly when the said act involves humans, environment and the nature on the whole, depends exclusively on whether or not such acts leads humans to a higher metaphysical (nirvāõic or non-nirvāõic spiritual) goal. This amounts to the position that the basis of ethicality is the metaphysical, aptly put as: "metaphysical-ethical" "ethico-metaphysical". This is no queerer to us as our higher cultures have lived happily with such presupposition. It has been argued with fervour that in so far as human-environment relationship is not soaked in metaphysical colours, imagining an environmental ethics is not possible. In fact, most of the philosophies of great "synthetic" nature sometimes prefer such things as in the synthetic web of things, the "metaphysical" spins threads of epistemology, ethics, religion and even logic. If you touch one string, the master 'metaphysical' is in way. You really cannot help this as the whole system is self-sufficient.

But there are questions, a number of them points towards the glaring misconception of deriving the "ethical" on the basis of "metaphysical" because the categories being obviously so different, does not leave any room for a category confusion. This is an old argument which is often met with so-called success that the categories are really not different. Then are they all the same? Can we conveniently substitute "ethical" for "metaphysical"? Can we for instance, write "environmental metaphysics" in place of "environmental ethics"? Or, is it proper to say that the "non-metaphysicians" are not "ethical"? Most disturbing point is that it encourages the argument that within the gamut of "ethics", there is perhaps little or no reason to justify the possibility of environmental ethics. For all this, you must take recourse to metaphysical arguments. Yes, ethicists are disturbed by such things no doubt but we have to consider the sense in which 'ethico-metaphysical' is used in this context. It may well be that a combination of ethical and metaphysical reasons constitute a different kind of argument to justify environmental ethics. But what is this combination like? And why such alchemy is needed? What I have got from these arguments is that the alchemy is essential for without it humans are not able to realize what is most needed, that is, they are not "pushed" enough for non-human considerations like those explained in environmental concerns. This is to say that if we do not consider the whole of existence to be extremely different from what it appears, that is, the mechanistic material, we are never "pushed" through an ethical life in environment. So it is not a point of confusing metaphysical with the ethical, it is putting things together for they cannot be isolated particularly when environmental concerns are supreme. In one sense then, the ethico-metaphysical explains without much logical hassle: Why should we be concerned about environment? And why do we (the habitual egoists), transcend ourselves to be concerned about the environment on the whole?

However, the logic inherent in such justification may be weak and not strong enough because non-metaphysicians argue that many among us conceive of the whole of existence or the ecological community as extremely valuable though not spiritual or sacred. We may not conceive of a trans-worldly end of ethical life, we may be welfarists, utilitarians, egoists, humanists, feminists, deontologists and so on proposing different ends of ethical life and thereby different basis of evaluating our concerns in environment. In fact, we are obsessionless and undogmatic ethicists having independent grounds to justify environmental ethics, we do not need spirits, gods, and eternal streams of consciousness etc. and yet extremely "charged up" people. To this, the first party retorts saying that teleological and deontological theories are really speaking staunch obsessions and dogmas in ethics which taken separately, have been proved to be not so full proof as we imagine them to be. Much problem about these theories is that devoid of metaphysical goal in life, nothing universal could be justified in ethics, which we know is the most important criterion for a satisfactory first principle in ethics. Yet, the non-metaphysical ethicist shows the incongruity in argument by saying that any conceivable metaphysical end in life is "teleological" and, therefore, it curses itself for the defects. Further, in proposing a unique "teleos" as "universal", the so-called ethicists forget its thorough batterings from several corners because non-metaphysicians, a large group throughout the world, disprove the illusory claim to be true. There are persistent disagreements about the universal status of nirvāõic and other goals of life as well as the spiritual status of the environment, and all antagonists claim to be pushed up to take ethical care about environment. A last ditch effort of the metaphysicians should be easily assumed. First, they fall back on insights, intuitions and personal realization than damp arguments with regard to what ought to be done. Second, they doubt laxity in environmental management on the part of non-metaphysicians as they have no sense of moral responsibility towards the environment on the whole as they fail to realize "empathy" and "oneness" rather, they are merely sympathetic people talking much about the whole of nature, but in reality are disguised anthropocentric egoists or altruists who put themselves first after using nature to its maximum. They are unclear about what they intend to do, and for these reasons they never agree about what ethically desirable environmental management should be like. The antagonists are not perturbed by these charges. They argue that insights, intuitions and realization are all extremely relative to us. If we really have such things as intuitions, first they have to be proved beyond doubt, for which, we need infinite number of intuitions giving us nothing at all.

Even if we accept that our intuitions is self proved or do they not need a proof, what it proposes as a sacrosanct truth is not at all of that nature, because different insights conflict and contradict. If non-metaphysical ethicists have their intuitions for constructing a first principle of morality or ethical end in life then such monolithic trends are always facing some challenges, which are put to rational scrutiny. In the field of ethics, the insightful people have been always challenged but some degree of agreement about the belligerent theories have cropped up through the years. It is perhaps due to this that monolithic trends in ethical philosophy have given way to pluralism when application of ethics in particular has come to the fore in modern times. Now this point is worth considering a little further. We have seen in an earlier chapter that the actual ground level ethical dynamics involving value-laden environmental problems needing solution projects several ethical theories as has been applied by people most unconsciously or deliberately. Or, even if some well read people were deliberate in such use, they were opposed. Yet, among everything, the theories were transcended and disagreements were resolved to give us something concrete about environmental management. Extremely interesting point is that among such views were the metaphysical views, which counted much because the proponents thought that their views were most pertinent in this matter. However, they too, merged in the ethical milieu. What difference did it really make for a metaphysician and a non-metaphysician when they all preached "welfare of the whole of ecological community"? There has always been a global tension among these parties as the one says that metaphysical overbearing in ethical matters make us really moral, whereas the others think that it makes us either unnecessarily dogmatic or becomes instrument to mass exploitation. What we found on the other hand is that both parties come together despite of any monolithic obsession in official ethics. Then do we not say that in reality, an applied ethical theory or some theories applied by some of us do not have much value? Really they do not have much value unless they help in decision-making. What an impartial and unobsessed ethicist finds in the 'post-corroborative venture' is that theories really emerge variously in a moral dialogue but they are not preferred by the public to be unilaterally applied. In a democracy, therefore, particularly that nurtures many castes, classes and faiths as also disparities in income and living as is in India, we just cannot think of metaphysicians creating moral islands for their decision-making rules or the other party doing the same. Hatred among ethical faiths does not help us, it nourishes unethical chauvinism. Countries having "official ethics" as they are having "official philosophies" rest on such hatred and also mistake because they prefer unilateralism and party-chauvinism in ethical matters. They are completely unaware that people may have strong ethical disagreements and that they were not taken into confidence. This will not help us in India—a red or a green or a saffron ethics is useless here. People's ethics is not coloured by intellectuals' opinions. So hats off to metaphysicians that they have come up to corroborate, but here is lesson for them, not everything must be right about them. It is due to this that deep realization of eco-community as divine and nirvāọa as the ideal does not prove that they will have any higher degree of moral attachment with nature. Equally true is that masked metaphysicians do plunder nature for they are having different socio-economic and political goals in life when their ethics is completely at the mercy of their whims. Those who think nature as sacrosanct need remain sacred themselves because high speaking welfarist often do not care for nature's welfare. What we think and what we do should *match* but for that we should come together because in isolation our "matching" is whimsical, a kind of self-deception. Then what do we say now, is metaphysical ethics true? There is no true or false ethics but there are either logical, or illogical ethics. I must say that our analysis reveals that metaphysical ethics is not completely sound but it has some value for it has proved to be pro-environment and pro-welfare in ethical dialogues.

Let us now weigh the ethic of technological and scientific progresses in the context of our ethical concern about environment. It is undeniable that human beings are qualitatively different from many other beings in nature. Don't read here 'supremacy' of our race as we realize it very well that amoebas and molluscs are unique and different, too. We mean to say that human beings have certain features which we now cannot help as not having them lest some factors playing in evolution does not take away something from us, which would amount to regress and not progress. The stark truth is that we now cannot help that we think about our good and good of other human beings and for that reason invent means available in and around us. There is no way that we shall continue using natural things and beings for our good such as vegetations, waterbodies, air, space, animals and even human beings. Yet, we shall talk about many things and beings whom we use for their good. Truly, we shall also continue not to care much for these things and beings. There is really no going back on our general progress for we are such beings who prefer progress and not regress. Equally true is that we cannot become hermits who live alone, does not need any thing or any being for him or her. The nature has been usually used by us and we shall continue using them for our existence. Still, in the midst of what we 'are' we continue to deliberate on what we 'should', and this marks our unique feature though not less unique is that we are dual-thinking and masked people often not caring to do something we thought, we 'should'. If we look around us, we do not find our parallel in the way we usually live in nature for we have extensively created or made 'a world within the world', that is, our created environment. Our creation has been huge, quantitatively numerous and qualitatively either gorgeous or ugly. We talk much about ant-colonies, beehives, nests of birds and so on but are they comparable to ours? Certainly not, because among other things, we have created an environment which needs not merely natural things but such things in plenty. Still, we are concerned about "judiciousness" of such needs which ants and bees might not care for they are believed to be judicious in their need and use by their standards. Our talk of "proper" and "improper" needs and uses of natural things and beings make us believe that something is wrong.

It may well be that we were not "judicious", something is running out of our control and we may be gradually slipping through a slope of calamity. Our optimism did lull us to danger while our pessimism may be now at best a warning but not a solution—we have to face it.

The enormous structural progress that human beings have made and the intricate patterns of living they have meshed were possible owing to the advents of science and technology. However, much talk about "judiciousness" hovers around such advents because they have been riding mainly for their success, on the waves of needs of people on the one hand and the extensive use of natural things and being on the other hand. It has been imagined that science and technology are the means for massive exploitation of our environment though, they were made scapegoats to cover up human follies. They were needed for noble purposes, not only for human beings but for the welfare of animals and even natural things. Hence, much blame is on people who have much need and greed but do not have balance and lack sense of rights of natural existential. They are thus charged as "injudicious" people, who are for many reasons, unethical (fools!), for among other things they were busy cutting the branch on which they were comfortably seated.

A glimpse of all this amounts to an ethical concern about scientific and technological progress man has made particularly in view of the states of environment, natural and artificial and our renewed ethical interests in the environment, the natural and the artificial. Weighing scientific and technological progress is a complex matter but here we have selected one parameter that is how far is the said progress ethically desirable or good particularly when an open eye could see what direct and indirect environmental impacts the progress had, which is already under our ethical scanner. Hence, any contemporary ethical analysis of environmental matters should include a correlative ethical analysis of scientific and technological progress man has made. If we look at this discussion starting from the seventeenth century, the advent of modern science, it has been "man centred" because the crux of it was to find out whether or not the "happiness" of man and his general "well-being" has been cared for. Hence, weighing the "progress" ethically was pivoted on utilitarianism and the anthropocentric welfare. Though it has been accepted that humans should be happy any way at the cost of using his powers in nature. There is perhaps no question of unethicality or undesirability here as human beings should show what they are capable of doing in nature and with nature. It would be also a kind of "disgrace" as Leibniz said to human nature which had every ability to make discoveries for need, satisfaction and happiness and yet, could not do all that. Perhaps it is due to this that we are so many but do so little for human folks. The ethic of scientific progresses we have made thus depends much on anthropocentric welfare and happiness maximization, which involves without any doubt, some degree of intervention in nature or even using up nature. However, by the nineteenth century the happiness balloon was punctured by several things. First was the endless rise of the need graph of man, among which was (though jokingly), use of our power to have birthdays throughout twelve months of a year. Second, with the need, we increased inventions, and with it we increased the use of nature's potentials. In any case 'human centredness' did not shift its grip in ethicality. The scientific and technological progresses were thoroughly dominated by gross hedonism; utilitarianism soon went this way and welfarism was constricted to the zone of economics. It was really looking like a brainchild of economics which had no ethics of "other centredness" owing to the growing capitalist cultures or in the cultures where colonies were made to exploit new resources, natural and human. The exploitation of man by man, growth of classes of the exploiters and the exploited and the battered environment really punctured the happiness ideal we constructed so dearly. Hence, what was the basis of the ethic of progress was itself under an ethical scanner when environmental well-being came to the fore. We were pressed further for our moral extension. Malthus and Darwin must have been particularly responsible for a moral extension in one way as renewed interests in ecology, stiff competition in nature and the moral ideal laid down by the evolutionary thesis were remarkable. Not only this, we came to realize that now as we move in a hedonic paradox, man must be against man for they have rivalry for the control of nature. The idea that scientific and technological progress enhances our happiness and well-being were questioned for ethical validity. This was noticed by many conscious philosophers as did Lotze who remarked that "each step of progress with the increase of strength it brings, brings also a corresponding increase of pressure".[1]

Nevertheless, the point is that we cannot easily trash scientific-technological progress as unethical just on the basis of anthropocentrism because "happiness" and "well-being" themselves did not have much to be blamed nor does human happiness and well-being have anything inherently evil. We are concerned about two things: maximization of happiness irrespective of minimization of the same of other beings and things. As such, even if the ethical basis of scientific progress is human happiness, we have to weigh this basis on its positive and the negative aspects, we cannot simply trust it. There is hardly anyone who doubts that human well-being was made possible by modern science and technology. The 'positive' contribution of science and technology amounts to some definite good, produced over some not so good or evil conditions prevailing erstwhile. Further such contributions are good or valuable in their own right and not for any other reason. Best instances come from the fields of medicine and information and communication technologies. Certainly we would not prefer deadly viruses killing innocent lives (or, shall we for the sake of rights of viruses?). Yet, some 'negative' aspects are clear as well, which aspects are and greater production of evil over good and they are on their own rights, disvalues. It is here that we may consider the case of using innocent monkeys for cosmetic tests and the case of using means of mass communication for ulterior political exploitations. In fact, most of the times science and technology comes out as admixtures of positive and negative values in so far as human happiness is concerned and as purely negative values only in so far as maximization of the ideal has been overplayed at the cost of larger well-being. In any case, it is our ways of using science and technology which comes under ethical scanner because it stands out as means to some end in our life. *Per se*, science and technology is neither moral nor immoral but the inventions and its users are moral or immoral in so far their goals might be intrinsically good or bad. The ethic of science and technology thus refers to humans because their positive and negative intuitions are weighed by us. Hence, it is we who should either encourage or discourage some technologies which are either good or evil. Potentially, certain scientific and technological discoveries, when they are made, does not necessarily unfold fully their inherent capacities (on their own right), to be called positive or negative or both. It may well be that though discoveries are for human well-being, which is good-itself, turns out to be negative as well as positive to some extent. For this, we may have to wait for some time. The point is that once its negative repercussions come out, should we still deliberately use them for the sake of yet another goal to be achieved? These negative repercussions, whether or not they outweigh the positive aspects, is not very important rather, most important is to know whether or not they contribute negatively to the happiness and well-being of humans only or they are so when we consider well-being of a larger whole. It is here that our moral extension is made possible and the ethic of science and technology could be weighed properly. Equally true is that while we consider the positive aspects of the same things, a moral extension is possible because we have every reason to find out whether or not these things enhances human happiness and well-being but does little or nothing for the well-being of existential. Both points made above make any sense if some scientific and technological discoveries are in reality capable of positive and negative contributions towards a larger ecological community and, therefore, not restricted to human beings. It is clear that complex ethical considerations about science and technology does not remain restricted to human centred talk rather, they should be extended when they are pertinent enough to cover other-centred talk. It is also clear that any talk about goodness of technologies as means towards an end, which is good and not bad, cannot be solely human-centred because such values are not restricted to things like 'human happiness' and 'human well-being'. We have also to understand that science and technology cannot be weighed as ethically good or bad by mere reference to what their inner capacities are but with reference to their use. Truly, these things cannot be valued overnight as good though they contributed negatively by inflicting much pain on human beings and others. It is due to this that though modern medicines have alleviated human miseries, we ask about the ethic of pain relieving drugs being tested on animals for their effectiveness. It is here that we tend to think broader in terms of ethic of science and technology because we feel that a human-centred-happiness-oriented-talk is too narrow. It is exactly due to our renewed weighing of the moral value of scientific and technological progress on the broader environmental or ecological contexts that we are seriously considering the obvious counterproductiveness of science. These counterproductiveness are too many for us now and they should be weighed not on non-ethical terms only because it is not only a matter of pragmatics and prudence rather, depends on what amount of evil it produces in the ecological community as such. However, as I have maintained thoroughly in this book, such resolution of ethical impasse regarding science and technology is a matter of moral dialogue.

Another point is that we have to weigh science and technology on ethical terms when not merely human well-being is kept in mind rather, when for that well-being many human beings are deprived or not. Most important is that many human beings on earth are deprived of scientific, technological and economic progresses. A large population is poor and needy. We should consider the ethic of reaching benefits of science and technology to a class of people at the cost of depriving the poor class. Often science and technology becomes sources of mass exploitation as we did see after the spread of British colonialism in today's commonwealth nations. Or, even today it may be seen in glorious democracies where a spurt of progress as planned by great people reach a class, not mass. Interestingly, many contemporary world organizations of great humanitarian repute make great environmental hullabaloo after they have licked top cream of technological progress because they now find the ideal of happiness to be reached by curbing science and technology for world disaster and preach ethics for the poor nations. Aptly Fidel Castro remarks: "The solution cannot be that which burn the development of those who need it the most; the fact is that everything that contributes to underdevelopment and poverty is an open violation of ecology".[2] An extreme view may be like this as well: "If I were a Brazilian without land or money or the means to feed my children, I would be burning the rainforest too". [3] Why should we wonder about such reactions? Are they grossly unethical? Granted that remaining poor is not an ethical ideal as is the ideal of remaining unhappy but being affluent by mass exploitation may be ethically still worse. Science and technology should be thus evaluated for ethical value if they have capacity to alleviate poverty and satisfy the most needy. It is here we have tough question to face when ethical concern in environment are posed, which we thought is a remarkable basis for our moral extension while evaluating the values of science and technology. The second quote projects this tough question. Is it proper to burn our rainforests to exist? Is it proper to cry over tiger projects when men are starving? Is it desirable to say that we must risk some green house effect to alleviate poverty? The last question is more pertinent here when we are considering the ethic of scientific and technological progress. The problem is that though human well-being and well-being of the poor and the needy are great ethical ideals, is it feasible if the means are slowly but surely counterproductive to people and to environment? Does it not amount to some kind of childish idea of allowing poor nations to use nature's bounties with the help of technologies available today in so far as they are not developed (if not rich), though they are sure to suffer in the longer run for their ruins? Is it not true that this opinion is extreme and anthropocentric welfarism? In fact, what the moral opinion stated here wants to stress is not that ecology is not worth caring about, what it stresses is that if poor do not exist, there is no status of an ecological hue and cry. It wants to say that an environmental ethic and an ethic of science and technology should not be unethically imposed on us who wants to exist decently first for we have all right to enjoy such existence when others did long ago. Still, they might not stress that we shall go on exploiting nature at whims for maximization of our goods in so far as rest of the cream has not been licked by none other than 'our' class. The poor and the developing are not unethical people who use technology for using up what has been left in nature and then cry over it. They know it well that this is a capitalist-mercantile ideal of happiness much touted in the name of gross utilitarianism. It is perhaps due to this ethical vision of science and technology in developing and underdeveloped countries (in the broader context of environment), that weighing the ethic of technological progress is possible, not merely on the basis of what are our preferences 'now' but what should be our preferences in the 'future' because we are not worried about the poor and the needy 'now' but for others as well who might still live under the oppression of economic exploitation. It is this ethical concern that enables them to weigh why some pseudo-scientific progress in the name of poverty alleviation are also ecologically (and thus humanly disastrous). Much of this point was focused in our discussions about multipurpose projects and ruthless urbanization owing to heavy industrialization. In reality, the poor and the needy are ecologically conscious for they fight against exploitative technologies which do not aim at poverty alleviation. The ethic of scientific and technological progress is thus closely related to our ethical concerns in environment and its ethical value should be weighed by keeping in mind several factors not any one.

Let us now switch over to consider weighing 'human interventions' in nature so that what has been maligned is at least patched up or substantially recovered. Richard Sylvan's essay on "Mucking with Nature" provides us a good insight. This is needed for no one would certainly like to live in a mucked nature, not even the poor and the needy, but what 'should' we do now as several things are gone out of control? More worse, a huge chunk of population holds the right to have decently and other natural things and beings too does enjoy rights to exist not really as they were but in a manner so that something we called "nature" is well recognized for days to come.

This assertion is not liked by many ethicists particularly those of the Heideggerian tradition remarks Sylvan because according to them we should not interfere with natural system in any case or change them according to our wish or exploit them for our needs. Hence, the ethical basis of ecological management is "Let Being Be", which is the guiding force for the deep ecologists of our times but not for the Leolpoldian naturalists. What I want to maintain about this should be preceded by weighing 'human non-intervention or non-inference' first and then by weighing 'human interference' in ecological management. Again the ethics-talk is predominantly "human" as it is we who likely interfere in nature though undeniably we rely heavily on the natural things and beings. And from our earlier discussion, we need not remain poor and needy.

First thing about the non-interference theory of ecological management is that we cannot swing the clock backward to find nature as it was, completely non-interfered because always forces within nature changed its face slowly or rapidly. The problem is when in addition to it, "we" interfere? "We" are in plenty, not only humans because other animals do interfere in the vast natural setting because they have their own needs and thus own environment to be built for living. It is also true that when we speak of the natural systems on the whole, we do conceive humans and animals within the system and not out of it. Hence, we cannot curse humans for using natural existential for their need and their living. Moreover, as other animals, we too, owing to the queerness of our "nature", do "use" nature and even create an environment for us. This is so natural within nature that there is hardly anything to be branded as "unethical". Yet, what the proponents of "non-interference" holds is that we do not have any ethical right to manipulate and change natural systems so much so for not merely our need but the greed of maximization of pleasures and happiness so that the Beings are unrecognizable for their identity or they simply fizzle out of existence. Strangely, we do not realize that proactive interference ruins the whole system, including man, so the message is not to become hermits or go for mass *hara-kiri* but do not tease natural systems for greed. Another point is the ethics of non-interference is that though it is foolish talking about swinging the clock back or to restore the nature as it were even in a small locality, we may nevertheless, seriously consider some kind of "nihilism" *now* in so far as any further ecological disturbances are concerned. This may amount to either complete nihilism in *extracting* something out of the natural systems any further or complete nihilism in *extracting as well as restoring* parts of the systems methodically or nihilism in *changing and manipulating* natural systems any further or nihilism regarding considerable positive changes and manipulation from external outputs by man like artificial settings to make-up for the loss. In any case, natural systems 'now' do *not* need any *therapy* to cope up with the various losses. The ethic of such "therapeutic nihilism" is supported by some of the Heideggerian supporters mostly championing deep ecology which is well explained by E. C. Hargrove (1989)[4]. These people build their ethic of nihilism on ontological basis as it is common with deep ecologists to hold that the *fundamental structure* of nature is much beyond our creation, manipulation, change and artificialities. Man cannot put any amount of effort to replace this structure and, therefore, should not enjoy any right to do so. At best they should put their hands off the natural systems to at least preserve the structure which has still not crumbled completely. Hence, the ethical upshot of the argument is that therapeutic nihilism is a desirable *ethical management* of environment as it restored the basic form of natural systems which is desirable in itself. Hargrove confuses such "nihilistic ethic" as environmental "management", which is perhaps owing to medical nihilistic management. This is a confusion because not doing anything or hands-off from environment cannot really "restore" or "heal" the mucked nature, it is really not "management". The idea may be at best nature "taking care of itself " and man has to do nothing at all. This may be an old fashioned idea as if wounds will heal as it does at times needing no medical intervention. To wait for some catastrophic change or disaster by virtue of natural forces as if nature "follows its own course" to fill up deeply dug coal quarries, may be the inherent idea. This in fact, ruled out our systematic and scientific efforts to facilitate *restoration* works in nature.

Well it is nothing unethical to nurture an ideal structure of nature but doing nothing now may be still unethical for different reasons. One is that active interference might result into stopping of nature's plunder as well help its restoration. The other is that we cannot stop fulfilling our urge to develop and needs. Those people who are equally vulnerable and nearly extinct also fight for basic subsistence. Hence, it is but natural for us to extract and also restore otherwise living in an ideal state of inaction does not count much. Moreover, when we are well-placed intellectually to understand the ways in which nature recovers its losses, we may go for some such means which may happily go along with it or help the healing up process to some extent. If our ethical basis is to care for the whole of the ecological community, including ourselves, why should inaction be good, is not known to us. At least 'inaction' is not ecological management.

Our discussion has revealed that though we cannot go back to an original natural system we can do something now for the recovery of and prevention of several losses. Equally true is that we have to use something for our existence. How can both things go together? The ethical basis of environmental management which calls for some action at least involving natural systems cannot be thus nihilistic because no ethical ideal remains valuable in inaction. Ideals are valuable or are known to be of worth if we try to achieve them. Truly, we cannot achieve well-being of the whole system immediately and fully but we can keep trying. Keeping these things in mind many ethicists criticize the severe *environmental pillage* including short-term 'using-up-to-satisfaction' and 'leaving-the-area-used for mercy'. These people are against the anthropocentric hedonists because it is grossly against ecological well-being, which we did not favour in our earlier discussion. There are also criticisms on ethical lines about environmental 'long-term-using-up' such as in agriculture or settlement, particularly when that is not backed up by restoration projects. It would be a bad idea to think that nothing in nature belonged to us and we have grabbed everything for our good. Hence, cleaning land for agriculture and using it up are said unethical practices. The grave error lies in thinking that one who is part and parcel of a natural system is unethical in using something he possesses. However, the reality is that we do not grab anything when we depend so much on those things for our existence. Naturally, we use natural things and beings for our basic needs. Similarly, we clear places for our settlement. We can ask the critics of use-theory that what should be done if we do not use what we already have. In fact, in an attempt to answer this we are further pushed to the ethical basis of restoration or ethical environmental management, but not inaction. Another criticism on ethical lines is about using up followed by filling the gaps or a combination of extraction and restoration because what has been used up is too substantial and too many or too much which can never be 'filled up' by 'us' in anyway causing mounting losses. There is suggestion about not using anything, therefore, and inaction as well. The problem is one of realizing the 'using up' dogma which has been misconceived for its meaning. Why should we think that using up is to unjustly grab natural things in its totality or to a maximum extent? Undoubtedly we have not stopped at extracting something where we should have done so, but that calls for action, some management on the part of human population, the evaluation of needs and the follow up judicious use. Further, granted that certain things cannot be 'filled up' now for their huge losses but we can still try to do so to some extent at least when it is physically possible. Or, we should leave no alternative left but to evaluate methods using nothing any more and wait for nature to heal its pains. If natural systems need an overall care for its well-being, doing something is a must. Environmental management thus calls for human interference to a certain degree. We have to understand that much useless criticism declaring using up thesis unethical is due to a misconception about what I call 'mercantile-capitalist' using up goals and correlate management *vis-à-vis* 'environmental conscious and people friendly' using up goals and correlate management. The former sets a goal of "class welfare", whereas the latter that of "mass welfare". The first thus takes up class-management or we may call it "covenantal management" whereas the latter calls for "people's management". When we go for a de-classed people's management of environmental losses most balancedly, we get results because people's management is essentially "feminine"—it is *caring and sharing*. What I feel is that the idea of nature taking care of itself is well understood when apart form the natural laws and movements, people as sons of nature support it as well. It is due to this that human interference in environmental management is ethical.

Many ethicists feel that a feminine people's management of natural systems is neither saving the system nor restoring the system, nor is it rehabilitating the system. It is nothing but "faking" the natural system, which is unethical because after incurring many losses for human good, nature is caricatured causing more bad than good. Such efforts amount to modern technological support, significant manipulations and replacing the natural with the artificial as well as dwarfing, constructing and narrowing down the natural things and set ups. The argument against the mass restoration management naturally amounts to either no use or no action. Though not all who are critics of such management, as is R. Elliot, suggests environmental inaction.[5] Though these people pretend to be with the environmentalists camouflaging arguments but certainly they trash environmental active management because they say that what the environmental engineers are propounding is that we accept a fake or a forgery instead of the real thing.[6] The problem with people like Elliot is that he or we or anyone else did not ever see the whole of the natural system in *originality* or more in its *basic form*. Given that impossibility, a 'fake' is really so-called when in relation to the 'original', it aims at taking the place of the 'original' but turns out to be laughing or inadequate or even bluffing. However, the wrongly called 'faking' is in reality *restoration* because the original has not been referred to or could not at all be referred to fake it. Isn't it possible that we very well know or have seen an original state of a part of the natural system which after much use has been abandoned by us? When we actively manage for recovery of that part by trees, lakes, brooks whatever were there, is it not "faking" and thus "joking" with nature? Is it proper to do such things? Elliot and others forget that there is little or no chance of a picture-perfect restoration of even a part of a natural system as if a great master's work of art has been replicated by someone for a great bluff. Moreover, unethicality of faking may be largely owing to wrong financial or other gains. Sometimes faking involves some sort of aesthetic end. In the case of environmental restoration none of these may happen. Greatest problem with Elliot's argument is that faking does not amount to restoration rather, replicating an original. When a work of art is restored, it is different. In any case, if environmental management amounts to not restoration but deliberate artificial manipulations of the states of nature as if now the "original" has been restored in totality, it is certainly unethical as that does not aim at nature's well-being rather well-being of some planners who invest hefty sums on such bogus things.

Some ethicists like Richard Sylvan[7] thinks that even the "restoration" programme in management may not be so desirable because it always follows in modern times after lot of extortion for what a classneeds. Hence, whatever steps are taken later on by that exploitative class or their henchmen, are at best *present value* assumptions involving unethical practice of *skimping* many relevant considerations. This practice is comparable to 'skimping' in ethical theories as well, particularly when we are obsessed with one moral theory for the resolution of environmental issues. Sylvan thus ends up saying that "even where some serious attempts were made at restoration, environmental value is significantly reduced".[8] This amounts to the acceptance of the truth that gross hedonic extraction or plunder of nature's wealth can never be 'restored' for the topsoil is lost, which is among other things, impossible to restore. Moreover, several losses of local species flora, fauna in more or less same number (eye estimated) cannot be restored. In fact, the concept of "eco-sustainable development" also faces great challenge in so far as the sense of development is not made clear. If the type of development envisioned is class-based and present-time-based maximization of human pleasures, eco-sustainability hardly makes any sense.

Sylvan is of the opinion that the ethically desirable alternative in environmental loss management is rehabilitation and not restoration. This is because successful rehabilitation programme show us that rather than restoring what was extracted, it is possible to increase the *natural value* of the damaged place. Hence, 'rehabilitation' is one way to manage damaged nature by increased value-inputs at different locales. It is due to this that a dead swamp may be cleaned, a pool of bad water may be thrown, dug afresh deeper, fresh water stored and so on. Such practices may be of limited nature and needs some technological interventions as well, but it does contribute to loss-make-up in most eco-friendly ways. The people of a locality may be often seen in such successful practices, most of the time not needing much capital investment. However, Sylvan reminds us that not all rehabilitation management be feminine or ethical. Some are very optimistic, involving huge technological inputs, money, manpower etc. These are vehemently opposed by E. Katz[9] who says that nature can never be wholly rehabilitated because any ultramodern technology and any amount of investment is to satisfy a class interest but not the interest of the mucked nature. But Sylvan thinks that though Katz may be partly right, he is not entirely right in bashing rehabilitation. Katz thinks wrongly that rehabilitation is an "artifactual enterprise", which it is not. In rehabilitating nature not necessarily we act as artifacts rather, we put some labour to give back some lost value so far as it is practicable. In fact, as Katz thinks, we do not create or recreate anything; we just help in regaining something lost.

Sylvan thinks that a nature-empathetic enterprise as this one at small scales may have many hurdles. Some are practical and some are about lack of information and lack of technology. To this I may add, lack of interest on part of our politicians and policy makers for they might form more domineering rehabilitation because they involve large investments. Sylvan rightly says that "Rehabilitation can be viewed as a cooperative venture, between the rehabilitators and Nature".[10] I would like to add that a lot of 'talking' about 'what ought to be done' among the restorers should go on continually, for which, not exploiters but empathesizers, who may be scientists, technologists and thinkers may be included. Equally true is that rehabilitation needs a lot of practical effort after thinking. Most important is to ward off dangerous exploiters if some chance of ecological management is there for us, any further. With this we conclude our Postscript, leaving out several things in the vast field of Environmental Ethics.

POSTSCRIPT

We now face the question of the prospect of people's participation in resolution of ethical issues in environment, which is also closely related to formulation of decision-making rules and ground level "rehabilitation" of battered environment. People's participation also raises the issue of possibility of the participation of some such people who believe that they are still fighting war against a system which is anti-people and anti-environment. Any ethical discourse makes little sense before people are liberated from the clutches of the wrong system or else the ethical project is another camouflaged intellectualism that helps the system. Both these issues shall take us to clearly formulate our views about *moral responsibility* and *moral action* with regard to environment. It may well be that in the final run readers shall come to know some of the *limits* of intersubjective corroboration as one way of people's participation in environmental ethics, though I have tried to overcome some limits by answering why the charge of structuralism and ethical chauvinism is not adequate.

Starting with the weighing of intersubjective corroboration first, let me recapitulate that its possibility as model of application of ethics and people's participation rested mainly on the possibility of ethical experience and ethical knowledge we have in this ecological community, which is certainly not the same as knowing ethical first principles. It is this knowledge that is applied by us in moral resolutions including environmental ethical impasse. The way we do it has been explained by taking three cases that have been resolved. The post-corroboration analysis of the discourse shows the way some of our known ethical first principles become prominent in the moral views expressed by those who are not theory-coopted. We shall now produce some more reasons in favour of intersubjective corroboration to show that the model itself could be easily 'weighed' as ethical and not unethical. The reasons are as follows: (i) application of ethics need not necessarily be application of ethical theories; (ii) application of ethics in environmental cases is firmly empirically based; (iii) an empirically based environmental ethical application shows what moral ends and means are worth considering; (iv) such an ethical application thus takes into account what people of a region as well as nation normally justify as morally justifiable ends and means; (v) ethical application by considering people's moral views of ends and means also proves that our conceived moral ideals are not merely intellectual manipulations rather, they are empirically justifiable generalizations of some trained people in ethics; (vi) a moral philosophy which is itself moral, should be empirically oriented; (vii) such moral philosophy should consist of an application of the moral views of people which are empirically based and thus moral itself; and finally (viii) intersubjective corroboration satisfies the conditions of being a part of a moral philosophy which is moral.

Readers should not worry too much about the logical consistency of the supposed conclusion in (viii) on the basis of premises (i) to (vii) because what I have produced here as reasons in favour of intersubjective corroboration as an ethical model of applying ethics in environmental cases are mainly matters of a *heuristic understanding* of application of ethics in one particular field and not strict logical construction to prove something. I will be precise in unfolding each component of the above mentioned reasons to show that (viii) holds good. In so far as (i) is concerned, I maintain this for the obvious reason that there is no point in finding a necessary relation between applying ethics and applying ethical theories because apart from the fact that they are obviously not one and the same thing, ethical theories are just one part of ethics as well as ethical application, it is mostly of second order inquiry than a first order necessity which ethicists feel in normativity subjugated unilateral use. Though some ethicists have formulated such academic theories, they were never successful in using one theory at a region or nation without any moral disagreement. If some cases are cited, most likely, they were done in isolation as a purely theoretical affair and then imposed on a state machinery to follow for resolving moral dilemma and for decision-making rules. It is interesting to know who were successfully doing so in history. I believe that they were moral chauvinists probably having least faith in the people's intellect, and people's opinion.

About (ii), it should be said that environmental ethical problems are those which we know from our experiences in the nature, we do not create them intellectually. An application of our ethical knowledge [not of any theory artificially as we now get from (i)], to resolve these problems, therefore, demands participation of such people who knows from experiences in nature that such and such things evoke questions of moral goodness and desirability. Those who do not care for such experiences in nature or do not feel that some problems of moral orientation are worth resolving, can never participate in their resolution. If they participate by having second hand information, they do not have any moral authority to resolve moral crises.

About (iii) we have to say that people having knowledge of what environmental problems are worth considering for value resolution, come together for they often find that it is not a matter of unitary resolution because that is impractical as well as strongly opposed by other people, though there is every likelihood of building a general opinion through dialogue. Again all this is known to those people who are "social", neither anti-social nor those who live in classes or groups obsessed by their ethics. These social people tending to converge among divergences applying their moral experiences, debate about what should be our acceptable moral ends and what should be the moral means to achieve these ends. In this context they are able to discuss about why such and such means are best suited for resolving environmental problems. When this is done, often in one region or in a nation, it is the people irrespective of their intellect, caste, class, creed, status, position and knowledge of ethical theories, come together by giving us an idea of what we want in general or what 'we' normally think to be "desirable" about environmental impasse. Such respect for people's power of value resolution eradicates misconceptions about ethical vision of ethics-gods as most powerful humans on earth. Such things evoke an idea among people that they are extremely important parties in moral resolution and it is their moral responsibility to contribute to the welfare of humans and non-human existence. The vision of ecological welfare and its valuable means makes us moral beings, moral theories do not make us people of moral characters.

In (v) we come to realize the value of those trained people in our societies who though de-classed, nevertheless take up important task to let theoretically interested people know in our societies about the fact that some constructed ideals of life, which are taken to be basis of moral worth, are not mere generalizations having no practical value rather, they are extremely close to our life and people. Those moral ideals are not intellectual edifices exploiting common people because they think that they need moral experts to help them out of moral crises, and that there are only few people who should be decision-makers in civil societies. From all this we realize that (vi) and (vii) holds good because any moral philosophy cannot be itself immoral and what is not so should be caring moral experiences and opinions of people for whom an environmental problem is so crucial to be resolved. Finally, in (viii) we come to realize that people's participation in ethics is also participation in applying ethics and not generalizing findings because that is theoretically unique venture of a few people. We also realize that by such participation we are able to resolve moral problem ourselves for we have the moral authority and responsibility to do so. To be noted here is that it is equally immoral to push ethicists and intellectuals caring for people's ethics out of a moral society for they have their part to play with other people.

If environmental ethics is based heavily on the conception of human and ecological well-being, then we cannot escape our "moral responsibility" to do so. One who still risks such neglect of responsibility should be termed 'immoral'. An empirically based environmental ethical application demands that this ideal should be in our mind, which is possible not by teaching and preaching of ethics. Again, this is matter of our experience here, in this world. Our experiences enable us to realize that we have a desirable way of life, which we should adopt seriously. It is not important to only intend well for the nature on the whole, important is to *do well*. Hence, people of any nation having such moral responsibility need not be immoral by escaping it. As such, a chunk of population remaining morally aloof is immoral.

On the basis of what has been said now, we may consider the final problem of whether or not people's ethics as envisioned here is really a nonsense because people of some nations may not be living a normal free life, they may be rather under oppression and fighting for freedom and for that reason a system looks them as 'out of mainstream'. Is ethical resolution of problems through people's participation feasible in states of bondage? The so-called 'bondage' may be of two types: bondage within a nation having definite territory and independent government without any foreign control over social, economic, political and legal decision-making, and bondage within a nation marked for a definite boundary but under some control of a foreign land over the matters stated above and over many other things. In the former case it may be that only a part of the entire population of the said free nation may feel the pangs of bondage or else, at times the number might outwit those who do not feel bonded. In modern times examples of such bondages are easy to find out. The bone of contention here is about people's 'moral responsibility' in any state of bondage. Hence, we shall not make too many distinctions between the states of bondage. I have mentioned earlier that people's moral responsibility is to live a way of life which is above several bondages imposed on them by internal as well as external factors, so that their well-being is guaranteed. People's moral responsibility is to lead a way of life which also guarantees well-being of the environment in which they thrive. Their responsibility is to lead a way of life in which created environments do not go too far to damage the natural environment. Equally true is that people's moral responsibility is to live a free life. Hence, their moral responsibility is to fight for freedom from bondage, when such things really exist. How can bondages of different types exist in a truly 'free nation'? It might exist because exploiters of people thrive within the nation, sometimes instigated by those from outside. The aim of exploiters is gross hedonic in essence and for that reason they are anti-people and anti-poor. In such nations developmental programmes and planning are essentially for a class, mostly for the exploiters or gross hedonists and, therefore, not for the poor and the needy. But power of a nation rests with the people, the common people who toil hard for living. They are largely not exploiters of human beings and the nature unless the exploiters do not use some of them as means for their good. It may well be that in some cases the group of exploited and the deprived people with or without the help of intelligentsia, wage a war of protest, non-violent and violent. When exploiters are too many and too strong like militia, people staging a violent war goes out of the so-called 'mainstream'. Indians know all this very well as all colonies know and those who at the post-independence era still suffer from within as well. The problem is, when these people have the same noble goal of good of the people of nation and its environment, and they feel so responsible for achieving the end even at the cost of pain and sacrifice, how can they talk about moral resolution of problems of emergency when the system (of exploiters) have compelled them to go out of the stream? For these people, ethics-talk is not needed now as that would not help achieve the highest goal and for those who run a 'system', ethics-talk should be without these people for they are 'anti-nationals'. In fact, when there is no exploitation in the system, a protest is unethical. The problem is worse when protestors grow in strength because a system truly exploits people, which is unethical. How can we disregard them? Yet, greatest problem is how do you reach these outcastes of society for their moral opinion or for consensus when they are "underground" and when they do not accept you as their but one of the system? In such cases, is it not true that intersubjective corroboration fails as a model of people's application of ethics in environmental issues? An ethicist having deep feeling for ethical consensus is here at the crossroads. He cannot accept intellectual bosses in moral resolution on the one hand and on the other hand, he is not accepted by his own people in moral discourses.

A careful reader would realize here that the model does not fail but a severe *limit* is put to it. When guns rage, there is no ethical talk really. When such is the state, you really do not have another way to broaden people's participation in ethics. The truth is that resolution of moral dilemma of the people of a nation happens in a relatively exploitation free states. If it has to be successful in India as a people's creed, our exploiters should be dominated along with the *masked* gun ragers. When we are still not in a casteless, creedless, exploitationless society (not only on paper but in reality), our democratic set-up only allows a *limited* intersubjective corroboration. Only happy end for me is perhaps that when people stop exploiting people and the nation is free, it may well be that they all shall agree to this model. Hopefully, even today those who are not in a democratic participatory life, feel this truth. If that be the case, the best place for the good of the nation is the arena for discussion.

I end this book with this limit and this optimism. Here is one problem for my readers over and under the ground for their perusal. A note-book called *'eco-buddy'* of the Navneet house in India reads **"Congratulations! You just saved a tree. Guess how? All *eco-buddy* books use paper made from *bagasse*, not trees. *Bagasse* comes from sugarcane. After the cane has been crushed to make sugar, the leftover pulp is turned into paper and your *eco-buddy* book uses this paper. Think about it, every time you buy an *eco-buddy* book which uses *bagasse* made paper; you make our planet a little greener. And every time you save a tree, you save a part of the planet earth. It's up to you now. Go ahead, create a revolution."** At the back it reads ***M.R.P Rs. 19.00 (80 pages).*** I used these pages for writing some chapters of this book. Would you? Take this real life story before answering: **'I had a friend and met a student who uses light pencils to write on coarse note books because they erased what they wrote after making notes or after learning them because they were not able to afford Rs. 5.00 for notebooks of more than 80 pages'.**

NOTES

1. Rescher, Nicholas, *Unpopular Essays on Technological Progress*, University of Pittsburgh Press, p. 5.
2. Rajagopalan, R. *Environmental Studies: From Crises to Cure*, O.U.P., Delhi, 2005, p. 287.
3. *Ibid.*, p. 287.
4. Hargrove, E.C., *Foundation of Environmental Ethics*, Englewood Cliffs, N. J., 1989.
5. Elliot, R., "Faking Nature", *Inquiry*, Vol. 25, 1982, pp. 81-95.
6. *Ibid.*, p. 144.
7. Sylvan, R., "Mucking with Nature" in Edgar Morcher et al. (Eds) *Applied Ethics in Troubled World*, Kluwer, London, 1998.
8. *Ibid.*, p. 71.
9. Katz, E., "The Big Lie: Human Restoration of Nature", *Research in Philosophy and Technology*, Vol. 12, pp. 231-241.
10. Sylvan, R., *op. cit.*, p. 79.

Bibliography

Agarwal, Anil and Sunita Narain, eds. *State of India's Environment—4, Fourth Citizens' Report*; Centre for Science and Environment, New Delhi, 1997.

Agarwal, Anil and Sunita Narain, eds. *State of India's Environment—5, Fifth Citizens' Report*; Centre for Science and Environment, New Delhi, 1999.

Barbour, I., ed., *Western Man and Environmental Ethics: Attitude Towards Nature and Technology*, Addison-Wesley Publishing Company, MA, 1973.

Beauchamp, Thomas, *Ethics and Public Policy*, Part 7, Prentice-Hall, Englewood Cliffs, NJ, 1975.

Botkin, Daniel and Edward Keller, *Environmental Science: Earth As A Living Planet*, John Wiley, NY, 1995.

Chappel, C. E. and Tucked, M. E., eds., *Hinduism and Ecology: The Interaction of Earth, Sky and Water*, O.U.P., New Delhi, 2001.

Crawford, Cromwell S., *Evolution of Hindu Ethical Ideals*, Firma K. L. Mukhopadhyay, Calcutta, 1974.

Daga, Mahesh, "Limits of Coercion: Tighten The Screw Till It Is Loose", *The Times of India*, 07.06.2002.

Davies, Donald, "Ecosophy: The Seduction of Sophia", *Environmental Ethics* 8, 1986.

Elliot, R., "Faking Nature", *Inquiry*, Vol. 25, 1982.

Fisher, William F., *Toward Sustainable Development? Struggling Over India's Narmada River*, M. E. Sharpe, NY, 1995.

Flader, Susan L. and J. Baird Callicot. eds., *The Rivers of the Mother of God and Other Essays by Aldo Leopold*, University of Wisconsin Press, 1991.

Fox, Warwick, *Ethics and the Built Environment*, Routledge, NY, 2000.

Gaard, Greta, ed., *Ecofeminism: Women, Animals, Nature*, Temple University Press, Philadelphia, 1993.

Gadgil, Madhav and Ramachandra Guha, *Ecology and Equity: The Use and Abuse of Nature in Contemporary India*, Routledge, NY, 1995.

Ghosh, Aurobindo, *The Life Divine*, Vol. I, Pondicherry, 1944.

Ghosh, Aurobindo, *The Life Divine*, Vol. II, Pondicherry, 1947.

Giddens, Anthony, *The Consequence of Modernity*, Stanford University Press, Stanford, 1990.

Gutiérrez, Gustavo, *The God of Life*, Trans. J. O'Connel, Orbis Books, Maryknoll, 1991.

Hargrove, Eugene C., ed., *Religion and the Environmental Crisis*, University of Georgia Press, Athens, 1986.

Hargrove, E. C., *Foundations of Environmental Ethics*, Englewood Cliffs, N. J., 1989.

Hopkins, E. W., *Ethics of India*, Yale University Press, New Haven, 1924.

James, George A., ed., *Ethical Perspective on Environmental Issues in India*, A.P.H. Publishing Corporation, New Delhi, 1999.

Katz, E., "The Big Lie: Human Restoration of Nature", *Research in Philosophy and Technology*, Vol. 12.

Kimpel, Ben, *Principles of Moral Philosophy*, Philosophical Library, NY, 1974.

Leopold, Arthur, *A Sand Country Almanac and Sketches Here and There*, O.U.P., NY, 1949.

Lobb, John, *Sustainability: Economics, Ecology and Justice*, Orbis Books, Maryknoll, 1992.

McCully, Patrick, *Silenced Rivers: The Ecology and Politics of Large Dams*, Zed Books, London, 1996.

McDaniel, J. B., *Earth, Sky, Gods and Mortals*, Mystic: Twenty-Third Publication, 1990.

McDonagh, Sean, *To Care for the Earth: A Call to a New Theology*, Bear & Co., Santa Fe, 1986.

McFague, Sallie, *Models of God: Theology for an Ecological Nuclear Age*, Fortress Press, Philadelphia, 1987.

Mc Fague, Sallie, *The Body of God: An Ecological Theology*, Fortress Press, Minneapolis, 1993.

McKenzie, J., *Hindu Ethics*, O.U.P., London, 1922.

Merchant, Carolyn, *Radical Ecology: The Search for a Livable World*, Routledge, NY, 1992.

Mies, Maria and Vandana Shiva, *Ecofeminism*, Zed Books, London, 1993.

Miller, Alan, *Gaia Connections: An Introduction to Ecology, Ecoethics and Economics*, Rowan and Littlefield, 1991.

Mookerjee, Asoke, Development Plans: Some Unreal Fears About New Technology, *The Statesman*, Kolkata Ed., 03.01.2001.

Morscher, Edgar *et al.*, eds., *Applied Ethics In A Troubled World*, Kluwer, London, 1998.

Naess, Arne, *Ecology, Community and Lifestyle*, Trans. by David Rothenberg, Cambridge University Press, NY, 1989.

Narayan, Shriman, eds., *The Selected Works of Mahatma Gandhi*, Vols. 1 & 2, Navjivan Publishing House, Ahmedabad, 1968.

Nash, Roderick Frazier, *The Rights of Nature: A History of Environmental Ethics*, University of Wisconsin Press, Madison, 1989.

Paul II, Pope John, *The Gospel of Life*, Random House, NY, 1995.

Purtill, Richard L., *Thinking About Ethics*, Prentice-Hall, NJ, 1976.

Rachels, James, *Moral Problems*, Part 3, Harper & Row, NY, 1975

Radhakrishnan, S., *Srimad Bhagavadgitā* (Trans and Commentary), Blackie & Son, Bombay, 1976.

Raghunathan Meena and Mamta Pandya, *The Green Reader: An Introduction to Environmental Concerns and Issues*, Centre for Environmental Education, Ahmedabad, 1999.

Rajagopalan, R., *Environmental Studies: From Crises To Cure*, O.U.P, New Delhi, 2005.

Rasmussen, Larry, *Earth Community and Earth Ethics*, Orbis Books, NY, 1996.

Rescher, Nicholas, *Unpopular Essays On Technological Progress*, University of Pittsburgh Press.

Ruether, R. R., *Gaia and God: An Ecofeminist Theology of Earth Healing*, Harper Collins, San Francisco, 1992.

Ruether, R. R., *Women Healing Earth*, Orbis Books, Maryknoll, 1996.

Rolston, Holmes III, *Environmental Ethics: Duties to and Values in the Natural World*, Temple University Press, Philadelphia, 1988.

Santmire, Paul H., *The Travail of Nature: The Ambiguous Ecological Promise of Christian Theology*, Fortress Press, Minneapolis, 1985.

Sessions, George, ed., *Deep Ecology For The 21st Century*, Shambhala, Boston, 1995.

Singer, Peter, *Animal Liberation: A New Ethics for our Treatment of Animals,* Random House, NY, 1990.

Singer, Peter, ed., *A Companion To Ethics*, O.U.P., Oxford, 1991.

Singer, Peter, ed., *Ethics*, O.U.P., NY, 1994.

Singh, I. P. and Tiwari, S. C., eds., *Man and His Environment*, Concept Publishing Co., New Delhi, 1980.

Singh, R. P. B., ed., *Environmental Ethics*, National Geographic Society, Varanasi, 1992.

Skolimowski, H., *Eco-Philosophy: Designing New Tactics for Living*, Marion Boyers, Boston, 1981.

Slicer, Deborah, "Is There an Ecofeminism-Deep Ecology Debate?", *Environmental Ethics* 17, 1995.

Smith, Pamela, *What Are They Saying About Environmental Ethics?* Paulist Press, NY, 1997.

Stone, Christopher, *Should Trees Have Standing? Towards Legal Rights For Natural Objects*, Kaufman, Los Altos, 1974.

Tallmadge, John, "Saying You to the Land", *Environmental Ethics* 3, 1981.

Taylor, Bron Raymond, ed, *Ecological Resistance Movements: The Global Emergence of Radical and Popular Environmentalism*, State University of New York Press, Albany, 1975.

Taylor, Paul W., "In Defense of Biocentrism", *Environmental Ethics 5*, 1983.

Warren, Karren J., "The Power and the Promise of the Ecological Feminism", *Environmental Ethics* 12, 1990.

Wenz, Peters, *Environmental Justice*, State University of New York Press, Albany.

Werhane, P. and Freeman, R. E., eds., *The Blackwell Encyclopedic Dictionary of Business Ethics*, Vol. XI, Blackwell, Cambridge, 1997.

Index

A POLITICAL HISTORY OF THE IMPERIAL GUPTAS

About the Author

Tej Ram Sharma is a Professor in the Department of History, University of Hyderabad. He has served the university for over two decades in various capacities such as Head of the History Department, Dean, School of Social Sciences, and the Member of the Executive Council. Also, Prof. Sharma has been the Co-ordinator of SAP-II, DSA-UGC, Department of History, University of Hyderabad.

Prof. Sharma has participated, organized and chaired many conferences and seminars of interdisciplinary nature. He has won wide acclaim for his earlier works. He has also won many prestigious awards.

A POLITICAL HISTORY OF THE IMPERIAL GUPTAS

TEJ RAM SHARMA

CONCEPT PUBLISHING COMPANY PVT. LTD.
NEW DELHI-110059

ISBN : 978-81-7022-251-4

Edition 2021

Published and Printed by

Concept Since 1974

Concept Publishing Company Pvt. Ltd.
A/15-16, Commercial Block, Mohan Garden,
New Delhi-110059 (INDIA)
T : +91 11 41101460
E : publishing@conceptpub.com, *W* : www.conceptpub.com

Cataloging in Publication Data--Courtesy: D.K. Agencies (P) Ltd. <docinfo@dkagencies.com>

Sharma, Tej Ram, 1941- **author.**
A political history of the imperial Guptas / Tej Ram Sharma.
pages cm
Includes bibliographical references.
ISBN 9788170222514

1. Gupta dynasty. 2. India--Politics and government--To 997. I. Title.

LCC DS450.S53 2021 | DDC 934.06 23

To

My Beloved Wife

BRIJ KISHORI SHARMA

Preface

Political History is the bedrock of historical writing, by an over emphasis on socio-economic history its importance cannot be minimised. There is an imperative need at times to reconsider the important points and settle the disputes or controversies among scholars in the light of fresh evidence or new interpretation whatever the case may be. The Gupta period of Indian History is such a mine of information that so much has been written and rewritten, created and recreated, discovered and rediscovered, that a review of the whole literature/new material is possible every decade. It has attracted the greater attention of scholars from time to time. We started with a scratch and now we have a plethora of information on it; still there remain some conflicting and confusing issues to be resolved and settled.

We have tried to consult all the available works printed both in English and Hindi and offer our respects to all of them namely V.A. Smith, Fleet, Allan, Rapson, H.C. Raychaudhuri, R.D. Banerji, R.K. Mookerji, R.C. Majumdar, A.S. Altekar, K.P. Jayaswal, A.L. Basham, R.G. Basak, R.N. Dandekar, S.K. Maity, Sudhakara Chattopadhyaya, P.L. Gupta, B.P. Sinha, S.R. Goyal, U.N. Roy and B.P. Panthari who have contributed to our understanding of the Gupta History.

There is much new in this book than can normally be expected in a new work on the Imperial Guptas. We have taken a balanced view of problematic issues leaving the final judgement to the future discoveries/researches, and have avoided any personal impositions on our part. The chronology of the earlier rulers of the Imperial Gupta Dynasty has been settled to the best of our satisfaction for the first time.

The present work mainly concentrates itself from the foundations of the Gupta Empire to the reign of Skandagupta.

Tej Ram Sharma

Acknowledgements

It gives me immense pleasure to acknowledge my gratitude to the authorities of the Central University of Hyderabad for providing me the Faculty Grant, which I have fruitfully utilised for the preparation of the present work. It will not be out of context, if I mention about Professor Bh. Krishnamurti, the esteemed Vice-Chancellor of the University, who praised my earlier works both publicly and privately and exhorted me to contribute more to the scholarly world. I am indebted to the world of scholars, whose works have been of great value for throwing new light on certain unknown vistas and oblique problems concerning the political history of the Guptas. The points raised by some of the eminent scholars helped me to resolve some of the intricate and knotty problems of the period by further surveillance and investigation. I must thank all of them as it is not possible to mention them individually. I am also thankful to the librarian and the staff, University of Hyderabad, for allowing me free access to the numerous books and helping in Xeroxing the material as and when required.

I am also thankful to my learned Gurus, colleagues, students and friends, who always reminded me to bring out some more scholarly presentations on the Guptas.

I am also indebted to my respected father Shri Kali Dass Sharma, who is a guiding spirit for this scholarly disposition and whose blessings always count very much for my judicious and balanced thinking and dispassionate attitude towards life. I cannot forget at this juncture the ever-living memory of my dear mother, Smt. Shahni Devi, who, though long ago went to the heavenly world, was very well-informed about Indian traditions and culture and inherently sowed the seeds in my mind to dedicate myself to the ideals cherished by her in my humble contributions towards Indian History and Culture.

I am also grateful to my elder brother Shri Sudarshan Sharma, younger brother Shri D.D. Sharma, my sisters Shrimati Nirmal Devi Sharma, Shrimati Swarnalata Sharma for shouldering my entire household responsibilities and keeping me free from worries and cares of a joint-family.

I shall be failing in my duty if I do not mention about my beloved wife Brij Kishori Sharma, who showered great love and affection and took all care for my concerted efforts when I deliberated over the subject having very long sittings and confinements to the table. I must also thank my daughter dear Richa Sharma and my darling son Ranjan Sharma for bearing with me, when I devoted myself entirely for the study leaving them to my wife's care.

I also acknowledge the affectionate regards shown to me by Shri Naurang Rai Ji and Shri Ashok Kumar Mittal of Concept Publishing Company, New Delhi, by bringing out this treatise in a record time.

Hyderabad

Dated : 1.6.1988

Dr Tej Ram Sharma
Professor of History,
School of Social Sciences,
University of Hyderabad,
Hyderabad - 500 134.

Contents

List of Abbreviations

ABORI — *Annals of the Bhandarkar Oriental Research Institute*, Poona.

AGE — *The Age of the Guptas and Other Essays* by R.N. Dandekar, Delhi, 1982.

AHG — *A History of the Guptas* by R.N. Dandekar, Poona, 1941.

AHI — 1. *A History of India*, Vol. I (Pelican) by Romila Thapar, England, 1972.
2. *A History of India* by R.S. Tripathi, Delhi, 1967.

AICL — *Ancient Indian Culture and Literature* edited by Mohan Chand, Delhi, 1980.

AIG — *The Age of the Imperial Guptas* by R.D. Banerji, Benaras, 1933.

ASAI — *A Study of Ancient Indian Inscriptions* (in Hindi) by Vasudeva Upadhyaya, Patna, 1970.

ASR — *Archaeological Survey of India*, Annual Reports.

BKSY — *Bharata Ka Swarna Yuga* by B.P. Panthari, Lucknow, 1974.

BSOAS — *Bulletin of the School of Oriental and African Studies*, London.

CCGD — *Catalogue of the Coins of the Gupta Dynasties and of Śaśāṅka, King of Gauḍa*, (in the British Museum) by J. Allan, London, 1914.

CII III — *Corpus Inscriptionum Indicarum*, Vol.III by J.F. Fleet, Second edition, Varanasi, 1963.

DC — *Dineśachandrikā*, Sri D.C. Sircar Festschrift, Delhi, 1983.

DHM — *Dynastic History of Magadha* by B.P. Sinha, Delhi, 1977.

DKA	—	*Dynasties of Kali Age* by F.E. Pargiter, London, 1913.
EI	—	*Epigraphia Indica*
ELNI	—	*Economic Life in Northern India in the Gupta Period* (Cir. A.D. 300-550) by S.K. Maity, Delhi, 1970.
GA	—	*Gupta Abhilekha* by Vasudeva Upadhyaya, Patna, 1974.
GGCBH	—	*The Gupta Gold Coins in the Bayana Hoard* by A.S. Altekar, Bombay, 1954.
GE	—	*The Gupta Empire* by R.K. Mookerji, Delhi, 1973.
GM	—	*Guptakālīna Mudrain* by A.S. Altekar, Patna, 1954.
GRY	—	*Gupta Rājavaṁśa Tathā Usakā Yuga* by U.N. Roy, Allahabad, 1977.
GS	—	*Gupta Sāmrājyā* by P.L. Gupta, Varanasi, 1970.
HI	—	*History of India* (150 A.D. - 350 A.D.) by K.P. Jayaswal, Lahore, 1933.
HIG	—	*A History of the Imperial Guptas* by S.R. Goyal, Allahabad, 1967.
HIPC	—	*Historians of India, Pakistan and Ceylon*, Edited by C.H. Philips, Oxford University Press, 1967.
HNEI	—	*The History of North Eastern India* by R.G. Basak, Calcutta, 1967.
IAB	—	*India of the Age of the Brāhmaṇas* by J.R. Basu, Calcutta, 1969.
IC	—	*Indian Culture*, Calcutta.
IG	—	*The Imperial Guptas*, Vol. I, by P.L. Gupta, Varanasi, 1974.
IGT	—	*The Imperial Guptas and Their Times* by S.K. Maity, Delhi, 1975.
IHQ	—	*Indian Historical Quarterly*
IA	—	*Indian Antiquary*, Bombay.
JBORS	—	*Journal of Bihar and Orissa Research Society*, Patna.
JBRS	—	*Journal of Bihar Research Society*, Patna.
JIH	—	*Journal of Indian History*, Madras, Trivandrum.
JMC	—	*Journal of Mythic Society*, Bangalore.
JNSI	—	*Journal of the Numismatic Society of India*, Bombay, Benaras.
JOI	—	*Journal of Oriental Institute*, Baroda.

JRAS	—	*Journal of Royal Asiatic Society of Great Britain and Ireland*, London.
JRASB	—	*Journal of Royal Asiatic Society of Bengal*, Calcutta.
MW	—	*A Sanskrit - English Dictionary* by Monier Williams, Delhi, 1963.
PBRI	—	*Prāchīna Bhārata Kā Rājanītika Ithihāsa Evam Samakālīna Rājavaṁśa* by S.R. Goyal, Allahabad, 1969.
PGNGI	—	*Personal And Geographical Names in the Gupta Inscriptions* by Tej Ram Sharma, Delhi, 1978.
PHAI	—	*Political History of Ancient India* by H.C. Raychaudhuri, 7th edn. Calcutta, 1972.
PIHC	—	*Proceedings of Indian History Congress*
RC	—	E.J. Rapson (ed.) *British Museum Catalogue of the Coins of the Andhra Dynasty, Western Kshatrapa Dynasty, The Traikūṭaka Dynasty and the Bodhi Dynasty*, London, 1908, Reprinted, Delhi, 1975.
RHC	—	*Readings in History and Culture* by B.P. Sinha, 1978.
RPHI	—	*Readings in Political History of India* edited by S.P. Gupta, Delhi, 1976.
SGAI	—	*State And Government in Ancient India* by A.S. Altekar, Delhi, 1972.
SHMI	—	*Some Historians of Medieval India* by B.N. Lunia, Agra, 1969.
SI	—	*Select Inscriptions* by D.C. Sircar, Bearing on Indian History And Civilization., Vol.I, 2nd edition Calcutta, 1965.
SIA	—	*Studies in Indian Art* by V.S. Agarwala, Varanasi, 1965.
SII	—	*Some Important Inscriptions of Aśoka, the Guptas, the Mukharis and Others*, by Sadhu Ram, Delhi.
SSI	—	*Selections From Sanskrit Inscriptions* by D.B. Diskalkar, Delhi, 1977.
VGA	—	*The Vākāṭaka —Gupta Age* edited by R.C. Majumdar and A.S. Altekar, Delhi, 1967.
WTWI	—	*The Wonder That Was India* by A.L. Basham, Bombay, 1963.

SYSTEM OF TRANSLITERATION

VOWELS

Sanskrit	English
अ	a
आ	ā
इ	i
ई	Ī
उ	u
ऊ	ū
ऋ	ṛi
लृ	lṛi
ए	e
ऐ	ai
ओ	O
औ	au
अं	aṁ
अः	aḥ

CONSONANTS

क	ka	छ	chha	द	da	ढ	ḍha	र	ra
ख	kha	ज	ja	ध	dha	ण	ṇa	ल	la
ग	ga	झ	jha	न	na	प	pa	व	va
घ	gha	ञ	ña	ह	ha	फ	pha	श	sha/śa
ङ	ṅa	स	sa	ज्ञ	jña	ब	ba		
ष	ṣa	त्र	tra	ट	ṭa	भ	bha		
क्ष	kṣa	त	ta	ठ	ṭha	म	ma		
च	cha	थ	tha	ड	ḍa	य	ya		

1

Historiography of the Gupta Period

It will be of some advantage to the lay reader as well as to a scholar if we draw out a line of origin and successive growth of literature on the writings concerning the Imperial Gupta monarchs of which we are concerned with the political aspect only right now. It would be better to review the Gupta historiography in general along with the writings on political history in particular.

We would first like to draw the attention of the readers to the strenuous efforts made by James Princep[1] in reading the Brāhmī script in 1837 without which the inscriptions of the Imperial Guptas would not be read and hence would have no meaning for us.

Goyal has drawn our attention to the inadequacies and errors committed by earlier scholars while reading the inscriptions of the Gupta rulers in 1834, in case they corrected one error they committed another due to uncertainty about reading the script.[2] In 1854 General Cunningham while writing on the Bhilsa Topes even though suggested A.D. 319 as the initial year of the era used in the Gupta inscriptions, he placed the accession of the Gupta in A.D. 319, of Ghaṭotakacha in 340, of Chandragupta-I in 360, of Samudragupta in 380, and of Chandragupta-II, Kumāragupta-I and Skandagupta in 400, 430, and 440[3] respectively.

Scholars even believed that the Gupta dynasty ended with the Skandagupta.[4] Fleet published his *Corpus Inscriptionum Indicarum*, Vol.III in 1888 and confirmed the beginning of the Gupta era in A.D. 319-320 but unfortunately Gereral Cunningham had changed his opinion about the initial year of the Gupta era which he accepted as A.D. 166-167 and accepted the A.D. 319 as the year of extinction of the Gupta rule and not its commencement[5] which now we know that his earlier opinion was the correct one that the Gupta era started in A.D. 319-320.

The knowledge of numismatics also remained similarly poor in the first half of the nineteenth century.[6] In 1832 James Princep or H.H. Wilson could not make much of the Gupta coins, however, by 1835 Princep had made considerable progress in reading their marginal inscriptions and showed that the Gupta coins followed the pattern of the "Indo-Scythic" coinage, but were purely Indian in execution.[7] The picture changed with the findings of a number of hoards of the Gupta gold coins in the later half of the nineteenth century followed by the early years of the twentieth century viz., at Bharsar (1851), Jessore (1852), Hugli (1883), Tanda (1885), Kotwa (1885), Basti (1887), Hajipur (1893), and Tekri Debra (1910). As a result Allan published his famous *Catalogue of the Coins of the Gupta Dynasties* in 1914. After which again a number of hoards have come to light - Kasarva (1914), Mithathal (1915), Sakori (1914), Kumrakhan (1953), and Bayana (1946), Bayana being the biggest of the hoards of Gupta gold coins so far discovered.[8] All these gold coins are published in a big volume entitled *Catalogue of the Gupta Gold Coins in the Bayana Hoard* by A. S. Altekar in 1954. Altekar summed up his results of the studies of the Gupta Coinage in his work Guptakālīna Mudrain (in Hindi) in 1954 which was followed by his *Coinage of the Gupta Empire* (in English) in 1957.

Another type of evidence for reconstruction of the history is the literary evidence of which we find many variants such as bardic legends, the *Vaṁśas*, the *Prabhandhas* and the *Charitas*. Apart from that we have the ancient *Itihāsa- Purāṇa* tradition and the *Kāvya* literature. Due care needs to be taken in the collection of the data as well as its interpretation for the purpose of historical reconstruction.[9]

Earlier works of the 19th century on Indian history either made a sundry reference to Gupta rulers like Samudragupta or did not care to mention them at all.[10]

The *Early History of India* by Vincent A. Smith which first appeared in 1904 is the first work which deals with the Gupta history in some details. In 1933 appeared the work of R.D. Banerji, *The Age of the Imperial Guptas*. The same year saw the publication of *History of India* A.D. 150 to A.D. 350 by K.P. Jayaswal. If the first work by V.A. Smith suffers from the imperialistic bias, the latter two works mentioned above reflect nationalistic bias.[11] Rather we may mention the work of H.C. Raychaudhuri, entitled the *Political History of India* first published in 1923 which is free from any bias in contrast to the two above quoted Indian authors and has been stated to have replaced Smith's *Early History of India* in the world of scholars.[12] R.G. Bhandarkar published his work *A Peep into the Early History of India*, in 1920. In 1932 Raghunandan Shastri published *Gupta Vaṁśa Kā Ithihāsa* (in Hindi). R.G. Basak's *History of North Eastern India* was published in 1934.[13] Basak has devoted the four chapters to a connected account of the Imperial Gupta rule adding some new points in the light of the discovery of the five Damodarpur Copperplate grants of the Gupta period, admitting his omission on the important question of Ramagupta which he deemed to be a controversial one at that time.[14] G.P. Mehta published his work Chandragupta Vikramaditya (in Hindi) in 1932. R.N. Dandekar's *A History of the Guptas* was published in 1941. R.N. Saletore's work *Life in the Gupta Age* was published in 1943.

In the same year was published *Select Inscriptions* by D.C. Sircar which covered the inscriptions of the Imperial Guptas. B.S. Upadhyaya published his work*Gupta Sāmrājya Kā Itihāsa* (in Hindi) in 2 volumes in 1939.

In 1942 appeared the *History of Ancient India* by R.S. Tripathi. The famous work, *The GuptaVākāṭaka Age* edited by R.C. Majumdar and A.S. Altekar appeared in 1946.[15] In 1948 came out *The Gupta Empire* by R.K. Mookerjee which was followed by his work *Ancient India* in 1956. In 1952 Dikshitar's work *The Gupta Polity* was published. In 1962 appeared a monograph on Samudragupta by B.G. Gokhale. In 1958 was published the *Early History of North India* by Sudhakar Chattopadhyaya which devoted one major chapter on the Guptas. In 1953, K.M. Shembavanakar's work *The Glamour About the Guptas* was published. In 1954 appeared the wonderful work of A.L. Basham *The Wonder That Was India* which gave an account of the political history of the Gupta period. In 1954 was also published *The Classical Age* edited by Majumdar and Pusalkar. In the same year appeared the monumental work of B.P. Sinha, the *Decline of the Kingdom of Magadha*. This followed the other works by Sinha, one as *Dynastic History of Magadha* (c.450-1200 A.D.) published in 1977 which is the revised and enlarged version of his earlier work on the *Decline of the Kingdom of Magadha* in 1954, the other *Readings in*

History and Culture published in 1978 is the collections of his articles some of which deal with some aspects of the Gupta history.

In 1963, Panthari's work *Mahān Gupta Rājavaṁśa* (in Hindi) was published, a revised version of which has been published entitled *Bhārata Kā Swarṇa Yūga* in 1974. In 1967 was published S.R. Goyal's, *A History of the Imperial Guptas* which is a good contribution on the Gupta history even though it has thrown some suggestions which are not acceptable to us. However, it has explored the new possibility of locating the original home of the Imperial Guptas in eastern Uttar Pradesh due to provenance of the coins of the Guptas in that area rather than in Bengal which we have endorsed by way of our interpretation of the connected passages from the Puranas. This was followed by his *Prachīna Bhārata Kā Rājanītika Itihāsa* published in 1969. We may also mention two works on the Hunas in India by Upendra Thakur and Atrayi Biswas published in 1967 and 1973 respectively. In 1977, *Gupta-Rājavaṁśa Tathā Usakā Yūga* (in Hindi) by U.N. Roy was published. *Economic Life in Northern India* by S.K. Maity was first published in 1957 republished as revised second edition in 1970 which was followed by his work *The Imperial Guptas and their Times*, in 1975. In 1970 was also published *Gupta Sāmrājya* (in Hindi) by P.L. Gupta which was followed by his publication *The Imperial Guptas*, Vol.I. in 1974, makes a good contribution to our understanding of the Gupta history. In 1974 Vasudeva Upadhyaya's *Gupta Abhilekha* was also published. In 1976, S.P. Gupta edited the articles by R.C. Majumdar in *Readings in Political History of India*. In 1976 Manabendu Banerjee's *A Study of Important Gupta Inscriptions* was published. In 1978 was published the *Personal and Geographical Names in the Gupta Inscriptions* by the author of the present work. Another work published in the same year is *India as Reflected in the Inscriptions of the Gupta Period* by Haldhar Pathak.

In 1982 Sarla Khosla's work entitled *Gupta Civilization* was published. The same year saw the publication of *The Age of the Guptas and Other Essays* (a compendium of his articles) by R.N. Dandekar. Bardwell L. Smith edited a work entitled *Essays on Gupta Culture* in 1983 which includes two bibliographical essays on "Gupta History and Literature" by Eleanor Zelliot and "Religion and Art in the Gupta Age" by Bardwell L. Smith which save our trouble for listing few more works on Gupta Culture, Art and Architecture.

Notes And References

1. Ravindra Vasistha, "Prāchīna Bhāratīya Abhilekhaon Kā Vāchana" in *AICL.*, pp.156-159.
2. S.R. Goyal, in *HIG.*, pp. 3-4; *JASB*, p. 119f. "Ghaṭotkacha" was read as "Yagnakacha".
3. *Ibid.*, pp. 4-5.
4. *Ibid.*, p.5; *JASB.*, VII, p.38.
5. *Ibid.* Also see f.n. 5, Fleet, Cii.III *Introduction*, p.125.
6. *Ibid.*, pp.22-23.
7 *Ibid.*, p.17.
8. *Ibid.*, pp.17-18.
9. *Ibid.*, pp.23-28.
10. *Ibid.*, pp.28-29.
A Comprehensive History of India, Vol. I by Henry Beveridge published in 1865 referred to Samudragupta as a "Fanatic" who changed the seat of Government and established a dynasty while Clark Marshman in his work *The History of India*, Vol. I. published in 1867 did not refer to the Guptas at all.
11. *Ibid.*, pp. 31-35.
12. A.L. Basham, " Modern Historians of Ancient India" in *HIPC.*, pp.284-85.
13. R.G. Basak, *The History of North-Eastern India*, Extending from the foundation of the Gupta Empire to the rise of the Pala dynasty of Bengal (A.D. 320-760), Calcutta, 1967, See *Preface to the Second Edition*, p.v.
14. *Ibid.* See *Preface to the First Edition*, pp.viii-ix.
15. S.K. Maity, *IGT.*, p.260

2

Sources for the Study of the History of the Imperial Guptas

Sources for the study of the history of the Imperial Guptas can be divided into two classes:

1. Literary sources;
2. Archaeological sources.

The Literary Sources are as follows:

The Smṛitis

The two *Smṛitis* namely the *Nārada Smṛiti* and the *Bṛihaspati Smṛitis* are believed to have been written during the third century A.D. and the sixth century A.D. [1] Both of them throw some light on the society and economy of the period.

The Kāmandkīya Nītisāra

It is a guide-book for kings and has been compiled along the lines of the *Arthaśāstra* of Kauṭilya with simple modifications here and there. In the opinion of some scholars the compiler was no other than Śikharasvāmin, the minister of Chandragupta-II who compiled this work under the pseudonym Kāmandaka. [2] It throws light on the polity and administration of the period.

The Purāṇas

The Purāṇas give the dynastic account of the royal dynasties after the Mahābhārata War. They follow a style to describe the events in future tense. The accounts of these dynasties are often brief and enigmatic. Out of the eighteen Purāṇas only five, namely, *Vāyu, Brahmāṇḍa, Matsya, Vishṇu,* and *Bhāgavatā* are more important for their historical narratives.[3] It is believed that a final shape was given to these Purāṇas during the Gupta period.[4] From these Purāṇas we briefly know about the origin and expansion of the Guptas which is corroborated by the archaeological sources.[4a]

Some scholars have drawn our attention to the *Kaliyuga-rāja-vṛittānta* of the *Bhavisyottara Purāṇa* which gives the vivid account of the Gupta kings but it is believed to be a modern forgery.[5]

Works of Kālidāsa

It has been generally accepted that the eminent poet Kālidāsa flourished during the Gupta period.[6] Hence the seven works of Kālidāsa namely *Mālavikāgnimitram, Vikramorvaśīy Abhijñāna-Śākuntalam, Ṛitusaṁhāra, Meghadūta, Raghuvaṁśa* and *Kumārasambhava* give us enough source material to know about the social, cultural and economic conditions of the period.

Apart from these seven works according to some scholars Kālidāsa seems to have composed two other works viz. the *Setu-bandha* and the *Kuntaleśvara dautyam* which give some allusions to the Gupta-Vakataka relations as mentioned in the literary works of the later dates.[7]

The Dramas of Viśākhadatta

Viśākhadatta, the known author of two plays - the *Mudrārākṣasa* and the *Devīchandraguptam* has been placed by scholars in the 4th-5th century A.D. and is believed to have been a contemporary of Chandragupta II.[8] The *Bharatavākya* in the play *Mudrārākṣasa* refers to a king named Chandragupta which equally applies to Chandragupta Maurya and Chandragupta-II and the Mlechchhas referred to are the Greeks in case of the former and the Śakas in the case of the latter.[9]

The second play *Devīchandraguptam* is lost to us and is available in fragments in the later literary works viz. in the '*Ṣriṅgāraprakāśa*' of Bhoja, *Nāṭya-Darpaṇa* of Ramachandra and Guṇachandra and the *Nāṭaka-Lakṣaṇa kośa* of Sāgaranandin as also in the *Abhinavabhāratī* of Abhinavagupta, a commentary on Nāṭyaśāstra.[10] An attempt has been made by P.L. Gupta to connect these passages to throw some light on the Ramagupta episode.[11]

The Kaumudī-Mahotsava of Vajjikā

At one time the *Kaumudī-Mahotsava* was deemed to be a source for knowing the political history of the times of Chandragupta-I. But in view of the dispute over the date and events, P.L. Gupta rules out the possibility of its being a source for the history of the Guptas.[12] The drama condemns Chaṇḍasena as an usurper of the throne of Magadha and belonging to low caste whom the citizens of Magadha could not tolerate and drove out to die in exile. Moreover, the description of the total annihilation of the dynasty after the death of Chaṇḍasena is not applicable to the dynasty of Chandragupta which ruled for several generations after him.[13]

Mṛichchhakaṭika of Śūdraka

The *Mṛichchhakaṭika* of Śūdraka is also generally believed to have been written during the Gupta period and supplements our knowledge by depicting the general condition of society which is rather different from that portrayed by Kālidāsa.[14]

Kāmasūtra of Vātsyāyana

The Kāmasūtra though mainly deals with erotics, gives us useful information about the social life of the people during the Gupta period.

Amarakośa

Though it is a lexicon and is believed to have been compiled during the Gupta period, it gives us a mine of information on the social and economic life of the period.

The Works of Varāhamihira

Texts on astronomy and astrology namely the *Bṛihatsaṁhitā* and *Bṛihajjātaka* of Varāhamihira throw a good deal of light on the geographical distribution of the Indian sub-continent and give us a general idea of the state of agriculture, cattle rearing, trade and commerce etc. about our period.[15]

Buddhist Sources

We get some information about the history of the Guptas from a Buddhist Mahāyāna text, the *Chandragarbha Paripṛichchhā*, [16] *the Life of Vasubandhu*[17] written by Paramārtha, a well known Buddhist philosopher and another Mahāyāna text the *Ārya-Manjuśrī-Mūla-Kalpa* [18] written in 7th-8th century A.D. gives us a graphic account of the Guptas.

Jain Sources

We gain some knowledge about the Guptas from the *Tiloya Paṇṇati*,[19] a Jain work by Yati Vṛishabha, the *Kuvalayamālā* [20] by Udyotana Suri written in Śaka 699 (A.D. 777) and the Jain *Harivaṁśa* Purāṇa [21] written by Jinasena Suri in the year 705 of the śaka era (A.D. 784). We know from these sources that the Guptas ruled from 231-255 years which is well corroborated by archaeological evidence.[22]

Kathāsaritsāgara

We get to know some stories of Vikramaditya tradition from the *Kathāsaritsāgara* of Somadeva written towards the end of the eleventh century A.D. Though it is claimed by the author to be a summary of an earlier work *Bṛihatkathā* compiled by Gunāḍhya, written in Paiśāchī dialect in the reign of some Sātavāhana ruler, the possibility of including some stories of the later rulers cannot be ruled out.[23]

Apart from these works mentioned above we get to know some information about the reign of Chandragupta-II from the *Vāsavadattā* [24] by Subhandu (5th-6th century A.D.).

The *Kāvya-Mīmāṁsā* [25] of Rājaśekhara (10th century A.D.) and *Āyurveda Dīpikā Ṭīkā* [26] of Chakrapāṇidatta (12th century A.D.) also gave us information about the Rāmagupta episode.

The Accounts of the Foreign Travellers :

Chinese Accounts

A number of Chinese pilgrims visited India and have left their memoirs about India which throw a good deal of light on the social, religious and political conditions of the country. We may mention the Memoir of Fa-Hsien known as *Fo-Kuo-Ki* who travelled over India for more than ten years (A.D. 400-411) during the reign of Chandragupta-II. [27] Wang-Hieun-Tse's (7th century A.D.) memoirs are preserved in *Fa-Yuan-Chulin*. Yuan-Chwang (also spelt as Hsuen-Tsang) who visited India during the reign of King Harshavardhana (A.D. 606-648) and stayed in India for fifteen years (A.D. 630-645) and covered many parts of India. His accounts are preserved in the *Si-Yu-Ki*. We also know of I'tsing who visited India after the death of King Harshavardhana i.e.. A.D. 671-672. In between the years A.D. 700-712 he compiled fifty-six distinct works into 250 chapters. One of these entitled : *Kau-Fa-Kao-Sang-Chu* is an account of fifty-five Buddhists who visited India during A.D. 627-672.[28]

Apart from the Chinese accounts we know about the geography, commerce and maritime relations of India with the known world from the '*Christian' Topography* (A.D. 535-547) written by Cosmas Indicopleustes and the works of Procopious of Caesarea, a contemporary Byzantine writer, the works known as "de Bello Persico".[29]

The Works of Arabian Scholars

(1). Al-Beruni's *Tahaquiq-ul Hind*. Al-Beruni, a resident of Ghazni who had come to India in the first part of the eleventh century A.D. (died in A.D. 1048) has left with us his notices about India in his *Tahaquiq-ul Hind* (Reality of Hindustan).[30] He informs us of the Eras used in India and states that the Gupta Era falls 241 years later than the Śaka era.[31]
(2). Abul Hasan Ali's *Majmal-ut Tawarikh* : Abul Hasan Ali wrote in the thirteenth century A.D. a book called *Majmal-ut-Tawarikh*. It is a translation of an Arabic work, which in turn may be the translation of

some text from India. This story corroborates the events of the Rāmagupta episode as we know from the *Devīchandraguptam.*[32]

The Archaeological Sources

"The Gupta period is regarded as the golden age of Indian history and its archaeology is sufficiently important to bear out this claim."[33] Its antiquities are spread all over Northern India. Excavations conducted in Assam, Bengal, Bihar, Madhya Pradesh, Uttar Pradesh, Punjab, Haryana, Kashmir, Rajasthan and Sindh throw light on the distinct town-planning, the various new forms of art and architecture and the beautiful forms of pottery which continued to dominate Indian Culture for centuries ahead.[34]

We may divide the archaeological source material into the following heads:

(A) Inscriptions
(B) Seals and Sealings
(C) Coins
(D) Monuments

Inscriptions

Inscriptions are one of the most authentic sources to know the history of any period. Fortunately enough we have good number of inscriptions of the Imperial Gupta rulers which throw a flood of light on the society, polity and economy of the period. In 1888 when J.F. Fleet published his *Corpus Inscriptionum Indicarum*, Vol. III on the inscriptions of the early Gupta kings and their successors, we knew about 20 inscriptions of the Imperial Guptas; by the passage of time the number has been more than doubled by now.[35]

Four inscriptions of Samudragupta have been discovered at Allahabad, Eran (Sagar District, Madhya Pradesh), Nalanda and Gaya. The latter two are the genuine copies of the original documents probably meant to replace the damaged original ones.[36] Six inscriptions of Chandragupta-II have been found at Mathurā, Udayagiri (North-West of Vidisha, Madhya Pradesh), Gadhwa (in Allahabad District), Sāñchī (Madhya Pradesh) and Mehrauli (near Kutub Minar, nine miles south of

Delhi).[37] Fourteen inscriptions of the period of Kumāragupta have been discovered at Bilṣad (Etah District, Uttar Pradesh), Gadhwa (Allahabad District), Udayagiri (Madhya Pradesh), Dhanaidah (Rajshahi District, East Bengal), Mathura, Tumain (Guna District, Madhya Pradesh), Mandasor (Madhya Pradesh), Karamdanda (Faizabad, Uttar Pradesh), Kulaikuri (near Nowgong in Bogra District, East Bengal) and at Mankuwar (Allahabad District).[38] Five inscriptions of Skandagupta have been discovered at Junagarh (Saurashtra), Kahaum (Deoria District, Uttar Pradesh), Supia (Rewa District, Madhya Pradesh), Indore (Bulandshahr District, Uttar Pradesh) and at Bhitari (Ghazipur District, Uttar Pradesh).[39] Eight inscriptions of the time of Budhagupta have been found at Sarnath (Varanasi, Uttar Pradesh), Paharpur (Rajshahi District, East Bengal), Rajghat (Varanasi, Uttar Pradesh), Damodarpur (Dinajpur District, East Bengal), Eran (Sagar District, Madhya Pradesh) and Nandanpur (Monghyr District, Bihar).[40] Apart from these inscriptions we have Mandasor inscription of Kumāragupta and Bandhuvarman the Mālava years 453 and 529, Eran posthumous pillar inscription of Goparāja and some other inscriptions of contemporary rulers like the Vākāṭakas, Kadambas, the Parivrājaka Mahārājas, Mandasor inscription of Yaśodharman and some later inscriptions containing traditions or references about the Guptas viz., the Rāshṭrakūṭa copper plate, Sanjan and Cambay copper plates.[41]

These inscriptions are inscribed on stone pillars made of stone, copper or iron, and copper plates. Some of them are praśastis or pūrvās, composed by the court poets or other state officials to eulogize their masters, while others are royal charters recording the gift of land or sale of land as the case may be.[42] A third category may be known as private records. These documents are mostly found engraved on the images of Gods and Goddesses and religious structures, and record pious donations.[43] Allahabad Pillar Inscription of Samudragupta composed by Hariṣeṇa and the Mandasor Inscription of the Malava years 453 and 529 composed by Vatsabhaṭṭi are well known Panegyrics. Allahabad Pillar Inscription of Samudragupta provides sufficient information regarding his conquests. Among the inscriptions of Chandragupta-II, Mathurā Stone Inscription, Sāñchī Stone Inscription and Udayagiri Cave Inscriptions are most important. Bihar Pillar Inscription, Bhitari Pillar Inscription, Junagarh Rock Inscription, Kahaum Pillar Inscription and Indore Copper Plate Inscriptions throw light on the reign of Skandagupta.

Seals and Sealings

In ancient times also the documents were secured by means of seals made of metal and ivory. The clay impressions of the seals were known as sealings which have also been found at all important ancient sites. The seals of the Gupta kings are very important as they supply us the genealogy of the rulers. Seals not only make the documents secure but also authentic.[44]

We may mention of Bhitarī Billion (made of silver copper alloy) seal, Basarh clay-sealings, sealings of Dhruvasvāminī, Queen of Chandragupta-II, sealings of Ghaṭotkachagupta and Nālandā clay sealings of Budhagupta, Vainygupta, Narasiṁhagupta, Kumāragupta-III and Visṇugupta.[45]

Coins

A host of scholars[46] have worked on the Gupta coins which have been discovered in large numbers which include the Kalighat (on the bank of Hugli near Calcutta) hoard, Bharsar (near Banaras) hoard, Bayana (Bharatpur, Rajasthan) hoard which is the biggest one bringing to light about 2100 gold coins of which 183 coins belong to Samudragupta, 983 coins to Chandragupta-II and 628 coins to Kumāragupta-I.[47] Gupta coins have been discovered in many hoards or sometimes as single coins from Bengal, Uttar Pradesh, Rajasthan, Madhya Pradesh, Haryana, Punjab and Orissa.[48] They are found made of gold, silver or copper and are distinguished for their weight, type and legends. The legends are sometimes metrical compositions which is very rare in the history of the numismatic world and is the first instance of its type where poetry was employed in its service.[49]

Originally the Gupta coinage is not free from foreign influence which is the Kushana or the Scythian and not Roman influence as propounded by Smith;[50] gradually it exhibits its creative talent towards Indianization. P.L. Gupta has also refuted the charge that the Gupta rulers adopted the weight standard of Kushaṇas or Romans or that it was replaced by the indigenous standard of *Suvarṇa* by showing the contrast in the weight of the Kushana coins (122-123 grains), Roman coins (122-123 grains or subsequently 112-113 grains) while the Gupta gold coins weighted between 118-120 grains, 121, 124, 127, 128, 130, 142-144 grains from Samudragupta to Skandagupta and after Skandagupta went on rising from 147-151 grains.[51] With the gradual rise in weight of the coins, the gold content was reduced and were made more and more alloyed viz. it was 80 percent or above gold content in the coins of Chandragupta-II which went down to 43 percent in coins of Vishṇugupta.[52]

There are almost twenty-one varieties of the Gupta coins viz. Archer type, Battle-axe type, standard type bearing Garuḍa standard in the left hand, Aśvamedha type, Tiger-slayer-type, Lion-slayer-type, Horseman type, Rhinoceros-slayer-type and Apratigha type, exhibit the qualities of valour, the Lyrist type, show proficiency in music and the coins with lesser gold content show the deteriorating conditions of the empire.[53]

The Guptas followed the model of later Kushāṇa coins in the reverse motifs of their earlier coins viz., the goddess Ardoksho seated on a high-backed throne, holding a cornucopiae [54] in the left hand and pāśa (noose) on the right, a well known reverse motif of the later-Kushāṇa coins is found adopted on some of the coins of Samudragupta and Chandragupta-II but on a number of other coins of these rulers certain modifications are made in the manner to represent the goddess Lakshmi in place of Ardoksho and thus an Indian form is given to the goddess when the cornucopiae in the left hand is replaced by a lotus and sometimes pāśa is eliminated from the right hand.[55]

Not only that on some coins Chandragupta-II, Ardoksho motif is fully transformed into the Indian goddess Lakshmī i.e., the goddess is shown seated on lotus holding lotus in her hand. But she still retains pāśa in her right hand. Of course, in some cases the right hand is empty or holding flower or scattering coins which we find in the coins of Kumāragupta. In some cases the goddess is shown seated on lion or standing on an elephant-headed fish (*makara*).[56] On the coins of Kumāragupta which is called Kārttikeya type or Peacock type, the war-god Kārttikeya is seen seated on his *vāhana* peacock, holding a *shakti* in his left hand.[57]

The coins usually bear the name of the king except the coins issued on behalf of the Lichchhavi republic where the name of both king and queen i.e., Chandragupta and Śrī Kumāradevī or Kumāradevī-Śrī are written. The coins also bear the legend in prose or verse which generally shows the king as invincible, the conqueror, valiant, performer of Aśvamedha, Paramabhāgavata or who after conquering the earth, wins the heaven by his meritorious deeds.[58]

Silver Coins

We find the beginnings of silver coins from Chandragupta-II which continued to the times of Skandagupta and were afterwards taken up by Budhagupta. These coins are almost imitations of silver coins of Western Kṣhatrapas who ruled over Kathiawar, Gujarat and Malwa regions for about two hundred years. They are about .5" in diameter and 24 to 26 grains in weight, most of the coins weigh near about 29 grains.[59] In the Sanund (Ahmedabad District) hoard 1395 silver coins were found in 1861 of which 1100 were the coins of Kumāragupta-I of Garuḍa type.[60]

Copper Coins

The copper coins of the Guptas are very rare. Scholars assign this rarity to the Kushāṇas who had issued it in northern India in such abundance that there was hardly any need for the Guptas to issue coins in this metal.[61] The coins of Rāmagupta are known exclusively in copper which seems to have followed local Nāga tradition rather than Mālava coins influencing them.[62] They have also striking similarity with the Gupta seals and coins. The lion seated to the left on these coins is exactly like the lion seen on the clay sealings of Dhruvasvāminī. The vase type is similar to the vase type coins of Chandragupta-II. The Garuḍa on these coins are similar to that found on the copper coins of Chandragupta-II and Kumāragupta-I and also found on the standard on the gold coins.[63]

Monuments

Monuments are also a source of both artistic and religious history. They illustrate different schools of art and architecture. Three different schools of Art are distinguished as (1) Mathurā, (2) Vārāṇasī, and (3) Nālandā. As in the case of coinage, the Gupta set the standard for Indian art free of foreign influence which are seen in Gāndhāra and Kushāṇa art. The image of the seated Buddha at Sarnath Museum is taken as the master-piece of Indian art and of its Vārāṇasī School, though by origin it is descended from the images of the Buddha and the Bodhisattva fashioned by the earlier Mathurā School. Examples of what may be called the Nālandā School of Art are seen at their best at Nālandā and at

Kurkihar from which typical specimens are gathered at the Patna Museum. As regards architecture, evolution of its different styles is seen in the various temples of its times.[64]

These temples throw light on religious history. They represent the main religions of the times and their deities of worship - Viṣṇu, Śiva, Durgā, Buddha or Bodhisattva and Jain Tīrthankaras. The temples of Udayagiri in Gwalior indicate worship of Viṣṇu and the emergence of Gaṅgā and Yamunā as goddesses while the famous temple at Deogarh in Jhansi (6th century A.D.) shows the worship of both Śiva and Viṣṇu or Kṛiṣṇa. The Durgā temple at Aihole of the fifth century and fine sculptures of the Badami caves also draw our attention. [65]

An Assessment or Evaluation of the Sources

We have to review our evidences in the light of recent researches based on 'the systematic study of society in its various facets' which helps us in adopting new approaches to the study of India's past. The study of archaeology also helps either to supplement the contemporary evidence thus sometimes helping us to fill in the gaps or corroborate literary evidence and provide statistical data.[66] As far as literary evidence is concerned we must be careful in knowing the possible data of the author, his style and motives if any for writing his works and finding out changes or modifications in the earlier sources or legends. For example the poet Viśākhadatta, author of the *Mudrā-Rākshasa* and the *Devīchandraguptam* was deeply concerned with the political disturbances in the empire which is evident in the *Bharatavākya* of the *Mudrā-Rākshasa* and that he gave a different colour to the Rāmagupta episode by showing his patron Chandragupta-II to be the original husband of Dhruvadevī through a *Svayaṁvara*, snatching her away from him by his elder brother Rāmagupta and later agreeing to surrender her to the Śaka Chief after suffering a defeat from the Śakas, the calamity which was averted by Chandragupta-II by hatching a plot against both the Śaka Chief and his brother Rāmagupta.[67]

We also need to be careful in looking into the different versions or interpolations in the Purāṇas. Similarly in the contemporary literary works as well as in the inscriptions known as *Praśastis* (panegyrical), we have to be conscious about the literary motifs of the periods such as marrying the ruler by the Goddess of Wealth (Royal fortune) of her own choice[68] and of the political ideals of a *Chakravartin* when there is mention of conquering of the whole earth or such other descriptions. We must see whether such a vast empire by the ruler is known from any other evidence viz., the numismatic evidence or its rule by the successors of the family.[69] We should also take into consideration the orthography and script of the inscriptions.

We should take with enough care any notices of the foreign travellers of India taking into consideration their assignment, period of stay and the basis of their narrative. It is true they reported faithfully whatever they were told by Indians without mincing any words but at the same time they could be easily misguided in their information as it is not necessary that they tapped the right sources of information and they were not fully familiar with the social institutions and different administrative patterns in vogue in various kingdoms. We need to study them before using them for our evidence.

With regard to the numismatic evidence we should not take for granted the domain of a particular ruler by finding a few coins from an area which may have travelled through by a person or businessman of the same period or even at a later date.

A careful analysis of the hoards may sometimes help us in fixing the chronology viz. the coins of Kachagupta are found only in those hoards which include the coins of Chandragupta-I, thereby, fixing his place with Chandragupta-I and Samudragupta.[70] The discovery of 1395 coins of Kumāragupta-I in the Satara district (P.L. Gupta states it to be Sahad in Ahmedabad District) and the issue of the *Vyāghra-Bala-Prākrama* type of coins may suggest the extension of the Gupta imperial boundaries even to the south of the Narmada which was the farthest limit of the Gupta empire in the south-west.[71]

Recently a new light has been thrown on the silver-plated copper-pieces bearing the name Kumāragupta found in Gujarat-Kathiawar region which were earlier believed to be so due to the atmospheric effects[72] or were attributed to a period of financial pressure[73] during his region by declaring them as contemporary forgeries[74] produced by some fradulent private parties, viz. traders or smiths, on the basis of a careful scrutiny of the data.

Secondary Sources

A word of caution is also essential for making use of the secondary sources for reconstruction of the history of the Imperial Guptas by which we mean the earlier historical writings in the form of books, monographs or articles brought out during the pre-independence period and the period after independence. Some of them are marked by imperialistic, nationalistic[75] or ideological[76] biases. We should not become prisoners to the earlier writings on the subject and must look afresh on them with our critical eye before assimilating or incorporating them in our material.

Notes And References

1. S.K. Maity, *ELNI.*, pp.6-7 , P.V. Kane, *HD.*, Vol.I, pp.477, 488.
2. P.L. Gupta, IG., p.155; Maity, *Op.cit.*, p.6.
3. P.L.Gupta, *Op.cit.*, p.114.
4. R.N. Dandekar, *AGE.*, pp.107-108.

4a. P.L. Gupta, *Op.cit.*, pp.114-117.

5. *Ibid.*, pp.117-121.
6. *Ibid.*, pp.150-151; Dandekar, Op.cit., pp.108-114.
7. *Ibid.*, pp.142-144.
8. T.R. Sharma, "Date of Viśākhadatta", *JOI* September, 1981, p.5; Dandekar, *Op.cit.*, pp.124-126.
9. *Ibid.*, p.6; P.L. Gupta, *Op.cit,* p.141.
10. P.L. Gupta, *Op.cit.*, p.135.
11. *Ibid.*, pp.135-141.
12. P.L. Gupta, *Op.cit.*, p.133-135.
13. T.R..Sharma, *PGNGI.*, p.21.
14. Maity, *ELNI.*, p.5; Dandekar, *Op.cit.*, pp.121-124.
15. *Ibid.*, pp.5-6.
16. P.L. Gupta, *Op.cit.*, pp. 154-155.
17. *Ibid.*, pp.145-147.
18. *Ibid.*, pp.121-129.
19. *Ibid.*, p.133.
20. *Ibid.*, pp.148-149.
21. *Ibid.*, pp.129-133.
22. *Ibid.*, pp.130, 133.
23. *Ibid.*, p.151.
24. *Ibid.*, pp.144-145.
25. *Ibid.*, pp.149-150.
26. *Ibid.*, p. 150.

27. *Ibid.*, pp.157-58. According to some other scholars Fa-Hsien visited India during A.D. 405-411; *See* Maity, *ELNI.*, Introduction, p.7.
28. *Ibid.*, pp. 162-163.
29. Maity, *ELNI.*, Introduction, pp.7-9.
30. P.L. Gupta, *Op.cit.*, pp. 157-157; B.N. Luniya, *SHMI.*, pp.11-35.
31. *Ibid.*, Luniya, *Op.cit.*, p.25.
32. *Ibid.*, pp.155-156.
33. K.N. Dikshit, "Archaeological Remains of the Gupta Period", *VGA.*, p.423.
34. *Ibid.*, pp. 423-441.
35. Fleet, *CII.* III, Contents, p.iv., T.R.Sharma, *PGNGI.*, pp.xxiii-xxvi.
36. P.L. Gupta, *Op.cit*, pp.5-14.
37. *Ibid.*, pp.14-22.
38. *Ibid.*, pp.23-31.
39. *Ibid.*, pp.31-41.
40. *Ibid.*, pp.44-47.
41. *Ibid.*, pp.50-56; Fleet, *CII.* III, pp.79-299; Vasudeva Upadhyaya, *GA.*, pp.121- 234.
42. Vasudeva Upadhyaya, *ASAI.*, pp.303-4.
43. *Ibid.*, pp.3-4
44. P.L. Gupta, *Op.cit.*, p.63.
45 *Ibid.*, pp.64-68.
46. J. Allan, *CCGD;* A.S. Altekar, *GM.*, Patna, 1954; P.L. Gupta, *Coins*, Delhi, 1969.
47. P.L. Gupta, *IG.*, p.93.
48. *Ibid.*, pp.89-95.
49. *Ibid.*, p.79.
50. Altekar, *GM.*, pp.9-10.
51. P.L. Gupta, *Op. cit.*, pp.69-70.
52. *Ibid.*, pp.70-71.
53. *Ibid.*, pp.70-76.
54. Ardoksho Goddess is believed to have possessed a horn full of fruits in her hands which was known as Cornucopiae.
55. P.L. Gupta, *IG.*, p.76; Atlekar, *Op.cit.*, pp.11, 15.
56. *Ibid.*, pp.76-79.
57. *Ibid.*, p.79.
58. *Ibid.*, pp.79, 88; Altekar, *Op.cit.*, pp. 22-199.
59. *Ibid.*, p. 97.
60. *Ibid.*, p.101.
61. *Ibid.*, pp.101-102.
62. *Ibid.*, p.105.
63. *Ibid.*

64. R.K. Mookerjee, *GE.*, p.1.
65. *Ibid.*, p.2.
66. Romila Thapar, *AHI.*, p.23.
67. T.R. Sharma, "Date of Viśākhadatta", *JOI*, xxxi, No.I pp. 1-6.
68. V.S. Pathak, *Ancient Historians of India*, Bombay, 1966, p.27.
69. S.R. Goyal, *HIG.*, pp.12-13.
70. P.L. Gupta., *Op.cit.*, p. 96.
71. B.P. Sinha, *DHM.*, p.1.
72. P.L. Gupta, *Op.cit.*, p.99.
73. Ajay Mitra Shastri, "Contemporary Forgeries of Silver Coins of Kumāragupta I " in *Dineśacandrikā Studies in Indology*, pp.97-102.
74. *Ibid.*, p.101.
75. C.H. Philips (ed.) *HIPC.*, pp.416-427, pp.260-293; S.R. Goyal, *Op.cit.*, pp. 29-40.
76. D.N.Jha, *Ancient India*, pp.101-103; Maity, *Op.cit*, p.5. The attention of Kālidasā was mostly confined to the Upper strata of society.

3

India Before the Rise of the Imperial Guptas

Political Conditions Regarding

With the disappearance of the the Kushāṇa empire in C.176 A.D. there arose many independent States in the form of monarchies and republics till the Imperial Guptas established a big empire in C.319 A.D. This period of interval has been named as the "Dark Age" in Indian History.[1] It has been rightly remarked by Romila Thapar that periods when large kingdoms flourished became the "Golden Ages" and those which saw the growth of smaller regional States became the "Dark Ages".[2] The history of the peninsula received so far less attention, except during those periods when it too could boast of empires.[3] We have scanty sources at our disposal to study the detailed history of these States. However, we can build up the story with the help of numismatics, inscriptions and the historical sections of the Purāṇas as well as taking help from the Allahabad Pillar Inscription of Samudragupta which helps us to provide enough clues for the political situation immediately before the rise of the Guptas. Not only that it supplies us additional information about some of States and republics about which we do not come across in other available sources.

Chattopadhyaya has pointed out that earlier some scholars tried to give the sole credit for overthrowing the Kushāṇa empire to the Bhara-sivas while the study of the contemporary coins and inscriptions brings home the point that the Yaudheyas, the Kuṇindas, the Mālavas, the Nāgas and the Maghas, and others played their own part in driving out the Kushāṇas.[4] We may enumerate in the following pages these States, monarchies, and republics in the following order:

NORTHERN INDIA

Monarchies

The Punjab, Sindh and Afghanistan

Due to paucity of sources it is possible to reconstruct the history of this period only in broad outline. There are no contemporary inscriptions to throw light on the political history of these provinces during the 3rd century A.D. The Purāṇas give us only scanty information in the form of referring to Śaka, Yavana, and Tuṣāra rulers in the North-West during this period. We largely depend on some foreign sources and coinage of the rulers of these provinces.[5]

After the death of Vasudeva-I (in C.180 A.D.) we come across the rule of Kaniṣka-III upto C.210 A.D. on the basis of numismatic evidence. It seems that Kaniṣka-III governed through the agency of Satraps or governors, the names of some of whom find mention in abbreviated forms in Brahmi Characters on the obverse side of his coins which are most Hindu names. Perhaps during the reign of next ruler Vasudeva-II (C.210 to C.230 A.D.) many of them got an opportunity to become independent during the troubled period of Sassanian invasions and brought the Kushāṇas under their suzerainty as evidenced by their coins bearing the title 'the King of the Kushāṇas' and 'the King of Kings of Kushāṇas'. However, during the period of Sassanian ascendancy in Afghanistan and the Indus Valley, some Śaka and Kuṣāṇa chiefs continued to rule as petty feudatories. We know of a feudatory Kuṣāṇa family ruling at Kabul, for the wife of a Sassanian emperor Hormuzd-II (A.D. 303-309) was a Kuṣāṇa princess belonging to that family.[6]

About Punjab proper we may state that the Sassanian conquest did not seem to reach that end. There the three Scythian houses, the Shāka (not Saka), Śhīlada and Gaḍahara dynasties ruled - the Shākas over Western Punjab, the Shīladas and Gaḍaharas over the Central Punjab who ruled after Vasudeva-II.[7] The rule of the Shākas and Shīladas was superceded by the little Kuṣāṇas also known as the Kidāra Kushāṇa known after their chief Kidāra in C. 340 A.D. It is he who seems to have sought help from Samudragupta who had extended his sphere of influence to the Punjab, to have an encounter with the Sassanian Emperor Shahpur-II in A.D. 367-368. After that Kidāra was successful in consolidating his empire in Gandhara, Kashmir, and the Western and Central Punjab.[8]

The Maghas of Kauśāmbī

The theories about the over-throw of the Kushāṇa power in the Gangetic plain by the Guptas [9] or the Bhāraśivas[10] have been examined by Altekar [11] and Chattopadhyaya [12] and are no longer held valid. The Allahabad inscription of Samudragupta makes it certain that there was no Kushāṇa rule in Magadha or U.P. when the Guptas were laying the foundation of their empire. Even if the Bhāraśivas had any connection with the Nāga rulers we do not find any coins of these rulers or the Nāgas in the Punjab, they are even rarely found beyond *Yamunā*.

The earliest challenge seems to have come from the Maghas of Kauśāmbī. The coins discovered at Kauśāmbī belonging to the centuries before the rise of the Imperial Guptas refer to Bhadramagha, Śivamagha, Satamagha, Bhīmavarman, Vijayamagha, Puramagha, Yugamagha, Pusvaśrī, Navika, Rudra, Dhanadeva etc. as the rulers of Kauśāmbī.[13] One of the rulers Bhīmasena had carved out an empire for himself as early as 130 A.D. as provided by his Ginja (40 miles to the south of Allahabad) record.[14] We do not know whether Śivamagha of the coins belonged to the dynasty of Bhīmasena as the forms of the names show the difference. The existence of a king named Bhīmavarman is proved by two Kosam inscriptions dated 130 and 139 (A.D.208 and 217). The last ruler of Kauśāmbī appears to have been Rudra who may be identified with Rudradeva mentioned in the Allahabad inscription of Samudragupta.[15]

The Muruṇḍas

Chattopadhyaya has shown on the basis of the numismatic evidence and the *Geographike* of *Ptolemy* in C.140 A.D. the Muruṇḍas had come to power in the Saketa or the Ayodhyā region and further there is no evidence that they in any way acknowledged the overlordship of the Kushāṇas.[16] The Jain traditions and the Chinese account quoted by Levi,[18] lay down that the Muruṇḍas also ruled over Pātaliputra. On the basis of the Puranic evidence Chattopadhyaya concludes that one ruler by name Viśvasphaṇi or Viśvasphurjhi is stated to have ruled over Magadha just before the rise of the Guptas who seems to be a foreigner, most probably belonging to this Muruṇḍa stock[19] and who might have extended his kingdom as far as Kānyakubja or Kanauj.

The Nāgas

We know from the Purāṇas that the Nāgas were very powerful in Vidiśā, Kāntipurī, Mathurā and Padmāvatī and that seven Nāga kings had ruled at Mathurā and nine at Padmāvatī (in modern Gwalior) when the Guptas were rising to power.[20] Altekar has tried to show on the basis of the numismatic evidence and the Vākāṭaka records that the Bhāraśiva rulers were no different from the nine Nāga rulers of Padmāvatī. One of the famous rulers named Bhavanāga had married his daughter to Pravarasena of the Vākāṭaka family. Prior to this matrimonial alliance the Bhāraśivas had been "anointed to sovereignty with the holy water of the Bhāgīrathī (Ganges) which they had acquired with their valour", and they had performed no less than ten Aśvamedha sacrifices.[21] They were thus mighty princes who flourished after the Kushāṇas and maintained their influence for a long time. We find the mention of one Gaṇapatināga along with other Nāga kings who suffered defeat at the hands of Samudragupta as mentioned in the Allahabad Pillar Inscription.[22]

Besides Mathurā and Padmāvatī Houses, we may assume the existence of other principalities of the Nāga rulers in Northern India immediately before the rise of the Guptas viz. Achyuta who rose to power in Ahichchhatra (Rohilkhand) may have been a Nāga king as also Nagadatta another Āryāvartta ruler; both of them find mention in the Allahabad Pillar Inscription of Samudragupta.[23]

The Lichchhavis

Samudragupta's father Chandragupta-I married a Lichchhavi princess and is said to have gained control over their territories with his empire as a result of the marriage. The Lichchhavis must be ruling in a contiguous region of the Guptas which seems to be located in North Bihar with Vaiśālī as its centre.[24] Majumdar is of the opinion that since the Lichchhavis were settled in Vaiśālī (modern Basārh in Muzaffarpur District) in the 6th century B.C. during the time of Gautama Buddha and continued to exist in India as an important clan, they probably founded later a kingdom in Nepal as early as first or second century

A.D. A separate reference to Nepal in the Allahabad Inscription, however, makes it clear that it was a separate kingdom which Samudragupta had inherited from her mother as he is vehemently named as 'daughter's son of the Lichchhavi' (Lichchhavi - dauhitra).[25] It seems that they followed a monarchical constitution.[26]

The Maukharis of Baḍvā

A small Maukhari principality is known in A.D. 239 from an inscription discovered at Baḍvā in Kotah State, about 150 miles to the west of Padmāvatī. It mentions one Mahāsenāpati Bala who had three grown-up sons belonging to the Maukhari stock to help him in administration. The Maukharis of Baḍvā probably had a feudatory status, owing allegiance either to the Western Kṣhatrapas of Ujjayinī or to the Nāgas of Padmāvatī.[27] It is very difficult,to state whether the Maukhari Bala was related to the Maukhari families that subsequently rose to power in South Bihar and Kanauj.[28]

THE REPUBLICS

The Republican States of Rajputana and the Eastern Punjab also took cudgels against the foreign rulers.

The Mālavas

The Mālavas originally belonged to Jhang district in Punjab (now in Pakistan) from where they spread all over Punjab and by the time of Samudragupta they had migrated to Rajasthan probably pushed by foreign invaders. The Mālavas had emerged in 250 B.C. as an independent tribal state. They came under the subservience of the Śakas in the first century B.C., to the Western Kṣhatrapas from the second to the fourth centuries A.D. and to Samudragupta in the fourth century A.D.[29] They were the immediate neighbours of the Ārjunayanas.[30]

Ārjunayanas

The tribe is stated to have drawn it name from the epic hero Arjuna. They fought with Alexander when he invaded India. Before the time of the Guptas they were occupying Agra-Jaipur area. They seemed to have formed a confederacy under the Yaudheyas to meet the Scythian threat. The Yaudheya coins bearing the terms dvi (two) and tri (three) are pointers towards this kind of loose confederation, the Kuṇindas being the third member, as during the third century A.D. they seem to have coalesced with the Yaudheyas.[32] The absence of the Kuṇinda coins after A.D. 250 and their not drawing any notice in the Allahabad Pillar Inscription supports this theory.[33]

The Yaudheyas

The Yaudheyas are stated to have drawn their name from the epic hero Yudhisthira.[34] They are the warrior-tribe of Punjab as the name indicates. Pāṇini mentions them among the *āyudhajīvī Saṅghas*.[35] Their coins are bearing the legend "Yaudheyānām Jayamantradharāṇām" which indicates that they always came out victorious as if they possessed some secret formula for victory.[36] The findings of the Yaudheya coins in large number at Saharanpur, Dehradun, Delhi, Rohtak and Kangra certify that they had driven out the Kushāṇas from these areas and had re-established themselves firmly, in the 3rd-4th century A.D.[37] Originally they seem to have occupied South-East Punjab but later extended their region to northern Rajasthan. They controlled the area lying on the banks of the river Sutlej upto the borders of the Bahawalpur State still known as Johiyawar which is the corrupt form of the Sanskrit Yaudheya.[38]

The Madrakas

They are known as descendants of an eponymous king Madraka, son of Śibi Auśīnara.[39] We know of a queen of Pāṇḍu known as Mādrī and Śalya, the king of the Madras from the Mahābhārata. During the period of Kauṭilya they were called as a corporation of warriors enjoying the title Rājan (king). Madra women were noted for their beauty while they are known as contemptuous and sinful people. The Madra country along with its capital Śākala (modern Sialkot in Pakistan) was one of the twenty ancient cities.[40] Though we have not come across any coins belonging to them, their mention in the Allahabad Pillar Inscription makes them one of the formidable powers of the times. Some scholars take them to be identical with the Mālavas,[41] but they seem to have their separate identity as proved by the literary evidence as well as their separate mention in the Allahabad Pillar Inscription. Possibly they formed neighbourly States.[42]

The Ābhīras

They are stated to have come to India shortly before or along with the Śakas from some part of eastern Iran and spread out from North-West Frontier provinces to other parts of India viz. Uttar Pradesh, Bihar and in the North-Western region of the Deccan including the Northern Konkan as far as Broach. They figure as generals under the Śaka Satraps of Western India.[43] The Principality of the tribe subdued by Samudragupta may be located in the Madhya Pradesh though according to the evidence from the Purāṇas they formed an Ābhira Kingdom in Northern Konkan and Maharashtra.[44]

The Prārjunas

They are mentioned as a tribe subdued by Samudragupta who are said to have obeyed his imperial commands and paid all kind of taxes.[45] They are placed in Narasimhapur or Narasiṁhagarh district of the Central Provinces.[46]

The Sanakānīkas

Another tribe subdued by Samudragupta is said to have lived in the neighbourhood of Bhīlsā, ancient Vidiśā. They were ruled by hereditary chiefs and were not tribal republics, as is generally supposed. The trends in their names indicate that though an aboriginal people they were coming under the influence of Sanskrit Culture.[47]

The Kākas

Mentioned as one of the tribes subjugated by Samudragupta in the Allahabad Pillar Inscription they are placed near Sāñchī at Kāhanāda in the Madhya Pradesh.[48]

The Kharaparikas

Another tribe subdued by Samudragupta mentioned along with the

Kākas is placed in Damoh district of Madhya Pradesh.[49] Apart from the tribes mentioned above there is an indication of some other tribes who were subdued by Samudragupta which do not find their mention obviously due to lack of space in the description which we may assume were cut off from the mainstream and were of no source of trouble to the emperor such as the Kulūtas and the Audambaras who lived in the hilly tracts of the Pubjab.

THE ŚAKA KṢATRAPAS OF WESTERN INDIA

We are not concerned here with the detailed history or genealogical lists of the Śaka Kṣatrapas. For our purpose we want to lay down the brief history of the regions which lay outside the domain of the Kushāṇas during the period under review.

The Western Kṣatrapas or Kṣatrapas of Ujjayinī who belonged to the house of Chaṣṭana had reached their height of power during the rule of Rudra-dāman-I who died about C.170 A.D.[50] He ruled over Malwa, Gujarat, Kathiawar, Northern Konkan, Western Rajputana and Sindh and has left with us one of the earliest important inscription in correct Sanskrit, a long Panegyric recording his material exploits around A.D. 150.[51] They appointed the Ābhīras as generals who sometimes asserted for their independent status during the troubled periods of the ruling family, the Mālavas also showed their concern for independence in the Udaipur-Ajmer tract.[52] The Ābhīras also worked as feudatory of the Satavahanas. Towards the last decade of the second century A.D. the region of Northern Konkan was retrieved from the Kṣatrapas by the contemporary Satavahana king Yajnasri Satakarni. Later, the Ābhīras ruled there after asserting their independence as well as in Maharashtra during the 3rd century A.D.[53]

The Western Kṣatrapas continued to enjoy the prestige of a great power with Ujjayinī as their capital earlier. They seem to go for matrimonial alliance by marrying one of the princess to Vīrapurusha-datta, the Ikshvāku king of Andhra country (A.D. 240-265) and another to the ruler of eastern India who ruled over a small principality in Magadha after the collapse of the Kushāṇa empire.[54]

During the first half of the 3rd century A.D. the Mālavas of Udaipur-Ajmer tract were successful in asserting their independence which is evidenced by the Nandsa Yupa Inscriptions in Udaipur State where a Mālava Chief (Śrī ?) Soma by name performed an important sacrifice to celebrate the liberation of his country.[55] During the same period the territory of Sindh also slipped away from the hands of Western Kṣatrapas later to be governed by the Sassanians.[56]

During the beginning of the fourth century (A.D. 304) the rule of the house of Chaṣṭana over Gujarat and Kathiawar came to an end after 175 years or more. Bhartṛidāman was its last ruler. After A.D. 304, the territory was ruled over by one Rudra-simha-II, son of Svami Jivadaman upto A.D. 316, and by his son Yaśo-Dāman-II upto A.D. 332. There is absence of any Kṣatrapa coinage during A.D. 332-348. Neither the Vākāṭakas nor the Sassanians seem to be responsible in overthrowing the Kṣatrapa power in the region, there is more likelihood of a struggle for power among the Kṣatrapas as Rudra-dāman-II, who ruled after Yaśo-dāman-II very probably was the scion of the house of Chaṣṭana which had been ousted by Rudra-siṁha-II in A.D. 304.[57] His successor Rudra-Sena-III ruled from A.D. 348 upto 380.[58]

The period from A.D. 351-364 seems to have been trouble-torn which is proved by lack of issue of any coinage and burying up hoards for safety by people both in the heart of kingdom as well as in its outlying cities.[59] Some scholars accredited this eclipse of the Kṣatrapas power for 14 years to the conquests of Samudragupta or to the Sassanian emperor Shahpur-II from his base in Sindh after conquering the Kidāra King of Punjab in A.D. 357,[60] but no definite evidence is forthcoming for either of the two possibilities as yet.

Soon after A.D. 380 when Rudra-sena-II died, we find four rulers within six years from A.D. 383-388. The last ruler is Rudrasiṁha-III whose coins are dated 310=A.D. 388. It is probably during his reign that the Gupta emperor Chandragupta-II put an end to the Śaka rule in Gujarat which is indicated by the absence of any further Kṣatrapa coins and inscription in the region.[61]

DECCAN AND SOUTH INDIA

After the death of Sri Yagna Śātakarṇi, the last great monarch of the dynasty in A.D. 199, the decline and dismemberment of the Sātavāhana empire started which came to an end around C. 225 A.D.[62] The rise of the Chutus in the Western and Southern districts, of the Ābhīras in the Nasik area and of the Ikshvākus in the East, and the relentless pressure of the Śakas of Ujjain gradually extirpated the branches of the Imperial Sātavāhana family. Thus the Deccan came to be partitioned among the

Ābhīras, the Chūṭus and the Ikshvākus.[63] The Ikshvākus, the Chūṭu-Śātakarnis and Ābhīras succeeded in carving out small principalities in its outlying province like central Andhra-deśa, southern Karnataka and Western Maharashtra.[64] In the South-East came the Pallavas.[65] The Western Kṣatrapas were no longer powerful to hold the politics or the destinies of the Deccan during the middle of the 3rd century A.D. due to a crushing defeat at the hands of the Mālavas around A.D. 225 who wrested several districts from their dominion who had reasserted their independence as had been done by the Yaudheyas, the Ārjunayanas and the Nāgas from the foreign rule of the Kuṣāṇas during the third century A.D. This gave a golden opportunity for the Vākāṭakas to found an empire reminiscent of the Sātavāhana empire covering the area of Berar and Madhya Pradesh.[66]

The Purāṇas as well as the Inscriptions of the dowager queen Prabhāvatīguptā, a daughter of Chandragupta-II (A.D. 375-414) who married Rudrasena-II, the sixth king in line of the Vākāṭaka dynasty, help us a great deal in determining the chronology of the Vākāṭaka kings. The chronology thus fixes Vidhyaśakti literally meaning who held power in the Vindhyan territory between A.D. 255-280, his son Pravarasena-I between A.D. 280-335 and his grandson Rudrasena-I A.D. 335-360 who seems to be a contemporary of Samudragupta.

The Kadambas

They were a Brahmana family who rose to power in the South-West of Deccan about the middle of the fourth century A.D. when the Pallava power had been weakened by Samudragupta's invasion.[67] The historical origin of the Kadambas is given in the Tālaguṇḍa (Shimoga District, Mysore) Pillar Inscription, of Śāntivarman. It is stated that Mayūraśarman (A.D. 345-370) wrested from the Pallavas the sovereignty of some territory along the west-coast round about Vaijayanti also known as Banavāsī (C. A.D. 345).[68] It is stated in the Tālaguṇḍa Pillar Inscription of the Kadamba king Śāntivarman that Kākutsthavarman, one of the great rulers of the line, gave his daughters in marriage to princes of many important royal families including the Guptas.[69]

EASTERN DECCAN

Andhra

The Ikshvākus

The Ikshvākus ruled over the Krishna-Guntur region, the heart of the Andhra country, immediately after the fall of the Sātavāhanas to whom they had been feudatories. They are called Śrīparvatīyas (rulers of Śrī-parvata) and Andhrabhṛityas (servants of the Andhras) in the Purāṇas which state that seven kings of the line ruled for fifty-two years in all.[70] Their capital seems to have been Vijayapuri in the Nagarjuna-konda valley of the Nallamallai range which was known in early times as Śrīparvata.[71] The independent rule of the Ikshvākus in the heart of Andhra country appears to have ended about the close of the third century.[72]

The Bṛihatphalayanas

The Ikshvākus were followed by the Bṛihatphalayanas (a gotra name) of whom only one king Jayavarman is known. The sudden rise of the Bṛihatphalayanas may have been at the expense of the Ikshvākus. But very soon both the dynastics appear to have been subdued by the Pallavas of Kāñchī.[73] Later the Śalankayanas of Vengi annexed some of the territories originally enjoyed by the Bṛihatphalayanas.[74]

The Śalankayanas

The Śalankayanas (name after their gotra) are known from the middle of the second century A.D. as mentioned by Ptolemy, a people called Salakenoi lived to the north of the Maiseloi, that is, the Masulipatam region.[75] All the Śalankayana charters so far available, were issued from Vengi or Vengipura. One of the southern kings subdued by Samudragupta in the Andhra country was Hastivarman of Vengi which has been located at the site of the modern village of Peddavegi near Ellore Taluka between the Krishna and the Godavari rivers.[76]

Kalinga

The Kalinga country roughly speaking the coastal land between the Mahānadī and the Godāvarī, earlier ruled by the great Khāravela of the Chedī dynasty, was now split up into a number of small principalities to which the Allahabad Pillar Inscription of Samudragupta stands a clear evidence.[77] Among the Dakshiṇāpatha rulers subdued by Samudragupta, four of them are usually assigned to this region namely, Svāmidatta of Koṭṭūra, Mahendragiri of Piṣṭhapura, Damana of Eraṇḍapalla and Kubera of Devarashtra. Koṭṭūra is identified with Kothur in the Ganjam District and Eraṇḍapalla is situated near Chicacole in the same district. Piṣṭhapura is the same as modern Pithapuram in the Godavari District and Devarashtra is identified with the Yellamanchili taluk of the Visakhapatnam District.[78]

Kosala (Dakshina-Kosala) and Mekala

Dakshina Kosala (literally South Kosala) comprising roughly the present Raipur, Bilaspur and Sambhalpur Districts of the Central Provinces and Orissa was one of the earliest Aryan settlements in the Deccan though its early history is wrapped in obscurity.[79] The Nasik Inscription gives an elaborate list of the countries ruled over by Gautamīputra Śātakarṇī (C.106 to 130 A.D.) but does not mention the name of Kosala. We come across a king named Mahendra of Kosala who ruled over the area around the middle of the fourth century A.D. when Samudragupta led his victorious campaign against the kings of Dakshiṇāpatha. The numismatic evidence of the coin-types of the Gupta era in their inscriptions bear testimony to the fact that kings of South Kosala became subordinate allies of Samudragupta and his immediate successors. The Arang (Raipur District, Central Province) plates of Mahārāja Bhīmasena-II were issued from Suvarna-nadī (the river Sona) in the year 282 of the Gupta era, corresponding to A.D. 601.[80]

FAR SOUTH

The region roughly comprises the peninsula to the South of the Tungabhadra and Krishna rivers.[81] The traditional division of Southern India was into three kingdoms:[82] (a) The Cheras or Keralas of the

Malabar coast, occupying what are at present known as States of Cochin and Travancore; (b) the Pāṇḍyas, whose territories included the modern District of Madurā and Tinnenvelly; (c) the Cholas, who ruled the tract to the north of the Pāṇḍya dominions upto the Pennar river along the east coast, called accordingly *Cholamaṇḍalam,* from which is derived the English name Coromandel. The boundaries of these rulers varied as their power waxed and waned in the course of their dynastic intrigues and internecine wars.

Though the Cholas, Cheras and the Pāṇḍyas find mention in the edicts of Ashoka, the Gupta inscriptions discovered so far are silent about them. The only mention we find is indirectly about the Pallavas in the Allahabad Inscription of Samudragupta whose ruler Viṣṇugopa of Kāñchī seem to have resisted Samudragupta's inroad into the Pallava territory along with his feudatory Ugrasena of Palakka. The earliest Pallava king known is Simhavarman whose stone inscription in Prakrit has been recently discovered in the Palnad Taluk of the Guntur District.[83] The characters of this inscription resemble those of the Ikshvāku inscriptions. Siṁhavarman belonged to the Bharadvaja gotra. His grandson Skandavarman performed the *agnishṭoma, vājapeya* and *aśvamedha* sacrifices and had Kāñchī for his capital. He ruled probably in the latter part of the third or early fourth century A.D. Next we know about his son name Buddhavarman and grandson Buddhyankura after whom we come across the name of Vishṇugopa of Kāñchī who was a contemporary of Samudragupta.[84]

Notes And References

1. Chattopadhyaya, *EHNI.*, p.138.
2. Romila Thapar, *AHI.*, p.21.
3. *Ibid.*
4. S. Chattopadhyaya, *Op.cit*, pp.138-140.
5. *VGA*, p.13.
6. *Ibid.*, p.19.
7. *Ibid.*, pp. 19-21.
8. *Ibid.*, p. 22.
9. R.D. Banerji, *AIG.*, p.5.
10. Jayaswal, *HI.*, pp.48ff, 174.
11. *VGA.*, pp.25-27.

12. S. Chattopadhyaya, *Op.cit*., pp.138-140.
13. *Ibid*., p.142.
14. *Ibid*., p.140.
15. *Ibid*., p.143.
16. *Ibid*., p.144.
17. Padalipta-Pravandha of the Prabhāvakacharita and the Āvaśyka-Bṛihadvṛitti as quoted by S. Chattopadhyaya, *Op.cit*., p.144.
18. Chattopadhyaya, *Op.cit*., p.145.
19. *Ibid*., pp.145-146.
20. F.E. Pargiter, *Dynasties of Kali Age*, London, 1913, p.53.
21. R.S. Tripathi, *HAI*, p.235
22. *CII* . III No.1, pp.1-17.
23. Altekar, *VGA*., pp-39-40; Chattopadhyaya, *Op.cit*., pp.147-148.
24. T.R. Sharma, *PGNGI*., pp.138-143.
25. *VGA*., p.130.
26. K.A. Nilakanta Sastri and G. Srinivasachari, *Advanced History of India*, Bombay, 1973, p.183.
27. *VGA*., pp.40-41.
28. *Ibid*.. Chattopadhyaya, *Op.cit*., p.151.
29. T.R Sharma, *PGNGI*., p.149.
30. Chattopadhyaya, *Op.cit*., p.150.
31. *VGA*., p.31; Chattopadhyaya, *Op.cit*., p.150.
32. *VGA*., pp.31-32.
33. *Ibid*., p.31.
34. T.R. Sharma, *Op.cit*., pp.171-172.
35. *V-3*., 117.
36. T.R. Sharma, *Op.cit*., p.174.
37. *Ibid*.
38. *Ibid*.
39. *Ibid*., p.143.
40. *Ibid*.
41. *Ibid*., p.147
42. *Ibid*.
43. K.A. Nilakanta Sastri & G. Srinivasachari, *Advanced History of India*, p.199; T.R. Sharma, *Op.cit*., p.125.
44. *Ibid*., p.200; *VGA*., pp.50-51.
45. *CII*. III, p.8, L.22.
46. T.R. Sharma, *Op.cit*., pp.154-155.
47. *Ibid*., pp.166-167.
48. *Ibid*., p.135.
50. *VGA*., p.47; Chattopadhyaya, *Op.cit*., p.152.
51. *Ibid*., A.L. Basham, *WTWI*., pp.61-62.

52. *VGA.*, pp.48-49.
53. *Ibid.*, pp.50-51.
55. *Ibid.*, p.52.
56. *Ibid.*, pp.53, 56.
57. *Ibid.*, pp.56-60.
58. *Ibid.*, p.61.
59. *Ibid.*
60. *Ibid.*, pp.61-62.
61. *Ibid.*, p.63; Chattopadhyaya, *Op.cit.*, p.157.
62. K.A. Nilakanta Sastri & Srinivasachari, *Advanced History of India,* p.198; *VGA.*, p.63.
63. *Ibid.*
64. *VGA.*, p.93.
65. Sastri, *Advanced History of India,* p.198.
66. Sastri, *Advanced History of India,* p.204; *VGA.*, pp.93-94.
67. Sastri, *Advanced History of India,* p.209.
68. *Ibid.*
69. *EI.* VIII, p.24, Vasudeva Upadhyaya, *ASAI.*, p.415; V.31.
70. Sastri, *Advanced History of India,* p.200.
71. *Ibid.*
72. *Ibid.*, p.67.
73. *Ibid.*, p.69
74. *Ibid.*
75. Ptolemy's Geography VII, i.79, as quoted in *VGA.*, p.73.
76. T.R. Sharma, *PGNGI.*, p.261.
77. *VGA.*, p.76.
78. *Ibid.*, pp.76-77.
79. *VGA.*, p.84.
80. *Ibid.*, p.85.
81. Tripathi, *AHI.*, p.438.
82. *Ibid.*, p.439.
83. Sastri, *Advanced History of India,* p.201.
84. *Ibid.*, p.202.

4

The Rise of the Imperial Guptas

As we have seen in the preceding chapter that after the eclipse of the Kushāṇa power in the Northern India and simultaneously in the South of the Sātavāhana empire in South India around the close of the second century A.D., the first half of the third century A.D. was full of small principalities like the Bharasivas, the Nāgas, the Mālavas, the Ārjunayanas, the Yaudheyas in the heart of India. The North-West was ruled by the Śāka rulers like the Śākas, the Shīldas and Gaḍaharas which was further seized by the Kidāra Kuṣāṇas who were for sometimes subordinated by the Sassanian rulers, who gave defeat to the Sassanian ruler Shāhpur-II around A.D. 366-367 seeking the help from the Guptas. Only the Śāka Kṣatrapas were an exception who continued their rule in Western India upto the close of the fourth century A.D. and were finally uprooted by Chandragupta-II around the same period from A.D. 388 onwards.

From the first quarter of the fourth century A.D. the Guptas rose to power who brought under their imperial umbrella all these smaller states both monarchical and republican in character except the Vākāṭakas who were reckoned as a great power in the Deccan.

The Guptas ruled the whole of Northern India as an imperial power for more than two hundred years upto A.D. 550. They are distinguished as the "Imperial Guptas" from the kings with 'Gupta' appendage to their names, who subsequently ruled in a circumscribed area in Magadha and the neighbouring regions and are designated as the "later Guptas"

ORIGINAL HOUSE OF THE GUPTAS

Mainly three theories have been given by the scholars to associate the Guptas as the founders of an empire which place their original home in:

(A) Bengal
(B) Magadha
(C) Uttar Pradesh

We shall examine each one of them. If we leave out other evidence for the time being, then the whole thing hinges around one problem, that is, the identification of a place called Mi-li-kia-po-no in vicinity of which a "Temple of China" meant for Chinese pilgrims was built by Mahārāja Che-li-ki-to (Sri - Gupta). I-tsing states that as mentioned by Hui-lun this temple was built 500 years before his time.[1] Other points are settled by the scholars as such the date will come around A.D. 175 which is short by one hundred years of the period of Śrīgupta i.e. which should be normally A.D. 275. This has been reconciled by scholars on the belief that this mention was based on rough estimate on hearsay by the old men. Some scholars who relied on exactly this date believed that there may be two Gupta rulers and former one may have been the grandfather who got this temple built.[2]

Ganguly,[3] Maity,[4] Chattopadhyaya,[5] Majumdar,[6] place this Mi-li-kia-si-kia-po-no (Mṛgaśikhāvana) in Bengal and subscribe to the theory of the original home of the Guptas in Bengal. Ganguly places this in Murshidabad, Bengal. Majumdar supports his theory by an illustrated manuscript at Cambridge which is dated A.D. 1015:

> There is a picture of a Stupa labelled as "Mṛigasthāpana Stūpa of Vārendrī" and that 'Mi-li-kia-si-kia-po-no, should be identified with Mṛigasthāpana."

This does not sound proper linguistically and may be quite a different place and cannot be equated with Mi-li-kia-si-kia-po-no which stands more closer to Mṛigaśikhāvana. Even if we agree to the view of Majumdar, it may be stated that Murshidabad is not in Varendri but in the Rāḍha Division.[7] Chattopadhyaya disagrees with Ganguli while counting the direction provided in the I-tsing's description and places the identity of Mi-li-kia-si-kia-po-no in Maldah. D.C. Sircar does not agree with the view about the identification of the home of the Imperial Guptas in Bengal even if Varendri was included within the dominions of 'Sri Gupta'. It does not necessarily mean that it was the original home of the Guptas.[8]

Identification in Magadha

Altekar while identifying Mṛigaśikhāvana in Varendra or northern Bengal remarked that the original patrimony of the Guptas was located in South-East Bihar and included a part of North-Western Bengal as well.[9] On the basis of the Puranic evidence he further states that obviously at the end of the reign of Chandragupta-I, the Gupta empire included Magadha, Sāketa (Ayodhyā) and the Gangetic valley upto Prayāga (Allahabad).[10] It is also confirmed by an analysis of the Allahabad Pillar Inscription which shows that Samudragupta has not conquered any portion of the territory in the above area. The inclusion of Ayodhyā and Allahabad - Benaras tract in the Gupta kingdom was not due only to peaceful diplomacy. Probably some battles had to be fought and the crown prince Samudragupta may well be presumed to have taken a prominent part in them.[11]

We should not ignore the corresponding passage in the *Vishṇu Purāṇa* in this context which states that the territory along the Ganges (upto) Prayaga will be enjoyed by the people of Magadha and the Guptas.[12]

We need to debate the issue whether the Guptas began their kingdom near Pāṭaliputra or near Allahabad for which we need to consider the evidence of the identification of the Mṛigaśikhāvana as well as the contemporary evidence of the availability of the coins and inscriptions of the Gupta rulers. After due consideration we shall see that it comes down to the eastern Uttar Pradesh of which the Mṛigaśikhāvana i.e. the Mṛigadāva or the Deer Park at Sarnath near Vārāṇasī formed a part of it and it was only at a later date that the Guptas expanded their empire to Pāṭaliputra and Bengal.

B.P. Sinha points out that there is some error in mentioning the direction by Hwui-lun. He should have mentioned the location of the Mṛigaśikhāvana temple forty stages west of the Ādityasena's temple rather than the east of this temple.[13] Majumdar objected to this view stating that it was based on the faulty and incomplete translations of Beal and seems to have ignored the authentic translation of I-tsing's Memoir by E. Chavannes.[14] In reply to this criticism Sinha has rightly drawn our attention to the identification of Mṛigaśikhāvana temple with the temple of the Deer Park in a footnote. Further he states that the Deer Park (Sarnath) was a famous and usual centre of pilgrimage for the Buddhists from China and so 'Śrī Gupta' of the 3rd century A.D., would have thought of building a China temple for the Chinese priests near the Deer Park, rather than somewhere in Bengal.[15] We may also notice that Majumdar has tried to evade the connected questions of the original home of territorial jurisdiction of the Imperial Guptas which cannot be avoided for a wholesome discussion in the matter.[16] A fallacy has been created that the original empire of the Guptas existed in Magadha while all evidences epigraphic, numismatic and the Pauranic go in favour of establishing the region of Eastern Uttar Pradesh as the original home of the Guptas. Earlier inscriptions belonging to the foremost Gupta rulers Samudragupta and Chandragupta-II have been located in Eastern Uttar Pradesh. Out of two in Gaya, one is stated to have been issued from Ayodhyā which is again in Eastern Uttar Pradesh. The inscriptions found in Bengal, belong to a later period i.e. of Kumāragupta-I.[17] The earliest and most important Gupta epigraph of Samudragupta, the Allahabad Pillar Inscription written by Hariṣeṇa is another pointer to the same direction that Prayāga region of the modern eastern Uttar Pradesh was the centre of power of the early Guptas.[18]

The largest number of hoards of Gupta coins as many as 14 namely Kaserwa in Ballia District, Devaitha (Dildarnagar) in Gazipur District, Bharsar near Benaras, Gopalpur in Gorakhpur, Kotwa in Bansgaon Tehsil of Gorakhpur District, Mauza Sarai and Rapti in Basti District, Jaichand Mahal and Madankola in Jaunpur District, Tebri Debra in Mirzapur District, Jhusi near Allahabad, Kusumbi District., Unnao and Kanauj,[19] have been found in Eastern Uttar Pradesh. While these hoards give evidence of the coins of early Gupta rulers, the coins of Kalighat hoard found in Bengal are the issues of the later Gupta emperors.[20] Similarly the Hugli hoard gives evidence of 5 coins of Chandragupta-II, 7 coins of Kumāragupta-I and only one coin of Samudragupta.

Not only that, the famous Chandragupta-Kumāradevī type of Gupta gold coins have also been discovered from Eastern Uttar Pradesh which include Mathurā, Ayodhyā, Lucknow, Sitapur, Tandi, Ghazipur, and Benaras.[21]

The Puranic evidence confirms the same in serial order while mentioning the jurisdiction of the Guptas: A verse from the *Vāyu Purāṇa* which is recorded as follows by Pargiter:

Anu Gangą Prayāgam Cha Sāketam Magadhāṁs tatha
Etān Janapadān Sarvān Bokṣyante Gupta Vaṁsajāḥ

i.e. "kings born of the Gupta race will enjoy all these territories namely, along the Ganges, Prayaga, Saketa, and the Magadhas". The corresponding prose passage in the *Viṣṇu Purāṇa* is as follows:

> Anu-Ganga Prayāgam Māgadha Guptāś -
> Cha Bhokshyanti

"The territory along the Ganges (upto) Prayāga will be enjoyed by the people of Magadha and the Guptas".[22] Even with variation in view on the above evidence there is no contest over the region as far as Eastern Uttar Pradesh is concerned. Patañjali has made use of this word *Anu-gaṅgaṁ* and cities the examples of Hastināpur and Vārāṇasī.[23]

Hence the cumulative evidence so far available supports the theory that some region in Uttar Pradesh most probably Eastern Uttar Pradesh was the original home of the Guptas.

The Capital of the Guptas

Some scholars have tried to conjecture the capital of the Guptas. U.N. Roy puts forward his views about Prayāga, Pāṭaliputra, Ayodhyā, Kośāmbī and Ujjayanī but is not certain about any one among them as being capitals. At the most they could be prosperous cities of the kingdom except Prayāga which developed into a city during the reign of Akbar.[24] Panthari points out to Mathurā, Kośāmbī, Pāṭaliputra, Ujjain and Eran.[25] In the Eran Stone Pillar Inscription of Samudragupta in line 25 we find epithet 'Sva-bhoga-nagara' 'the city of his own enjoyment' for it,[26] which indicates Eran to be an important principality or a capital town at the time of Samudragupta. We may assume the continuation of the Mauryan system of administration through viceroyalties at strategically important provinces of the empire.[27]

Caste of the Guptas

All the names of Gupta kings excepting Ghaṭotkacha have the suffix 'gupta' at the end. Does it signify the family (a vaiśya family) or the predecessor of the family?

In the inscriptions of the family Śrī Gupta appears as the founder of the dynasty. His name is always given first in the dynastic table.[28]

Moreover, we find in the Udayagiri Cave Inscription, of the year 106 [29] the wording "Guptānvyānām ṇripasattamānām rājye" (in the reign of the best of kings, belonging to the Gupta lineage) which shows that all these kings belonged to a family which was founded by Gupta, hence they were called Gupta.[30]

Here the prefix 'Śrī' is an honorific as in the case of other Gupta emperors mentioned in the inscription.[31] Had the name of the first king been 'ŚrīGupta', it would have been mentioned as Śrī śrī Gupta [32] as we find in the case of the name Śrīmatī in the Deo-Barnark Inscription of Jīvitagupta-II.[33] If we accept that Gupta was the name of the first King of the family we may dismiss the possibility of Gupta ending being a vaiśya suffix.

Now, the question arises why the family got named after Gupta? In many cases, families are named after some important person in them, and when once a family is so named, the tradition is maintained even though the successors may reach much higher positions. Before Gupta this family might not have attained any significant status. For the first time Gupta got the status of a *Mahārāja* as is mentioned in the Gupta inscriptions,[34] the status remained unchanged in the second generation and from the third generation, the Gupta kings became *Mahārāja-dhirājas*. Literally *Mahārāja* means 'a great king'. But the apparent and deliberate differentiation in the status of Gupta was not much high in his own times.[35] Probably he was only a feudal chief and not an independent king.

The name Gupta is so short that it looks suspiciously queer. But we must point out that first part has not been lost or damaged in the inscription. Palaeographically it is quite categorical that the name is Gupta only, there is no loss or damage of syllables.

In ordinary life in all societies we find the convenient tendency to drop one part of the name. We address a person by the pūrvapada or the uttarapada whichever is convenient to us. By the passage of time that name becomes his popular name. In some cases even his original name may be forgotten. In our own case Gupta may have been the uttarapada of the name of the first king by which he may have been generally known. The name Gupta was probably very popular, so much so that the dynasty itself was named after it.

The practice of shortening the names is not known in the Vedic times; it is noticed by Pāṇini and seems to have been fashionable in the times of Kātyāyana and Patañjali.[36] Several examples of it are also met within the Buddhist literature.

In modern historical usage Sri has become so much associated with the name of the first king of the Gupta dynasty that it has become a real part of his name generally written as Sri Gupta.

The psychology behind it may be that the use of the smaller name sounds queer and it is brought at par with other names in the dynasty, e.g. Chandragupta, Samudragupta, Kumāragupta, etc.

Smith [37] suggests that this name was not simply Gupta, but Śrīgupta, implying thereby that Śrī is an integral part of his name, not an honorific prefix. Fleet has thoroughly refuted all his arguments and we may not discuss them here.[38]

Some corroborative evidence for the historicity of Śrī Gupta is afforded by two seals of which one is in Prakrit and gives the legend 'Gautasya'. It is most probable that these seals belong to the founder of the Gupta dynasty, especially the Sanskrit seal.[39]

The word 'Gupta' is derived from 'gup' to protect.[40] The *Viṣṇu Purāṇa* [41] says "(The termination) Śarman is prescribed for a Brāhmaṇa; Varman belongs to a Kṣatriya (and) a name characterised by Gupta and Dāsa is approved of in the case of (respectively) a Vaiśya and a Śūdra". The commentary in the Bombay edition cites as examples, Soma-śarman, Indravarman, Chandragupta, Śivadāsa.[42] The *Mānavadharma-śāstra* [43] also lays down a similar rule without specifying terminations. On the basis of these authorities, it has been suggested that the early Guptas were not of a high caste, being at best Vaiśyas, and hence felt pride in their matrimonial alliance with the Lichchhavis.[44]

But we find that the rules regarding the naming of persons pre-scribed in the Dharmaśāstras were not always strictly followed. For example, the name of the well known astronomer, Brahmagupta, a Brāhmaṇa, in line 36 of the Nerur grant of Vijayāditya (dated Śaka-Saṁvat 627).[45]

We know of the names of the kings ending in Gupta as early as second century B.C. from the records of the excavations and explorations conducted in central India.[46]

It may be noted from the Tālaguṇḍa Pillar Inscription of the time of Śāntivarman (C. 455-70 A.D.) [47] that the grandson of a Brāhmaṇa king Mayūraśarman was named as Kāku (ut) sthavarman. Thus on linguistic consideration no weightage can be given to the word 'gupta' denoting a Vaiśya class.

In this context, we must note that Prabhāvatīguptā, the daughter of Chandragupta-II and the chief queen of Vākāṭaka king Rudrasena-II describes herself as belonging to the Dhāraṇa gotra in her Poona and Rithpur copper plate inscriptions.[48] Dhāraṇa is clearly the gotra of her father, as the gotra of her husband is specifically mentioned as 'Vishuṇvṛiddha' in the Chammak copper plate inscription of Pravarasena-II.[49]

This Dhāraṇa gotra has been variously interpreted by scholars.

Jayaswal takes it to stand for Dhanrī, a Jāṭ clan found in Amritsar, and on the basis of the *Kaumudīmahotsava* he concludes that Chandragupta-I was a Kāraskara or Kakkar Jāṭ.[50] This view has been supported by Gokhale.[51]

Jayaswal emphasises the similarity between the name of the Dhāranīya Jāṭs in Ganganagar District of Rajasthan and Dhāraṇa gotra of the Guptas.[52] Chandragomin's grammatical illustration"ajayat jarto Hūnān" (The jatra or Jāṭ king defeated Hūṇas) has been interpreted by Jayaswal to refer to the Gupta ruler Skandagupta's victory over the Hūṇas.[53] Thus the Jāṭ origin of the Guptas has been a favourite thesis of Jayaswal.

According to Raychaudhuri the Dhāraṇa gotra of the Guptas suggests that they were related to Dhāriṇī, the chief queen of Agnimitra Suṅga.[54] This view is untenable. The similarity in the two names is not sufficient to establish the origin of the Dhāraṇa gotra.

On the basis of the evidence of the *Skandapurāṇa*, Dasarath Sharma[55] says that the Dhāraṇa was a gotra of the Brāhmaṇas of Dharmāraṇya, a tract in the present Mirzapur District of Eastern Uttar Pradesh. But Sharma is not ready to accept that the Guptas were Brāhmaṇas, he considers them to be either Kṣatriyas or Vaiśyas who adopted the gotra of their gurus, without violating the laws of Smṛitis and the Dharmaśāstras.[56]

On the other hand Goyal[57] considers the Guptas to be Brāhmaṇas. He relies on the evidence of their matrimonial alliance:

We find that Prabhāvatīguptā, the daughter of Chandragupta-II was married to Brāhmaṇa King Rudrasena-II.[58] Kadamba King Kāku (ut) sthavarman who was a Brāhmaṇa says that he married one of his daughters to a Gupta King.[59] Buddhist scholar Paramārtha (A.D. 600) says that Bālāditya, the Gupta King married his sister to Vasurāta, a Brāhmaṇa by caste.[60] According to the Mandasor inscription of Yaśodharman, Bhānuguptā (most probably a daughter of the Gupta

King Bhānugupta) was the 'wife of a certain Ravikīrtti, evidently a Brāhmaṇa, who was the grandfather of Dharmadosha, the minister of Yaśodharman.[61]

Thus, we see that three of the Gupta princesses were married to Brāhmaṇas.[62] There is only one instance of the Gupta prince marrying a daughter of a Kadamba King, who was a Brāhmaṇa.[63]

It is to be noted that matrimonial alliances played a significant part in the foreign policy of the Guptas. Chandragupta-I rose to power by marrying the Lichchhavi princess Kumāradevī and Samudragupta accepted the offers of daughters from his feudatories. Thus, marriages with the most powerful and distinguished royal families in different parts of India continued to be an important policy of the Guptas.[64]

The matrimonial alliances of the Guptas seem to have sprung from political considerations.[65] Politically the Kadambas were no match for the Guptas. It may be inferred that it was on account of political pressures or a matter of pride for the Kadambas that they had married their daughter to the Gupta King. We can explain all the matrimonial alliances of the Guptas even without bringing political reasons in the picture. As we know intercaste marriages; especially of the *anuloma* type, has been permitted by the Smṛitis. In three out of the four cases Gupta princesses were married to Brāhmaṇa bridegrooms. If these are taken to have been *anuloma* marriages Guptas could have belonged to any of the remaining three varṇas. It is only the marriage of a Kadamba princess with a Gupta King which requires the Guptas to have been Brāhmaṇas, otherwise it will be a case of *pratiloma* marriage.

The Guptas do not mention their caste in any of their records. Had they been Brāhmaṇas they must have been proud to refer it, especially because they were staunch supporters of Hinduism. We find a parallel in the case of Pāla Kings of Bengal who are silent about their caste since they were Buddhists.

Finally Chandragupta-I agreed to have a joint coinage with the Lichchhavi princess Kumāradevī. Had the Gupta kings been Brāhmaṇas, they would not have agreed to have a joint coinage (bearing the legend 'Lichchhavayah', the Lichchhavis) with the Lichchhavis who were *Vrātya* Kṣatriyas.[66] Even if the Guptas had agreed for a joint coinage as a political matter, they might have objected the word '*Lichchhavayah*' on the coins. More astonishing is the fact that even the name of the Guptas is not linked with the legend '*Lichchhavayah* '. Above that, Samudragupta was ready to be called *Lichchhavi-dauhitra* and seems to have mentioned this epithet in his records as a matter of pride. [67] If Guptas could do such acts as political matters, we do not insist them to be orthodox Brāhmaṇas, and are not ready to give any weightage to their matrimonial alliances as Goyal has done for the consideration of their caste. They were kings, for them all such matters were first political and then social.[68]

Thus we can conclude that the question of the caste of the Guptas cannot be said to have been finally settled. If their Dhāraṇa gotra was not borrowed from the gotra of their purohita then they must be described as Brāhmaṇas. We will have to wait for some more weighty and specific evidence to give the final verdict.

THE GUPTA ERA

Thanks to the statement of Alberuni that 'the Gupta era was separate from 'Śāka era by an interval of 241 years',[69] and is substantiated by the Verawal inscription of Valabhī-Saṁvat 945.[70] This means that the Gupta era begins at 319A.D. as the Śāka era was started in A.D. 78. The views about A.D. 200-201 to be the initial year of the Gupta era held by Shama Sastry, and Govinda Pai's theory about the origin of the Gupta era in A.D. 272-73 are no longer accepted as correct.[71] Similar is the fate of other dates[72] viz. A.D. 166-67, 190-91, 194-95, 278, 284-85, which need not be discussed here.

We do not find the use of the Gupta era till the time of Chandragupta-II whose Mathurā inscription is dated in the year 61 which is the fifth year of his reign.[73] This is the earliest mention of the Gupta era where regnal year as well as the Gupta era have been mentioned. We may quote other few references which are as follows:

(i) Mahārājadhirāja - 'Śrī Kumāraguptasyābhivardhamāna - vijaya-rājya Samvatsare Shaṇ-navate[74]

(ii) Saṁvatsarāṇām adhike śate tu triṁśadbhi anyair api shaḍbhir- eva rātrau dine praushthapadasya shaṣṭhe Gupta prakāle gaṇanām vidhāya.[75]

The above reference makes it quite clear by mentioning "Gupta-Prakāle gaṇanām Vidhāya" that is "Making the calculation in the reckoning of the Guptas" that there was the Gupta era practically in vogue and is no longer a matter of speculation.

Alberuni has also mentioned that the Valabhī era and the Gupta era are one and the same.

The Mandasor Inscription of Kumāragupta and Bandhuvarman dated Mālava year A.D. 493-436 falls within the period of Kumāragupta C. 414-455 A.D. both on the basis of epigraphic and numismatic evidence. The dates of Kumāragupta do not tally with the date in the Mandasor record if we think that Gupta era started in A.D. 200-201 or A.D. 272-73 or any other dates mentioned earlier.[76] This also shows that Mālava era was the same as Vikrama era which has further been proved beyond doubt on the basis of inscriptional evidence from Chahamāna dynasty. They bear the dates closer to each other, while in one of them it is called Mālava era, in the other two it is named as Vikrama era. [77] We may quote V.A. Smith in this context:

> "The popular belief which associates the Vikrama era of 58-57 B.C. with a Rājā Vikramāditya or Bikram of Ujjain at that date may be erroneous. No such person is known. It is, however, true that the earliest known use of the era was in Malwa and it may have been invented by the astronomers of Ujjain. An early name of it was the Malwa era. The term Vikramakāla used in later times may refer to one or other of the many kings with the title Vikramāditya or Vikrama, who was believed to have established the era. The king referred to may be presumed to be Chandragupta-II Vikramāditya, who conquered Ujjain about A.D. 390. The Gupta and Śāka eras changed their names. Similarly becoming known in after ages as the Valabhī and the Śālivāhana eras respectively".[78]

Who is the founder of the Gupta Era?

Again different views have been expressed by scholars in this context, starting from the founder of the dynasty Gupta to Chandragupta-II which are as follows:-

Chandragupta-I as the founder of the Gupta Era

H.C. Raychaudhuri mentions that the first independent sovereign (Mahārājadhirāja) of the line was Chandragupta-I, son of Ghaṭotkacha, who may have ascended the throne in A.D. 320, the initial date of the Gupta era[79] but states that the identity of Gupta king with whom the era of A.D. 320 originated, is by no means clear. The claims of Mahārāja Gupta or even (less plausibly) of Samudragupta cannot be altogether disregarded.[80]

R.C. Majumdar remarks that it is generally assumed that Chandragupta-I founded the era dating from his coronation which took place on December 20, A.D. 318 or February 26, A.D.320 according to different calculations of the exact epoch of the Gupta era. It merely rests on the belief that he was a powerful king and his two predecessors were too insignificant to found an era.[81]

Samudragupta as the founder of the era?

R.C. Majumdar also puts forth another view that possibility of Samudragupta founding the era from the date of his accession cannot be ruled out.[82] In his support he quotes the evidence of the two copper plates grants of Samudragupta found at Nālandā and Gayā dated respectively in years 5 and 9. The genuineness of these two plates have been doubted but it has been asserted that they were not acts of complete forgery.[83] If therefore, we believe that it was issued in the year 5 of Samudragupta, "it would be more reasonable to regard the Gupta era as commemorating the year of Samudragupta's accession, rather than that of his father".[84] Majumdar is more favourably inclined to this view even if we regard the Nālandā charter of the 5th year as a late copy of genuine grant.[85] The ground that a period of 136 years (as Kumāra-gupta-I, the grandson of Samudragupta, was on the throne in the year 136 of the Gupta era) is very high for three generations, does not come on the way which is no impossibility as three generations of Western Chālukyan kings from Vikramāditya-V to Someśwara-III ruled for 118 years.[86]

This question, of course, is directly related to the chronology of the Guptas which we shall take up later. Suffice it to mention here that the dates given in the Nālandā and the Gayā records are the regnal years of Samudragupta while in the Mathurā inscription of Chandragupta-II, both the regnal year 5 and the Gupta era 61 have been mentioned.[87]

Now the only hypothesis which is left with us is that the Gupta era must have been started at a later date most probably by Chandragupta-II Vikramāditya beginning from the date of accession of the first most important sovereign, Chandragupta-I.

Chandragupta-II is the founder of the Gupta era

As we have seen that if we believe that the Gupta era begins with the date of accession of the first ruler Gupta, then the period of 56 years appears too short for 4 generations to Chandragupta-II in whose Mathurā inscription of G.E. 61, the G.E. 61 was his fifth regnal year. Similarly it is also not possible that Chandragupta-I's son Samudragupta could have started the era in A.D.319-320 as it does not fit in the chronology as we shall see later that his accession to the throne took place around this date. It is more likely that Chandragupta-I married Kumāra-devī., the Lichchhavi queen around this date and became a paramount sovereign, the first upholder of the title *Mahārājadhirāja* in the family, to his accession to the throne the Gupta era might have been allocated. As P.L. Gupta holds the opinion "If the era began with the reign of Chandragupta-I, it does not necessarily mean that he was the founder of the era". It should not be ignored that the dates found in the records of all the early kings of India, refer to the regnal years of an individual king and never to an era.[88]

The use of eras, in fact, has been started by the foreign rulers in India that is, the Kushāṇas on their inscriptions based on the dating by the regnal years of Kaniska and by the Western Kṣathrapas on their coins. The Guptas certainly adopted a foreign tradition. As such, the idea of establishing an era would not have occurred to Chandragupta-I. It would only be Samudragupta or his son Chandragupta-II, who could borrow the idea from the Śākas and the Kushāṇas, as only they had come in direct touch with the territories occupied by the foreigners.[89] We have already ruled out the possibility of Samudragupta having founded an era.

Earlier to Chandragupta-II, no ruler has ever made mention of the Gupta era in his records. If we add 319 years in the date of the Gupta era 61 which is mentioned in the Mathurā Pillar Inscription it comes to A.D. 380 which is the fifth regnal year of Chandragupta-I, a period of 56 years can easily take care of the reigns of Chandragupta-I and Samudragupta also including the brief period of Rāmagupta being the Gupta ruler. If we believe that the Gupta era started with the accession of Samudragupta, then it would give a very long reign to him i.e. A.D. 320-375.

Secondly, it comes in the fold of Vikramāditya tradition the legendary king Vikramāditya who is said to have founded an era in the

57th year of the paramount status of his family. Fifth regnal year of Chandragupta-II in the G.E. 61 indicates the 57th year of the paramount status of the Gupta family, becomes also the year of accession to throne by Chandragupta-II. The Vikramāditya had assumed the said title after his conquest over the Śākas. Chandragupta-II might have assumed the title of *Vikramāditya* around 380 A.D. only after the extermination of the Śāka Satraps of Western India so far as the evidence goes now by the first mention of the Gupta era in the Mathurā inscription. It tallies with the Vikramāditya tradition in this context. We have numismatic evidence in our support for this theory. The last known date of the Śāka dynasty is 304 as found on the coins and is referred to the Śāka era it gives the date 382 A.D. which comes very closer to 380 A.D. the date of the Mathurā record, which is the fifth victorious regnal year of his reign meaning truely that by this date he had exterminated the Śākas from that region.

It is more plausible to believe that Chandragupta-II thought of starting it from the accession of Chandragupta-I as an emperor, the first paramount sovereign and the real founder of the Gupta empire. This dynastic era started with a retrospective reckoning from some event but had occurred 56 years anterior to the accession of Chandragupta-II. Such a practice is not unknown in Indian history.[90] On this basis of numismatic evidence we can say that he was the first Gupta ruler to have assumed the title of *Vikramāditya.* [91]

The Gupta Chronology of the Earlier Rulers

So far the chronology of the Imperial Guptas has remained in utmost confusion and is far from settled. It looks no serious attempt has ever been made in this direction except making some statements. We want to hopefully settle this when we have got a definite clue of the date of accession of Chandragupta-II in the Mathurā Pillar Inscription of the Gupta Era 61 i.e. A.D. 480 which is the fifth regnal year of Chandragupta-II. That means that Chandragupta-II ascended the throne in A.D. 376.

We may have a look on a few chronologies suggested earlier:

(i) R.G. Basak states that following the view of late V.A. Smith and other schclars we may place the rule of Mahārāja Gupta conjecturally between A.D.275 and 300 and that of Ghaṭotkacha between A.D. 300 and 320 respectively.[92]

(ii) R.K. Mookerji places the reign of king Gupta in A.D. 270-280, giving to his son Ghaṭotkacha the period A.D. 280-319 for his reign.[93]

(iii) R. N. Saletore assigns the reign of king Gupta to A.D. 245-270.[94]

(iv) Smith taking 319 (the epoch of the Gupta era) as the initial date of Chandragupta-I, places Mahārāja Gupta between A.D. 275-300 and his dates have been accepted by almost all other scholars.[95]

(v) P.L. Gupta states that the terminal date of Ghaṭotkacha must have been A.D. 319, which is the epoch of the Gupta era and indicates the rise of Chandragupta-I. The beginning of the reign of Ghaṭotkacha may be placed in circa A.D. 300.[96]

(vi) Altekar[97] places Śrīgupta between A.D. 270-290, Ghaṭotkacha between A.D. 290-305,[98] Chandragupta-I A.D. 305-325,[99] Samudragupta A.D. 330-370,[100] Rāmagupta A.D. 370-375, [101] and Chandragupta-II A.D. 375-414.[102] So far we shall concern ourselves with the dates upto the reign of Chandragupta-II.

(vii) Goyal states that the chronology of the predecessors of Chandragupta-II is also far from settled.[103] The date of the establishment of the dynasty is not yet known,[104] the identity of the founder of the Guptas era is still a debatable question and the initial year of the reign of Samudragupta variously placed from A.D. 319 to C.350 A.D., is yet to be satisfactorily determined.[105] Goyal puts the probable date of Samudragupta's accession as A.D. 350 and he opines to construct the chronology of his predecessors on this basis[106] rather than catching the thread from the Mathurā inscription of Chandragupta-II, G.E. 61 i.e. A.D. 380. He suggests the dates C.295 to C.300 A.D. for gupta and his son Ghaṭotkacha from C.300 to 319 A.D.[107]

We propose to start our chronological reconstruction of the early imperial Guptas on the basis of Mathurā inscription of Chandragupta-II, Gupta Era 61 i.e. A.D. 380 which is beyond any dispute by just applying our common sense on the factual events and the possibilities of other related happenings.

Chandragupta-II is stated to have proposed a matrimonial alliance to the Vākāṭaka king Pṛithvīshēṇa by offering his daughter Prabhāvatī-Guptā in marriage to the Vākāṭaka crown prince Rudra-Sena. The offer was accepted by Pṛithvīshēṇa and the marriage was celebrated with great pomp, probably at Pāṭaliputra, in C.380 A.D.[108] which is further supported by epigraphic evidence[109] as well as a literary tradition. A literary tradition states Pravara-shena-II, the youngest son of this union, was spending early youth in the pursuits of pleasure in C.410 A.D. as his maternal grandfather Chandragupta-II could be relied upon to look after his administration. At the time of the death of Chandragupta-II in C.414 A.D., his grand son would be about 25, his birth, therefore, may be placed in C.390 A.D. He was atleast the second if not the third or fourth[110] child of his parents and so the marriage may be placed in C. 380 A.D.

Pṛithvīshēṇa died about five years after the marriage in C.385 A.D. and Rudra-sena-II ascended the throne in C.385 A.D. He suddenly died in C.390 A.D. at the premature age of about 30 after a short reign of five years. Prabhāvatīguptā was left with his two sons namely Divākara-sena and Dāmodara-sena (as known from the Rithpur plates), the former about 5 years in age and the latter about 2 years. As a result she became the Regent Queen. Divākara-sena, her eldest son and the minor king died soon after the 13th year of her regency. Dāmodara-sena, her younger son, was installed on the throne, and Prabhāvatīguptā continued to be the regent for another five years or six years. Eventually the regency terminated in C.410 A.D., when Dāmodara-sena ascended the throne assuming the coronation name of Pravara-sena-II.[111] Thus we have Vākāṭaka chronology also at our disposal to compare with the contemporary Gupta chronology.

Now we know for granted that Chandragupta married his daughter Prabhāvatīguptā to Vākāṭaka crown prince Rudra-sena in A.D. 380, which means he must be married at least twenty years before i.e. A.D. 360 when he must have been roughly twenty years. If we go back at least twenty years we clearly arrive at Chandragupta-II's date of birth i.e. A.D. 340. That takes us back roughly another twenty years i.e. A.D. 320 when Samudragupta is born while a son is born to him in A.D. 340. Pushing another twenty years back i.e. A.D. 300 we can arrive at the date of birth of Chandragupta-I who might have ascended the throne in A.D. 319-320. He must have married the Lichchhavi queen Kumāradevī around that date and assumed the title *Mahārājadhirāja*, 'paramount sovereign' as Samudragupta seems to have born around this date. His father Ghaṭotkacha must have been born atleast twenty years before i.e. A.D. 280. Therefore, Gupta, the first ruler of the family might have born around A.D. 260 or even a little earlier i.e. A.D.250 leaving some margin for his assuming the status *Mahārāja*, an independent ruler. We may place the reign period of Gupta when he assumed the age of atleast 25 years i.e. A.D. 275 -A.D. 295; of Ghaṭotkacha A.D. 295-A.D.319; Chandragupta-I A.D. 319-345 and after a few years, and of Samudragupta A.D. 346 or after to A.D. 375. According to our chronology we have stated that Chandragupta-II is born around A.D. 340 and must be of 18 years before he acted as a Dūtaka (executor) of the Nālandā grant of Samudragupta as his name is mentioned in the last line of the record as Kumāra Śrī Chandragupta who is no other than Chandragupta-II[112] which means that this record was executed around 358 A.D. The above record mentions year 5 which obviously seems to be the fifth regnal year of Samudragupta which brings us to the date of accession of Samudragupta five years earlier i.e. A.D. 353.

We have arrived at the date of birth of Chandragupta-I on A.D. 300. It shows that Chandragupta had a long reign of 33 years and must have thought of abdicating his throne in favour of his son Samudragupta as has been assumed on the basis of the evidence provided by the Allahabad Pillar Inscription of Samudragupta:

"Verily, you are the worthy...rule this earth".[113]

Thus we see that Samudragupta may have ruled for 20 or 21 years upto A.D. 373-374 giving margin of one or two years to the reign of elder brother of Chandragupta-II i.e. Rāmagupta before Chandragupta-II took the reigns of administration in his hands in A.D. 376.

We have no justification of Majumdar's theory of Samudragupta's ascending the throne in 325 or 335 A.D.[114] or of A.L. Basham's dating of Samudragupta C. A.D. 335-376 [115] as we have reasonably put his date of accession to the throne around A.D. 353.

Another method of verifying the chronology of early Imperial Gupta rulers is to compare them with the contemporary Vākāṭaka and Nāga kings.

Vākāṭaka Chronology

Vindhyaśakti	C. 255-275 A.D.
Pravarasena-I	C. 275- C. 335 A.D.
Gautamīputra	300 A.D. (crown prince married the daughter of Bhava-Nāga, the Nāga ruler and who predeceased his father and was succeeded by his son Rudrasena-I).
Rudrasena-I	(335 A.D. to 360 A.D.)
Prithvī-sheṇa-I	(360 to C. 385 A.D.)
Rudra-sena-II	(385 to C.390 A.D.)
Regency of Prabhāvatīguptā	(390 - C. 410 A.D.)
Pravara-sena-II	(410- 440 A.D.)
Narendra-sena	(440 - 460 A.D.)
Prithvīsheṇa-II	(460- 480 A.D.)

Vākāṭaka Chronology [117]

(Basim Branch)

Sarva-sena	(younger son of Rudrasena-I) 330 - 350 A.D.
Vindhya-sena	350 - 400 A.D.
Pravarasena-II	400 - 415 A.D.

(from C. 410 - C.415 A.D. rulers of the main Vākāṭaka line and its Basim branch bore the same name).

Name of the king	415-455 A.D.
Deva-sena	455-475 A.D.
Harisheṇa	475-510 A.D.

According to the Purāṇas seven kings had ruled at Mathurā and nine at Padmāvatī when the Guptas were rising to power by C. 325 A.D.[118] It can be safely assumed that these houses started their career in the latter half of the second century and they must have been the feudatories of the Kushāṇas. [119] It is said about the Nāgas that they were the distinguished heroes who got the Ganges water by their valour for coronation and performed ten horse-sacrifices (Prākramādhigata - Bhāgīrathy a (a) mala - jala - mūrddhan (rddh) - ābhishiktānām - daśā śvamedha-avabhṛitha - snānānām = Bhāraśivā - nām..... .[120] It appears that the Nāgas got the upper Gangetic plain freed from the yoke of the Kushāṇas during the first half of the 3rd century A.D. They flourished after the Kushāṇas and exerted their influence for a long time. Later traces of Nāga rule may be found in the Allahabad Pillar Inscription [121] which mentions the defeat of Ganapatinaga and other Nāga Kings at the hands of Samudragupta. This epigraph gives us an idea of the political condition of India about the middle of the fourth century A.D. Coins disclose the existence of ten Nāga rulers, Bhīma-Nāga, Vibhu-Nāga, Prabhākara-Nāga, Skanda-Nāga, Bṛihaspati-Nāga, Vyāghra-Nāga, Vasu-Nāga, Deva-Nāga, Bhava-Nāga and Gaṇapati-Nāga; the *Harshacharita* refers to 11th ruler named Nāgasena who also figures in the Allahabad inscription of Samudragupta, which possibly mentions a further 12th Nāga king Nāgadatta.[122] Among the Nāga rulers of Padmāvatī definite historic information is available only of Bhava-Nāga, who was ruling from C. 305 to 340 A.D. His daughter was married to the Vākāṭaka crown prince Gautamīputra in C. 300 A.D.[123] The Vākāṭaka records never fail to mention that Bhavanāga was the maternal grandfather of Rudra-sena-I. Bhavanāga's son-in-law Gautamī-putra predeceased his father and so the crown passed on to the latter's son Rudrasena-I. Bhavanāga died in C. 340 A.D. After whom Nāgasena followed by Gaṇapatināga became the rulers.[124] Thus a comparison of the Vākāṭaka chronology with those of the Guptas also shows the Samudragupta rose to power and exterminated these Nāga rulers around the middle of the fourth century A.D. The Guptas claim to have forcibly uprooted all the Nāga families. But some members of the old houses continued to exist either as Gupta feudatories or officers down to the decline of the Gupta empire. Samudragupta himself married his son Chandragupta-II to the daughter of a Nāga Chief around C. 360 A.D.[125] We may not agree with the date given by Altekar i.e. C. 370 A.D.[126] as the marriage of his daughter born of the union with the Vākāṭaka crown prince Rudra-sena II in C. 380 A.D. would not be possible in that case. Secondly as shown earlier when he was born around A.D. 340 and acted as a Dūtaka of the Nālandā record around A.D. 358, it is likely he may have been married atleast a few years later not as late as A.D. 370.

Another test we can put to our chronology is to compare the life span of 4 generations with those of the Vākāṭakas which comes to a total of 130 years upto A.D. 385 - beginning with Vindhyaśakti to Pṛithvīseṇa-I. The same in case of the Imperial Guptas comes to approximately 120 years upto A.D. 375 - beginning with Gupta to Rāmagupta. The Vākāṭakas had one crown prince Gautamīputra in the third generation who predeceased his father Pravarasena-I and among the Guptas we had the fifth ruler who ruled for a short interval of 2-3 years at the most between Samudragupta and accession to the throne of Chandragupta-II.

The margin for one ruler on an average comes to about 30 years. We find Majumdar a bit confused to the date of accession to throne of Samudragupta who mentions it in A.D. 320 or C. 350 A.D.[127] While ruling out any justification for the dates A.D. 325 - 335.

Our analysis has made it clear that Samudragupta ascended the throne around A.D. 353. We have tried to clear all doubts in this regard by reviewing the evidence of chronology, reconstructing it on the basis of a definite clue provided by the Mathurā record of Chandragupta-II, by collaborating the data provided by the coins, inscriptions, purāṇas etc. seeking corroboration of our evidence by the clues provided by the Vākāṭaka and Nāga-chronological frame-work at our disposal.

So we get the following chronology of the early Imperial Guptas.

1.	Gupta	A.D. 250 - 295 A.D. A.D. 275 - 295 A.D. (Reign period)
2.	Ghaṭotkacha	A.D. 295 - 319 A.D. (Reign period)
3.	Chandragupta-I	A.D. 319 - 353 A.D. (Reign period)
4.	Samudragupta	A.D. 353 - 373 A.D. (Reign period)
5.	Rāmagupta	A.D. 373 - 375 A.D. (Reign period) (Eastern Malwa region)
6.	Kāchagupta	A.D. 373 - 378 A.D. (Reign Period)?[128] (Pāṭaliputra)
7.	Chandragupta-II	A.D. 376 - 414 A.D. (Pāṭaliputra)

Notes And References

1. S.P. Gupta (ed.) *RPIII.*, p.76.
2. R.N. Dandekar, *AIIG.*, p.21.
   H.C. Raychaudhuri, *PIIAI.*, pp. 467-68.

3. *IHQ*. XIV, pp.532-535.
4. *IGT*., pp.17-20.
5. *EHNI*., pp.166-168.
6. Majumdar, *VGA*., p.130; *RPHI*., p.78.
7. Chattopadhyaya, *EHNI*., p.168.
8. *Ibid*. pp.18-19.
9. *GGCBH*., *Introduction*, pp.xi-xii.
10. *Ibid*., p.xiv. Anugaṇgā Prayāgaṁcha Sāketam Magadhāms, Tathā, Etān Janapadān Sarvān bhokṣyante Guptavaṁśajāh. *Vāyu*, p.99-383.
11. *Ibid*., pp.xiv-xv.
12. Anugangā Prayagām Māgadhā Guptāścha Bhokṣyanti.
13. B.P. Sinha, "Original Home of Imperial Guptas", *RHC*., p.92.
14. R.C. Majumdar, "Original Home of Imperial Guptas", *RPHI*., p.75.
15. Sinha, *RHC*, pp.97-98.
16. *RPHI*, p.78; "Whether this Śrī Gupta was the founder of the Imperial Guptas; and, if so, whether that portion of North Bengal, which was comprised in his kingdom, was his home territory or conquered land - these are questions which need not be discussed on the present occasion".
17. P.L. Gupta, *IG*, pp.5-14.; S.R. Goyal, *HIG*., pp.47-48.
We are unable to trace eight inscriptions of early Gupta rulers prior to Kumāragupta-I as mentioned by Goyal. They appear to be not more than three in number. On verification, his work *Prāchīna Bhāratavarsa Kā Itihāsa*., p.50 f.n.3. substantiates our view.
18. Goyal, *Op.cit*., p.49.
19. P.L. Gupta, *Op.cit*., pp.91-93.
20. *Ibid*., p.89.
21. S.R. Goyal, *Op.cit*., p.45.
22. Majumdar, *VGA*., p.134; Pargiter, *DKA*., p.53 f.n. 8 as quoted by Majumdar.
23. Anu-Gangaṁ Hastināpuraṁ, Anu-Gangaṁ Vārānasī Anu-Śoṇam Pātaliputraṁ - Patañjali, II. 1.2.
24. U.N. Roy, *GRY*, pp.58-64.
25. B.P. Panthari, *BKSY*., pp.48-53.
26. *CII*. III. pp.20-21.
27. A.S. Altekar, *State and Government in Ancient India*, Delhi, 1972, pp 320- 321.
28. Allahabad Pillar Inscription of Samudragupta, *CII*. III L.29, p.28.
29. *CII*. III, p.258, L.1.
30 Sarnath Buddhist Stone Image Inscription of Buddha Gupta, G.Y. 157, *ASR*., 1914-15, pp.124-125, L.I. Sarnath Buddhist Stone Inscription of Kumāragupta-II, G.Y, 154, *ASR*, 1914- 15, p.124, L.I.

31. Cf. Allahabad Pillar Inscriptions of Samudragupta, *CII*. III. LL.28-29, p.8.
32. It may be noted that here also the epithet Śrī indicates that the founder's name was 'Gupta'. Cf. D.C. Sircar, *IHQ*, XIX, p.19; "The first known king of the Gupta dynasty was Gupta whose son was Ghaṭotkacha; but when the latter's son Chandragupta-I founded an empire, his descendants always stuck to the name-ending Gupta and soon the family became known as the Gupta dynasty". Cf. Dashrath Sharma, *JBRAS*. XXXIX, p.265.
33. Śrī Śrīmatyam : *CII* . III p.215, L.2.
34. Allahabad Pillar Inscription of Samudragupta, *CII*. III L.28, p.8.
35. B.G. Gokhale, *Samudragupta, Life and Times*, p.28.
36. For details of abbreviated names, places, see V.S. Agarwala, *India as Known to Panini*, pp.190-192.
37. *JASB*., Vol.LIII, Part I p.119 and note.
38. *CII*. III, p.8. note 3.
39. *EI*. XV/4/ pp.2-43.
40. See, F.W. Thomas, "The Root Gup and the Guptas", *JRAS*, 1909, pp.740-743.
41. Book III, Chap. 10, Verse 9.
42. See also F.E. Hall's edition of H.H. Wilson's Translation, Vol.III, p.99f.
43. Mānavadharmaśāstra, II, 31. Brunell's translation, p.20.
44. This is shown by the appearance of the name of Kumāradevī and her father's family on some coins of Chandragupta-I, and by the regular use of the epithet, "daughter's son of Lichchhavi (or of a Lichchhavi King)" for Samudragupta in the Genealogical passages in the inscriptions of the Gupta dynasty.
45. *IA*., Vol.IX, p.131.
46. K.D. Bajpai, *S.K. Bhuyan Commemoration Volume*, p.119; A circular lead piece bearing the seal mark of Indragupta 'Rano Indagustasa' inscribed in the Mauryan Brahmi script was recovered which gave the name of a king who ruled over Eran about 200 B.C.
47. D.C. Sircar, *SI*., p.475.
48. *Ibid*., p.436.

    See Ibid., pp.436-37, f.n.9. The queen refers to her paternal gotra rather than that of her husband's family and thus contradicts the injunctions of the Smritis, p.439.
49. *Ibid*, p.443; *CII*. III, pp.245, 248: Rithpur Copper Plate Inscription of Prābhāvatīguptā of the time of Pravarasena-II, lines 7-8 and Poona Copper Plate Inscription of Prābhāvatīguptā lines 7-8.
50. K.P. Jayaswal, *HI*., pp.115-16.

51. B.G. Gokhale, *Samudragupta, Life and Times*, pp.25-26.
52. *JBORS.*, 1934, p.235.
53. *Ibid.*, XIX, pp.115-6.
Majumdar disagrees with this surmise. Some scholars have given the emendation 'Gupta' for the original Jarto, Jato, or Japto which also is not however, acceptable (Majumdar, *VGA*, 197, f.n.1) Hoernle while identifying the people with jats interprets the passage, as referring to the defeat of the Hūṇas by Yaśodharman.
54. H.C. Raychaudhuri, *PHAI.*, 5th ed., p.526, f.n.1. P.L. Gupta takes it to be gotra for Vaiśyas, see *IG.*, p.234.
55. Purāna, Vol.VII, No.1 (Jan.1965), pp.183-85; Skanda Purāna, Dharmāranya Khaṇḍa 2, Chap.9, Vs.27-31. Out of 24 gotras mentioned, Dhārana is the twelfth.
56. *Ibid.*, p.185, fn.8; (Mitaksara, N. 5th ed. p.151). "Rājnyaviśām prātisvik gotrābhavāt pravarabhavastathāpi purohitapravarau veditavyau". This is the view also of Baudhāyana, Āpastamba and Laugākṣi
57. S.R. Goyal, *HIG.*, pp.78-81.
58. *Ibid.*, p.78.
59. *Ibid.*
60. *Ibid*, p.80.
61. *Ibid.*, p.81.
62. These three marriages are not of much importance as a Brāhmaṇa can marry from any caste according to the provisions of the Smritis. So even as of lower class than Brāhmanas, Guptas could have married their daughters to the Brahmaṇas.
63. R.C. Majumdar, *VGA.*, p.170;
It has been suggested that Chandragupta-II arranged a marriage between his son and the daughter of Kākusthavarman, the most powerful ruler of the Kadamba family who was the ruler of Kuntala, Kanarese country in the Bombay Presidency.
64. Cf. *Ibid.*, p.170.
65. *Ibid.*, p.169.
66. R.C. Majumdar states that "We may reasonably assume that the marriage of Chandragupta and Kumāradevī led to the amalgamation of the Gupta principality with the Lichchhavi state, and the epithet 'Lichchhavi-dauhitra' was deliberately given to Samudragupta to emphasize his right of succession to the dual monarchy;"
R.C. Majumdar, *VGA.*, p.129, Cf. V.S. Pathak, *JNSI*, XIX, Pt.II, pp.140-141. Pathak takes the meaning Dauhitra in the technical sense of the Smritis, i.e. "a person having dual parentage (dvāmuṣyāyana)".
67. Allahabad Pillar Inscription of Samudragupta, line 29 *CII.* III, p.8.

68. We know that Seleucus married his daughter to Chandragupta Maurya eventhough the Greeks used the word barbarian for non-Greek. H.W. Fowler and F.G. Folwer *The Concise Oxford Dictionary*, p.91 and were not in favour of mixing with them. Marriages among Kings attached more significance to political than to social considerations.
69. J.F. Fleet, *CII*. III, pp.23-25, 61-64.
70. *Ibid.*, p.69, pp.84-93. Vasudeva Upadhyay, *GA.*, pp.104-105.
71. S. Chattopadhyaya, *EHNI.*, p.176.
72. P.L. Gupta, *IG.*, p.210.
73. V. Upadhyaya, *GA*, p.130.
74. *CII*. III, p.43, Bilsad Stone Pillar Inscription of Kumāragupta. The Year 96, L.6.
75. *CII*. III., p.60. Junagarh Rock Inscription of Skandagupta, L.15.
76. See, f.n. 72; S. Chattopadhyaya, *EHNI*, p.176.
77. P.L. Gupta, *IG.*, pp.212-213.
78. V.A. Smith, *The Oxford Students History of India,* OUP, 1933, pp.80-81.
79. H.C. Raychaudhuri, *PHAI.*, p. 468.
80. *Ibid.*, f.n. 2.
81. *VGA.*, p.131.
82. *Ibid.*, p.132.
83. S.P. Gupta (ed.), *RPHI.*, pp.84-88.
84. R.C. Majumdar, *VGA.*, p.132.
85. *Ibid.*, p.159.
86. *Ibid.*, pp.159-160.
87. Goyal, *HIG.*, pp.105-106.
88. P.L. Gupta, *IG.*, p.207.
89. *Ibid.*, p.208.
90. *Ibid.*, p.209.
91. *Ibid.*, p.83.
92. R.G. Basak, *HNEI.*, p.7.
93. R.K. Mookerji, *GE.*, pp.11-12.
94. R.N. Saletore, *Life in the Gupta Age,* Bombay, 1943, p.6.
95. P.L. Gupta, *IG.*, p.239.
96. *Ibid.*, p.240.
97. A.S. Altekar, *GGCBH.*, p.xi.
98. *Ibid.*, p. xiii.
99. *Ibid.*, p.xiv.
100. *Ibid.*, p.xvi.
101. *Ibid.*, p.xxii.
102. *Ibid.*, p.xxvi.
103. S.R. Goyal, *HIG.*, p.103.

104. *Ibid.*,
105. *Ibid.*, pp.103-104.
106. *Ibid.*, p.109.
107. *Ibid.*, p.110.
108. R.C. Majumdar, *VGA.*, p.110.
109. *Ibid.*, p.11, f.n.1.
110. *Ibid.*, p.110, f.n.1.
111. *Ibid.*, pp.110-113.
112. V. Upadhyaya, *GA.*, p.126;
     Nālandā Inscription, L.12;
     P.L. Gupta, *IG.*, p.208.
113. Fleet, *CII.* III, p.6, LL.7-8 in the Inscription, Verse 4.
114. R.C. Majumdar, *VGA.*, p.159.
115. A.L. Basham, *WTWI.*, p.63.
116. A.S. Altekar, *VGA.*, pp.96-119.
117. *Ibid.*, pp.119-125.
118. *Ibid.*, p. 36.
119. *Ibid.*
120. Fleet, *CII.* III, p.237, Chammak Copper Plate Inscription of the Mahārāja Pravarasena-II, Lines 6-7; Also see *CII.* III, p.245, Swani Copper Plate Inscription of the Mahārāja Pravarasena-II, Lines 5-6.
121. Fleet, *CII.* III, No.1, pp.1-17.
122. A.S. Altekar, *VGA.*, p.37.
123. *Ibid.*, p.38; Chattopadhyaya, *EHNI*, p.147.
124. *Ibid.*, p.39; *Ibid.*
125. *Ibid.*, p.40.
126. *Ibid.*
127. R.C. Majumdar, *VGA.*, p.160.
128. See pp.70-73 and Chapter 7.

5

Evolution of the Gupta Empire

R.K. Mookerji puts the dates C. A.D. 240-280 for Śrī Gupta, the founder of the Gupta dynasty.[1] Though the beginnings of the Gupta empire are obscure we have tentatively arrived at the dates C. A.D. 275-295 for the rule of Śrī Gupta[2] who had achieved the status of a *Mahārāja* which continued with the next ruler Ghaṭotkacha until and unless Chandragupta-I, in the third generation acquired the status of a *Mahārājādhirāja* as is evidenced from the Gupta inscriptions. All the Gupta inscriptions unequivocally testify Mahārāja Śrī Gupta to be the founder of the dynasty whenever the genealogy of the former rulers is mentioned.[3] Mookerjee wrongly quotes from the Poona plates of Prabhāvatīguptā Vākāṭaka stating that Mahārāja Śrī Gupta has been aptly called Ādirāja.[4] We may state that we should give full credence to the Gupta inscriptions rather than believe the version of Rithpur copper plate inscription of Prabhāvatīguptā of the time of Pravarasena-II, Royal year 19 and the Poona copper plate inscription of Prabhāvatī-guptā, Regnal year 13 where in the first two lines Mahārāja Śrī Ghaṭotkacha has been called the Ādirāja of the Guptas while narrating the genealogy of her illustrious father śrī Chandragupta-II.[5] Historians have accepted the testimony of I-tsing's Mahārāja Che-li-ki-to with Mahārāja Śrī Gupta who is said to have built a temple near Mṛigaśikhā-vana for some pious Chinese pilgrims giving some margin of error in the matter of chronology as he merely stated 'the tradition handed down from ancient times by old man.'[6] I-tsing (673-695 A.D.) notes that the building of the temple began 500 years before his travels.[7] This goes against the dates proposed above for Gupta as it brings his dates to 173-195 A.D. So we have to give a margin of 100 years to this evidence which is based on mere tradition while accepting our dates for

Gupta as A.D. 275-295. We mention that Sri is merely an honorific before the original name Gupta.[8]

Gupta was succeeded by his son, Ghaṭotkacha who is also styled *Mahārāja* . We know almost nothing about him except his mention in the genealogical list in the inscriptions of the Guptas and the inscriptions of Prabhāvatīguptā. He may have ruled from C. A.D. 295-319. When the throne passed on to his son Chandragupta-I who styled himself as the *Mahārājādhirāja* and seems to have raised the status of the family as a paramount sovereign. Even if we assume the title *Mahārāja* to be the status of an independent ruler if we do not know the first two rulers as vassals of any other ruler; the title *Mahārājādhirāja* certainly denotes that Chandragupta-I might have subdued some neighbouring kings or chieftains.

CHANDRAGUPTA-I (C.A.D. 319 - 353)

The evolution of the Gupta empire started with the reign of Chandragupta-I who is described in the Gupta inscriptions as *Mahārājādhirāja* 'Kings of Kings' or Emperor. The title signifies his conquests and the extension of territory but unfortunately we do not come across about his conquests in any source except the version of the Purāṇas that 'the kings born of the Gupta family will rule over the territories (janapadas) situated along the Ganges (anu-gaṅgā) such as Prayāga, Sāketa (Oudh) and Magadha' which most of the scholars believe that they describe the territories ruled over by the Guptas prior to Samudragupta's conquests i.e. of Chandragupta-I.[9] Majumdar, however, quotes the differing views of other scholars in this context that these territories may belong to Samudragupta as well at a certain stage of his victorious career, may denote the Gupta domination in the period of decline after Skandagupta.[10] He is of the opinion that it is difficult to rely upon the puranic evidence in deducing the extent of territory ruled over by Chandragupta-I.[11]

The theories of K.P. Jayaswal and D. Sharma based on the *Kaliyugarāja-Vṛittānta* of the Bhaviṣyottara Purāṇa and the story of the *Kaumudī-Mahotsava* of comparing Chandragupta-I with Chaṇḍasena and the events described in the drama leading to the usurpation of the throne of Magadha by a relative of the ruler, who had a marriage alliance with the Lichchhavis are no longer accepted[12] and we need not discuss them here.

Chandragupta-I was the first Hindu king who issued gold coins probably soon after assuming the imperial title.[13] We have the names and figures of the king and queen on the obverse and the name of the Lichchhavis (Lichchhavayaḥ) on the reverse of these coins.[14] Inspite of the growing greatness of the Guptas, the Lichchhavis insisted on maintaining their separate identity in the dual kingdom, we have not yet got a single coin of Chandragupta-I where the name of the Lichchhavi clan and princes does not appear on the reverse and obverse respectively. The recently published standard type coin of Chandragupta is now settled to be an issue of Chandragupta-II and not of Chandragupta-I.[15]

Significance of Chandragupta-I Kumāradevī type of Coins

The coins described above of Chandragupta Kumāradevī bearing the legend 'The Lichchhavis' (Lichchhavyah) become more significant when we notice the epitect 'Lichchhavi-dauhitra' (grandson of the Lichchhavis) in all the Gupta inscriptions which describe the genealogical list to raise certain basic questions before we discuss the various views of scholars on the subject.

(i) Whether the word 'dauhitra' has been used just to take pride as the Lichchhavis were a great power or meant for the purpose of inheritance?

(ii) The Lichchhavis were a republic as it appears from the coin legend, 'Lichchhavyah' even the word 'Gupta' is missing. If the Lichchhavis were a republic, how did they (the people) allow for its amalgamation by the Guptas?

(iii) When Chandragupta-I chose Samudragupta as his successor as evidenced by the Allahabad Pillar Inscription, he was himself alive. If so whether the coins were commemorative or not as opined by some scholars and issued by Samudragupta?

(iv) There is no mention of Sāketa and Magadha in his conquests by Samudragupta in his Allahabad Pillar Inscription which means he inherited this empire from his father Chandragupta-I.

(v) Whether Samudragupta inherited the empire of the Lichchhavis by virtue of his right as a *dauhitra* of the

Lichchhavis, the status as such being presented by the Smṛitis or Chandragupta-I himself amalgamated the Lichchhavi empire with the Guptas during his lifetime?

We may answer the above questions one by one giving the opinion of different scholars relevant to them.

(i) R.C. Majumdar states that the marriage of Chandragupta and Kumāradevī led to the amalgamation of the Gupta principality with the Lichchhavi state, and the epithet *Lichchhavi-dauhitra* was deliberately given to Samudragupta to emphasise his right of succession to the dual monarchy.[16]

(ii) Majumdar has prefaced the above opinion by stating that the use of the tribal name in plural number sems to imply that a vestige of the old republican Constitution still persisted among the Lichchhavis, but the position of Kumāradevī indicates something like a hereditary monarchical constitution in actual practice.[17] We may not agree with the view of Chattopadhyaya in this connection that Samudragupta was grateful and indebted to the Lichchhavi people and not to the Lichchhavi royal house as evidenced by the scant respect to the Lichchhavis in his Allahabad record by mentioning that Nepal, which was at this time ruled over by the Lichchhavis, paid respects and bore allegiance to him.[18] He attributes these coins to Samudragupta stating that Kumāradevī may or may not have belonged to the Lichchhavi royal house. "She was taken in respect more for her being a Lichchhavi by nationality, than for any thing else."[19]

(iii) Allan suggested that these coins were issued by Samudragupta to commemorate the marriage of hisparents[20] which is supported by Chattopadhyaya.[21]

This cannot be accepted. Firstly, it is against Indian tradition that sons celebrate the marriage of their parents. Secondly, if the issuer of these coins was Samudragupta, he must have mentioned his name on these coins.

Goyal,[22] however, puts forth some arguments to support the theory first proposed by Allan that Samudragupta was the issuer of these coins speculating a different purpose of claiming the Lichchhavi empire as well as his right of succession to throne of Chandragupta-I being thus gainful for the Gupta empire rather than for commemorative purposes as allocated by Allan. He also takes into account the provenance of these coins in eastern U.P., Lucknow and Bharatpur state and their complete absence in Bihar where the Lichchhavis actually ruled,[23] for they were issued to publicize that he had a better title to rule over the amalgamated kingdom, it was only natural for him to circulate them in the region where the rebellious princess 'of equal birth' could hope to find some support. He tries to build up his argument on the assumed identity of the issuer as conjectured by Pathak[24] and Sohoni stating that 'there was enough indication left by Samudragupta on Chandragupta-I Kumāradevī coin type to indicate a reference to him, viz., names of his parents and of a community which had helped him.[25] We could agree with Goyal's assertion in this regard that the Chandragupta - Kumāradevī type, the earliest of the Gupta gold coin series, was issued 'not by Chandragupta-I but by Samudragupta in the initial years of his reign to show that he, being a dvāmushyāyana, had a better claim to the throne than his rival brothers,[26] if these coins were bearing the legend 'Lichchhavi dauhitra' along with Samudragupta's name as is shown by the Gupta inscriptions.

Sohoni, on the other hand, states that the marriage of Chandragupta-I with Kumāradevī was dictated at the battle field.[27] If it were so the Guptas will not take pride in the Lichchhavi blood specially a son born of such marriages never enjoyed himself stating that he belonged to one whom he or his father or fore-father had humiliated on the battle-field. In such case, Samudragupta would have taken pride by calling himself as belonging to one who had exterminated or defeated that power.[28] P.L. Gupta opines that if the Guptas were proud of their Lichchhavi descent there can be one and the only inference that the marriage of Chandragupta-I with the Lichchhavi princess was a normal one; and the Lichchhavi matrimonial alliance assisted the rise of the Guptas.[29]

About the finds of these coins in the eastern Uttar Pradesh and adjoining regions and their absence so far in Bihar we cannot definitely say anything in the matter and have to wait for future discoveries.

(iv) We may assume that Chandragupta-I himself was a great conqueror who might have added more territories to the boundaries of his empire by his conquests before assuming the title *Mahārājādhirāja*. One thing is very clear that these coins reflect that the prosperity of Chandragupta was due to his Lichchhavi alliance.[30] From the records of Samudragupta's conquests it has been deduced that his father's rule was confined to Magadha and the adjoining territories.

Some scholars have expressed the view that these coins were issued by Chandragupta jointly with the Lichchhavis and their princess Kumāradevī was a queen by her own right, and the proud Lichchhavis, to whose stock she belonged, must have been anxious to retain their individuality in the new imperial state.[31] V.A. Smith expressed the view that Kumāradevī brought to her husband as her dowry valuable influence, which in course of a few years secured to him a paramount position in Magadha and the neighbouring countries. He even went so far as to suggest that 'the Lichchhavis were makers of Pāṭaliputra, and Chandragupta, by means of his matrimonial alliance, succeeded to the power previously held by his wife's relatives.[32]

It is possible that the Lichchhavi king died without a male successor and Chandragupta helped the administration of the kingdom by Kumāradevī who was a queen in her own right. These coins might have been issued to mark this opportunity for circulation in the Gupta as well as the Lichchhavi Kingdom or even more likely for circulation only in the Lichchhavi Kingdom which might have adopted a mixed constitution by the time, having a hereditary monarchy but actually governed by a Lichchhavi Council.[33]

After a son was born out of the union he became the rightful successor. Chandragupta and Kumāradevī may have continued their conjoint rule over the Lichchhavi Kingdom till Samudragupta became their successor. This explains the fact that these coins had wider circulation as according to Altekar "their recorded findspots are Mathurā, Ayodhyā, Lucknow, Sitapur, Tanda, Ghazipur, Benaras in Uttar Pradesh and Bayana in the Bharatpur State". According to P.L. Gupta, the legend as such (the Lichchhavayaḥ) seems to suggest that the coin was issued on behalf of the Lichchhavi tribe as a political body.[34] D.C. Sircar is of the view that these coins were issued by Samudragupta in memory of his parents and may have been circulated in the territory of the Lichchhavis.[35]

The greatest objection to the commemoration medal theory of Allan also in a way supported by Sircar is the absence of the name of the commemorator on these coins. It appears Allan had in his mind the parallel of Indo-Bactrian rulers Agathocles, Antimachos and Eucratides issued a number of commemorative medals, commemorating their predecessors or parents.[36]

There is least possibility of Samudragupta issuing these commemorative medals as in that case he would have given the names and figures of his parents on the obverse; and his own name or atleast his *biruda* on the reverse like his Aśvamedha type of coins where he has given his *biruda Aśvamedha Prākramaḥ* that enables us to identify the issuer.

> "If we are to assume that like the Aśvamedha coins, the Chandragupta-Kumāradevī coins were also commemorative medals issued by Samudragupta, it is indeed strange that he should not have put atleast his *biruda* on these coins, as he did on the Aśvamedha type".[37]

Thus we may conclude that these coins were in no case commemorative medals issued by Chandragupta-I himself during his own reign.[37a]

SAMUDRAGUPTA (C. 353 - A.D. 373)

We have many sources in the form of inscriptions and coins to throw light on the career and achievements of Samudragupta. While finding ample material in the Allahabad Pillar Inscription of Samudragupta, scholars have usually ignored the evidence of Eran Stone Inscription of Samudragupta which is obviously earlier than the Allahabad Pillar Inscription and the Mathurā Stone Inscription of Samudragupta. Before taking up a discussion on the Allahabad Pillar Inscription let us take into account the information provided by these two inscriptions cited above found at Eran and Mathurā.

The Eran Stone Inscriptior[38] states that Samudragupta excelled Pṛithu, Rāghava and other kings in giving gold, who was equal to gods Kubera and Yama in pleasure and anger and who had brought under his suzerainty the whole tribe of kings; whose enemies are terrified when they think, "even in the intervals of dreams - in a place in Airikiṇa, the city of his own enjoyment....". This shows that even earlier to the period of the Allahabad Inscription Samudragupta had subdued a number of kings who are not named as the focus was installation of a temple with the colossal figure of Viṣṇu to which most probably this inscription belonged.[39] Another information which we get is that Eran formed a capital town of the Guptas if not the chief capital, atleast a provincial capital which is the import of the word 'Sva-Bhoganagara."

The Mathurā Stone Inscription of Chandragupta-II[40] brings to light the epithet of Samudragupta as one "who was the restorer of the Aśvamedha sacrifice that had been long in abeyance (Chirotsana-aśvamedha - āharrttā)[41] which we find in the later inscriptions of the Imperial Guptas which describe genealogy of the earlier kings. This inscription also describes Samudragupta as an Exterminator of all kings (Sarvvarājochchhettā) who had no antagonist of equal power (Apratiratha) in the world, whose fame was tasted by the waters of the four oceans, who was equal to the gods, Kubera, Varuṇa, Indra and Yama, who was giver of millions of cows and gold....."[42]

Allahabad Pillar Inscription of Samudragupta

This panegyric composed by Harishena who was himself a minister for peace and war, Kumārāmātyā and Mahādaṇḍanāyaka in 33 lines in a mixed style verse and prose which is known as *Champukāvya* in literature, is inscribed on some stone pillar which carries the famous edicts of Ashoka on it presently preserved in the Allahabad Fort.[43] Originally this pillar was set up at the ancient Kośāmbī, about 28 miles west by south from Allahabad and being still at that place when the present inscription was engraved; the executor of the inscription is Mahādaṇḍanāyaka Tilabhaṭṭaka. Later it was moved from there to Allahabad by one of the early Muslim rulers of Delhi.[44] This argument finds support as the column contains a short Asokan edict addressed to the rulers of Kośāmbī and further the absence of any mention of the above column in his account of Po-lo-ye-kia i.e. Prayāga or Allahabad by the Chinese pilgrim Hieun Tsang.[45]

Whether Posthumous or Not?

Fleet declares this Allahabad Pillar Inscription as Posthomous.[46] He goes so far to state that it belongs to the time of his son and successor Chandragupta-II and must have been engraved soon after the accession of the latter.[47] We cannot agree to this statement as in that case some reference to Chandragupta-II must have been there as usual in the Gupta Inscriptions.

Fleet states that it is not dated as it describes Samudragupta as deceased.[48] He forcibly alludes the reference to the term "Paramabhaṭṭāraka" in line 33 of the Allahabad Inscription for Chandragupta-II[49] He seems to have drawn the inference of Samudragupta being deceased from lines 29-30 of the inscription which states that "his fame resulting from his world conquest had spread over the whole universe and even upto heaven." But the fame of a man on earth may go upto heaven without the man himself going up to heaven. This expression is quite common in his coins: "The invincible (king) who had won victories on a hundred battle fields and conquered the enemies wins the heaven" (Samara-śata-vijata-vijayojita-ripurajito divaṁ jayati)[50] is the legend on the obverse in his standard type of coins. "Having conquered the earth, the invincible one (or the lord of the earth) wins the heaven by meritorious deeds" (Apratiratho vijitya Kshitiṁ Sucharitair (or avanīśo divaṁ jayanti)[51] is the obverse legend on the Archer type of his coins. Similarly we find the legend "the king of kings, who had performed the Vājimedha (horse-sacrifice), wins heaven after protecting (or conquering the earth) (Rāj(a,¯)dhirājaḥ prithviī - mavitvā (or vijitya) divaṁ jayaytāhṛita =vājimedhaḥ)" on the obverse of his Aśvamedha type of coins.[52] Moreover, it was a poetic way of describing his fame in all the three worlds. Fleet had not taken into account the lines 30-31 which state that his fame purifies the three worlds like the water of the Ganges. Secondly the inscription is not posthumous on the ground that it might have been issued before Samudragupta had performed Aśvamedha sacrifice which is evident from his coins and the inscriptions of his successors. Samudragupta evidently could perform the Aśvamedha after completion of all his conquests mentioned in the Allahabad Pillar Inscription and their consolidation.[53]

In this connection Buhler and Chattopadhyaya also hold the same view as ours.[54]

There are two other records [55] of Samudragupta one from Nālandā of the year 5 and another from Gayā of the year 9. Scholars have no unanimous opinion about the genuineness of these records but we cannot definitely declare them to be spurious or forged.[56] Both have been written by the orders of one Gopaswāmin who is described as the Chief Controller of Records (Mahābalādhikṛita) and Dyūta (Messenger, Dūtaka, or Executor) in the Gayā record and as Chief Controller of Records, Chief of Elephantry (Mahāpīlupati) and Chief Commander of Armed Forces (Mahābalādhikṛita) in the Nālandā record which also mentions the name of Chandragupta-II as a Kumāra.[57] Curiously enough in the introductory portion of both the inscriptions the epithets used for Samudragupta are almost the same as mentioned in the Mathurā Stone Inscription of Chandragupta-II discussed earlier.[58] The reference to the epithet "the restorer of Aśvamedha sacrifice which was for long in abeyance". (Chirotsanna-asvāmedhāharttā) is also[59] common with all these three records. It is possible that the two records of Samudragupta might have been damaged and restored again in the time of.his son and successor Chandragupta-II probably with the addition of the above epithet which is not very certain otherwise it would create doubts with regard to its genuineness that Samudragupta had performed his Aśvamedha sacrifice even before his regnal year 5 mentioned on the Nālandā record which creates strong doubts about the non-mention of Aśvamedha sacrifice in the Allahabad Pillar Inscription which is deemed to be a much later record of Samudragupta. The performance of Aśvamedha sacrifice by Samudragupta is a valid fact attested by his coins and specific mention of it in all the records of his successors which describe the genealogy of their predecessors.

We should not forget that the poet Harisheṇa the author of the Allahabad Pillar Inscription was a minister of peace and war (Sāndhivigrahika), Kumārāmātya and Mahādaṇḍanāyaka. He carefully drafted the panegyric by putting general things in the form of introduction in poetic form (eight verses), for the middle portion he turned to prose to give a graphic description of his heroic exploits, again turned to poetry just to give one verse before the Epilogue.

Samudragupta's accession to throne and Kācha problem

Scholars have expressed different opinions based on verse 4 of the Allahabad Pillar Inscriptions and Kācha type of coins. While some scholars mention this nomination of Samudragupta for succession to throne by Chandragupta-I was followed by his abdication of throne and renunciation. As a result Samudragupta became the king, Kācha being another name of Samudragupta while others allude to a war of succession accepting the separate identity of kācha being the elder brother of Samudragupta.

The verse 4 in the inscription states that Chandragupta-I nominated Samudragupta as his successor with the words "verily thou are worthy, protect this earth" and embraced him with the hairs of his body standing erect out of pleasure. This describes a Court-scene whereas the courtiers were rest assured with this selection of Samudragupta. Others of equal birth (other princes) looked at (this) with melancholy faces.[60] S.V. Sohoni is of the view that Kāchagupta was an earlier name of Samudragupta himself and he gives no weightage to the theory that there was no smooth transfer of power in favour of Samudragupta or it led to a civil war.[61] S. Chattopadhyaya is also of the same view and accepts Kācha to be another name of Samudragupta stating that it is not unlikely that Kācha acquired the name Samudragupta after his conquests extended up to ocean.[62] Allan had also given the same view about it.[63] Majumdar holds the view that the circumstances described in the inscription suit more with abdication and final leave-taking than merely a formal announcement of his successor. Though it is possible to detect some political disturbances at the beginning of his reign by the description followed in the fragmentary verses (5 and 6) which describe how some were attracted to him by his extraordinary deeds of valour, and others submitted after being afflicted by his powers.[64]

We have earlier stated that 'the identification of Kācha with Samudragupta has been rightly opposed by scholars'.[65]

Gokhale is of the opinion that Kācha was a person different from Samudragupta as the inscription clearly hints at a disputed succession, the coins of Kāchagupta cannot be ascribed to Samudragupta, especially in the matter of standard and the small number and there is possibility of the occurrence of a rebellion against the claim of Samudragupta. 'Presumably the rebellion led by Kācha was short-lived and was early crushed by Samudragupta.'[66]

Altekar,[67] Rapson[68] and Heras[69] are all of the view that the selection of Samudragupta as his successor by Chandragupta-I in a full Darbar did cause a heart-burning to those of his brothers, who had aspired for the imperial throne, and it would appear that one of them,

Kācha, successfully contested the throne and ruled for sometime. Altekar has withdrawn his opinion about the identification of Kācha with Rāmagupta after the discovery of the copper coins of Rāmagupta in Malwa.[70] R.K. Mookerjee calls Kācha an usurper and this explains why his name is not mentioned in the genealogical list of the inscriptions which also do not generally mention the name of a king who does not come in direct line of succession.[71] Further he stated that the reference to Chandragupta's selection of Samudragupta as his successor in preference to his other brothers is also indicated in the Riddhapur inscription in the expression 'tatpāda-parigṛhīta'.[72]

Goyal[73] and Gupta[74] drew on the evidence of an eighth century work *Ārya-Mañju-Śrī-Mūla-kalpa* which speaks of a Buddhist tradition mentioning the names of anti-Buddhist kings namely Bhasma (Kācha). Soma (Śaśāṇka) and Graha (Mihirakula).[75] It also mentioned that Samudragupta had a brother named Bhasma who contested the throne against him and ruled for a short period. That this intruder was no other than Kācha which according to Lexicographers means Bhasma 'alkaline ashes'.[76] Goyal even stretches the evidence to state that it was fight between two factions - one led by Kācha who was anti-Buddhist and belonged to a more conservative and rather intolerant ideological faction while Samudragupta and his supporters represented the liberal aspect of Brahmanical revival which was not unsympathetic to other faiths.[77] Evidently Goyal's hypothesis is based on the testimony of Yuan Chuang that a king of Srāvastī did not give proper hearing to a Buddhist scholar Manoratha who sent an account of the happenings to his disciple Vasubandhu before he died. Later on, this king lost his kingdom and was succeeded by a king who patronised Vasubandhu. Goyal identified this king with Samudragupta.[78] Majumdar is sceptical about this identification and states that it is not impossible that Vasubandhu's patron belonged to a local dynasty of Ayodhyā.[79]

While taking care of the cumulative evidence discussed above, we are of the opinion that Buddhist tradition mentioning the anti-Buddhist kings by name as well as the numismatic and epigraphic evidence gives more support to the theory that Kācha cannot be equated with Samudragupta.[79a] There might have been rise of some trouble as soon as Samudragupta succeeded the throne or even earlier which Samudragupta successfully brought under his control.

Samudragupta's date of accession to the throne

As we have discussed earlier that Chandragupta-I ruled for a long period from A.D. 319-353 i.e. for about 34 years and must have grown old enough to deem it fit to nominate his successor and go for renunciation. Apart from his heroic qualities described in the Allahabad Inscription being a successful heir to the Lichchhavi kingdom, his being a 'Lichchhavi - dauhitra' might have gone in his favour, for Chandragupta-I while taking this decision in favour of Samudragupta among the princes who were equally aspiring for the throne when they might have thought a time ripe for such an event as he has been inferred by scholars from the Raghuvaṁśa Mahākāvya of Kālidāsa where they see an echo of the Gupta Kings who abdicated the kingdom when they grew old in age in favour of their sons.[80] The prevailing atmosphere in verse 4 of the Allahabad Inscription is self-explanatory that Chandra-gupta-I thought it necessary to nominate his successor rejecting other claimants to the throne who felt jealous and dejected after his choice went in favour of Samudragupta who got appointed to the throne around A.D. 353. He might have faced some turmoil soon after his nomination or thereafter which he got over in no time as envisaged by the following fragmentary verses 5 and 6 of the Allahabad record which describe how some were attracted to Samudragupta by his extra-ordinary deeds of valour, and others submitted after being afflicted by his prowess.[81]

Samudragupta's conquests

It seems in the first eight verses which cover sixteen lines, the poet Harisheṇa covers in general the career of Samudragupta and comes to prose for describing his achievements in particular which cover the rest of Allahabad Pillar Inscription except one verse before the end.

Verse 7 in the inscription covering lines 13-19 has often been misinterpreted by scholars to stand for his Āryāvartta campaign.[82] which certainly deals with his achievements as a Yuvarāja which is explicit in the expression 'Puṣpāhavaye Krīḍatā' which is to be read with 'yena' in line 13 meaning thereby 'By him while he played in Pāṭaliputra,'. "By whom, rising with full force by the prowess of his arm were uprooted Achyuta, Nāgasena, Gaṇapatināga, 'ga' with mutilated portion ahead is restored to stand for Gaṇapatināga mentioned later among the Āryāvartta rulers [83] and other kings in warfare and (by whom) a prince born of the

Kota family was taken as prisoner through his army, while he played in the city named Puṣpa, i.e. Pāṭaliputra'. Nāgasena was king of Padmāvati (Padam Pawaya, 25 miles north-west of Narwar, near Gwalior), Gaṇapatināga of Mathurā, Achyuta of Ahichchhatra (near Bareilly) and the Kotas whose coins have been found in East Punjab and Delhi probably ruled in the Upper Gangetic valley.[84] In the *Harsha-charita* of Bāṇa (7th century A.D.) we know of a king Nāgasena, born in the Nāga family who got destroyed himself due to announcement of his policy by a Sārikā bird.[85] According to Rapson, Nāgasena mentioned in the Allahabad Inscription and in the *Harsha-charita* are identical.[86]

Aims of Samudragupta's conquests

Scholars have dilated upon the motives of Samudragupta's campaigns. According to Goyal, he was inspired by the Hindu ideal of Chakravartitva or universal sovereignty which was very popular in the Gupta age which, in practice, usually meant the establishment of one's overlordship over the whole of Bhāratavarsha.[87]

According to Gokhale, Samudragupta may have been influenced by the reputation of the Nāgas for the performance of the coveted vedic rite of horse sacrifice. That Samudragupta thought of performing the horse sacrifice after his encounter with the Nāgas and "that his South Indian campaign was in the nature of an 'Aśvamedha campaign'.[88] Buddha Prakash has stated that in his southern expedition Samudragupta was led by economic considerations of the newly rising demands of trade and commerce between India and South East Asia which could only be met through increased maritime activities catering to such needs.[89] Recently one scholar while accepting the progress of South East Asian trade and commerce as the motivating factor for southern campaign has suggested that Samudragupta could not think of achieving his objective through direct conquest because its military-political apparatus was not strong enough to embark on such a project outside Āryāvartta and that it showed the relative powerlessness of the Gupta state machinery when compared to that of the Mauryan Empire. Further, he was interested in the Aryanisation or Sanskritization suggestive of social and economic development of South India.[90]

Chattopadhyaya accepting Samudragupta's strong desire to bring under his control many good ports newly sprung up on the eastern coast of South India also states that Samudragupta wanted to keep the

Vākāṭakas under control who were now becoming the dominant power in the Maharashtra region.[91] Some scholars[92] have tried to support this theory by identifying an Āryāvartta ruler Rudradeva with Vākāṭaka ruler Rudrasena-I who was a contemporary of Samudragupta which is not likely. In that case, the above fact would have been more forcefully mentioned, as they were equally powerful rulers, and would not have been summarily dismissed by the mere mention of him among other Āryāvartta rulers. Secondly Rudrasena of the Vākāṭaka family was a king of Deccan.[93] Thirdly, Vākāṭaka King Rudrasena's grandson Rudrasena-II married Prabhāvatīguptā, the daughter of Chandragupta-II. It stands to reason that Pṛithvīsheṇa-I would never have married his son to the grand-daughter of one who killed his father. It is more likely that both the powers refrained from directly confronting each other and ultimately connected their relations by a matrimonial alliance in the time of Chandragupta-II.

Before concluding the nature of his campaign we have to take into consideration the total evidence about the campaign as referred to in the Allahabad Pillar Inscription. In lines 17-18, it is stated in general that Samudragupta had to fight hundred battles variously where he reflected upon himself hundreds of (numerous) wounds which looked like his marks of glory and who assumed for himself the title of *Prākramāṇka*. After enumeration of his conquests from lines 19-24, in line 24 itself the poet Harisheṇa concludes that thus the king had realised the ideal of "Dharaṇi-bandha" (Chakravartin). "Thus was the whole world bound by means of the amplitude of the vigour of (his arm)"[94]

This clearly shows that Samudragupta was motivated by his desire for conquering the whole earth (i.e. Bhāratavarsha) which was a stage prior to the performance of Aśvamedha sacrifice after successful completion of which he could become a Chakravartin king. This is supported by lack of any mention about the performance of Aśvamedha sacrifice (which is proved beyond doubt by his coinage). The mention of which the poet would have in no case avoided if the panegyric was written after its performance. As regards other motives alluded to by the scholar, it may be remarked that a king would always keep some material advantages before him while taking up a victorious campaign. Thus the fundamental motive was ideological one to realise the ideal of a Chakravartin ruler while economic, maritime or any other motives were subsidiary to it. He had thus become a "Apratiratha on earth" (i.e. who is not subordinate to anybody on earth, or the invincible).[95]

Order of Samudragupta's campaigns

There is some controversy among the scholars whether Samudragupta's conquests are arranged in a chronological order, in order of geographical distribution or based on his different policies of conquests and are categorised accordingly.

Mookerji states that this inscription may be taken to be faithful record of events it narrates and also of the order of their happenings. The order in which these are mentioned in the inscription is important for the history it records.[96] He, however, holds his view in favour of two Āryāvartta campaigns - one before his southern campaign and the other soon after it,[97] like some other scholars who subscribe to this view.[98] But the use of the word 'unmulya' (having uprooted) in connection with these rulers in the 7th verse itself goes against this explanation. It may also not be easy to believe that a ruler of the calibre of Samudragupta would have gone for his southern campaign without first consolidating his position in Āryāvartta.[99]

We have already explained that verse 7 describes Samudragupta's achievements as a Yuvarāja keeping in view the long reign of Chandragupta-I. Their territories had become a part and parcel of the Gupta empire and these rulers were completely exterminated. The complete absence of any mention of their territories clarifies the issue. We cannot talk about one campaign or two campaigns in this regard as it could have been the result of several campaigns. Even if the names of eight kings are mentioned in the Āryāvartta conquests, the expression in the inscription alludes to many more kings.[100] It is likely that poet Harishena in order to extol the greatness of his patron as also to cover the geographical region of Āryāvartta thought it necessary to reiterate again these rulers with little more details than described in verse 7 especially when he has come to prose which practically provides a better scope for it. We do not agree with Goyal when he states that the scheme of Allahabad *Praśasti* is neither geographical and nor chronological.[101] It is more plausible to believe that the author of the inscription needed to strike a synthesis of the details of the conquests. It is not possible that any king will expand his territory to one side only and it is more likely that he may conquer his enemies of contiguous states in all directions. In case Harishena described the victories merely geographically he would miss the sequence chronologically. So while admitting that the scheme of description in the panegyric is partly geographical and partly chronological we are of the opinion that the poet is more particular about geographical distribution or location which is clear from the various terms used in the inscription in this context viz. Dakshiṇāpatha, Āryāvartta, Āṭavika-rāja, Pratyanta and Dvīpa.[102]

So far the scholars have generally emphasized that the portrayal of these conquets, is cheifly based on Samudragupta's relations with the states[103] which some scholars divide into 4, while others into 5 categories. If we believe in relations theory to be the sequence of the description then Harisheṇa should have mentioned the Āryāvartta rulers first who were completely uprooted and next in precedence will come the relations of the rulers of Dakshiṇāpatha who were captured, released and reinstated in their kingdoms (after they accepted his suzerainty).

Majumdar[104] and Goyal[105] divide Smudragupta's description of conquests into 4 categories while Dikskalkar [106] and Narayanan [107] divide them into 5 categories. The categorization is important by itself but it does not seem to be done in hierarchical order of nature of relationship but apparently taking due care of the geographical location. It is likely that while putting the sequence of events in some order the poet had given more importance to Samudragupta's southern campaign which might have been successfully completed immediately before the writing of the inscription. Usually immediate happenings get more precedence over others which may be described even later with due weightage as might have been the case in the present inscription by describing southern list first and the Āryāvartta list later.

Obviously the conquests may be divided into the following five categories:

1. Twelve rulers of Dakshiṇāpatha (Deccan and South India) with the names and their territories who were captured, liberated and reinstated in their kingdoms after they admitted Samudragupta's suzerainty.
2. Eight rulers of Āryāvartta by name only and others who were completely exterminated.
3. Kings of forest kingdoms who were reduced to his servanthood.
4. Kings of frontier states with names and others along with several tribes by name and others who paid him tribute and obeyed his orders and performed obeisance to him.
5. Some rulers of the North-West and the people of Siṁhala and other islands who paid him obeisance and offered the presents in the form

of maidens or married their daughters and solicited charters bearing the Garuḍa Seal for governance of their territories.

The details follow:

Southern Kings and their Kingdoms

It is stated in the lines 19-20 of the Allahabad Inscription that he had added to his fortunes and glory by the first capturing them (grahaṇa), liberating them (Moksha) and then favouring them by reinstating them in their kingdoms (Anugraha). According to Narayanan, the term "anugraha" is used in the technical sense to mean financial aid given to an ally as described in the Arthaśāstra of Kauṭilya, Book XIII, *Labdha-praśamana.*[108] On verification we have found that it has been used in a different context, to earn the goodwill, the king should help (anugraha) the poor, orphaned and afflicted sections of the conquered people and liberate (Moksha) all those taken prisoners.[109] The word 'anugraha' means 'blessing' or 'favouring' and has been rightly used by scholars in the sense of blessing the kings with their empires in the context of the inscription. It would look ridiculous to accept the interpretation of Narayanan that "Samudragupta gave protection and aid to the Southern States in order to secure alliance which in the long run brought him great support in treasury and allied forces".[110]

In relation to the Southern conquests described above Samudragupta has been termed as a king who believed in conquest as ordained by Dharma (Rājadharma) which means reinstating the defeated kings by reducing them to tributary status, as stated in the *Raghuvaṁśa Mahākāvyam* of Kālidāsa.[111]

We may mention that the term "Dakshiṇāpatha" has been earlier used in the Nānāghāṭ Cave Inscription (second half of first century B.C. in the Junagarh Rock Inscription of Rudradāman A.D. 150) and also in the Nasik Cave Inscription of Vasiṣṭhīputra Pulumāvī (A.D. 149) while its literally references go back to the Baudhāyana Dharmasūtra (where it is used for Deccan) and as far as the Ṛigveda where it refers to a place where exiles are expelled. In short it denoted the territory of South beyond the limits of the recognised Aryan world.[112] In Allahabad Pillar Inscription it covers the kingdoms of Kosala, Mahākāntāra, Kurāla, Piṣṭapura, Koṭṭura, Eraṇḍapalla, Kāñchī, Avamukta, Vengī, Pālakka, Devarāṣṭra and Kusthalpura.[113]

Identification of the Southern Kingdoms

We have discussed the different views of scholars about the identification of the Southern kingdoms mentioned in the inscription in our earlier work *Personal and Geographical Names in the Gupta Inscriptions*[114] and need not repeat them here. A mention of our preferred view in each case will serve the purpose on the table given below:

	The King	His Kingdom	Identification
1.	Mahendra	Kosala	Dakṣhiṇa Kosala comprising the districts of Raipur, Bilaspur of Madhya Pradesh and Sambhalpur in Orissa.
2.	Vyāghrarāja	Mahākāntarā	Mahāvana, present Jeypore forest region in Orissa.
3.	Maṇṭarāja	Kurāla	Kolada near Russelkonda in Ganjam.
4.	Mahendragiri[115]	Piṣṭapur	The modern Pithapuram in Godavari district of Andhra Pradesh.
5.	Svāmidatta	Koṭṭura	Kothoor in the Ganjam District.
6.	Damana	Eraṇḍapalla	Eraṇḍapalli near Chiccole on the coast of Orissa.
7.	Viṣṇugopa	Kāñchī	Modern Conjeevaram in Chingliput district, Tamilnadu.

	The King	His Kingdom	Identification
8.	Nīlarāja	Avamukta	On the bank of Gautamī i.e. Godāvarī according to *Brahma Purāna* Chapter 113, 22, there is an Avimukta Ksetra which may be located between Kāñchī and Vengī mentioned next.
9.	Hastivarman	Vengī	Vegi or Peddavegi, a village near Ellora Taluka between the Krishna and Godavari rivers.
10.	Ugrasena	Pālakka	Palakkada, the capital of a Pallava viceroyalty in Nellore district.
11.	Kubera	Devarāṣṭra	The province of Devarāṣṭra (= Yellamanchili tract) mentioned in a copper plate grant of the Eastern Chālukya Bhīma-I, found in the district of Visakhapatnam.
12.	Dhanañjaya	Kusthalapur	A tract around river Kuśasthalī near Dakshiṇakosala.

Route of Southern Campaign

It seems that Samudragupta's military expedition to the South was confined to the eastern part of Deccan along the coast of the Bay of Bengal, passing through the forest tracts of Madhya Pradesh he proceeded to the Orissa Coast, and then marching through Ganjam, Vizagapatam (Viśā)khāpaṭṭanam), Godāvarī, Krishṇā and Nellore districts, he reached as far as Kāñchī modern Conjeevaram, South of Madras.[116]

Category - The Āryāvartta Rulers

In line 21 of the Allahabad Inscription it is stated that Samudragupta had added to his glory in great magnitude by forcibly uprooting many rulers of Āryāvartta which include Rudradeva, Matila, Nāgadatta, Chandravarman, Gaṇapatināga, Nāgasena, Achyutanandin and Balavarman. Their territories are not mentioned as they had become part and parcel of the Gupta empire. Out of them we have already identified Achyta, Gaṇapatināga and Nāgasena mentioned in the earlier portion of the inscription which is in poetic form where the poet may have avoided to give full name of Achyutanandin and only wrote the name Achyuta probably to meet the needs of metre.

The identification of the rest of the kings follows:

Rudradeva : Some scholars had earlier tried to identify Rudradeva with Rudrasena of the Vākāṭaka family[117] which we have refuted earlier while discussing the motives behind Samudragupta's campaigns. Identification of Rudradeva with Rudra whose coin has been found at Kauśāmbī and who may be placed in the fourth century A.D. the paleography of the coin being similar to that of the pillar inscription as suggested by Gupta[118] is more acceptable.

Matila : The name Matila may be explained as a diminutive of Matideva, Matidāsa or Matidatta like the formation of Devila from Devadatta.[119] He is not satisfactorily identified. Some scholars identified him with Mattila of a clay seal discovered in Bulandhsahr (Uttar Pradesh) while other scholars have opposed it on the ground that since there is no honorific before the name the seal may belong to some trader or petty chief.[120]

Nāgadatta : Most probably a Nāga King who has been differently identified by scholars. Jayaswal identified him with the father of

Mahārāja Maheśvaranāga[121] and D.C. Sircar identifies him with a king of North Bengal, an ancestor of the Datta family of Viceroys who served the Guptas.[122] The first identification cannot be accepted as the text of the seal gives the name of the father of Maheśvaranāga as Nāgabhatta[123] which is quite different from Nāgadatta. The second identification by D.C. Sircar may be accepted for the time being till we get some other clues to arrive at a final verdict in the matter. He may be originally a ruler of a Nāga kingdom whose successors later on were sent as Viceroys in Bengal after they lost their status as independent rulers.

Chandravarman : He has been identified with Chandravarman of the Susunia hill inscription near Bankura district in Bengal[124] where he is stated to be a son of Siṁhavarman and the king of Pushkarana which is identified with modern Pokharan, about 25 miles away from Susunia hill.[125] This identification has, however, been questioned by P.L. Gupta on the ground that most of Bengal was already part of the Gupta empire and no king of Bengal appears to be mentioned in the inscription.[126] We cannot agree to this view as only "Samataṭa" i.e. South-Eastern part of Bengal has been mentioned in the Allahabad record to be under his sphere of influence and not the whole of Bengal and the first epigraphic reference to Samataṭa is to be found in this inscription.[127] Gupta's suggestion that Chandravarman may be indentified with the Mandasor ruler, a brother of Naravarman and Siṁhavarman[128] as proposed by some scholars earlier[129] is not acceptable as even Samataṭa is described as a frontier state which is outside the pale of Aryandom. Secondly, all the Āryāvartta rulers were exterminated and their territories were made part of the Gupta empire where as according to this evidence Naravarman appears to have ruled as a feudatory king in Western Malwa in A.D. 404 i.e. during the reign of Chandragupta-II, son of Samudragupta.[130] Thirdly, the father's name being the same can be possible even by coincidence.

Gaṇapatināga : Already identified as a ruler of Mathurā.

Nāgasena : Already identified as a king of Padmāvatī (Gwalior Region).

Achyutanandin : Already identified as a ruler of Ahichchatra (ancient Northern Pāñchāla modern Ramnagar in Bareilly district). Some scholars were confused by the name Achyuta mentioned in line 13 of the inscription, and took Achyuta and Nandin as two separate names here. Even P.L. Gupta tried to take Nandin as a separate name but reverted back with his suggestion finding it difficult.[131] As explained earlier the name Achyuta may have been used without the surname in V.7 (lines 13-14) to meet the exigencies of metre.[132]

Balavarman : He has been been generally identified with the king of Assam who was a predecessor of Bhāskaravarman, the contemporary of Harshavardhana but this cannot be accepted as Assam has been separately mentioned as one of the frontier states who came under Samudragupta's sphere of influence and certainly existed outside the Āryāvartta kingdoms.[133] U.N. Roy has, however, suggested that he may be one of the predecessors of the Maukharis whose names have been mentioned with Varman endings and who worked as feudatories to the Guptas in the fifth century A.D.[134] We need to wait for a final settlement in the matter.

In the Allahabad record only names of eight kings of Āryāvartta have been given but this does not exhaust the list and more conquests are hinted at in the region according to the inscription. Some scholars have tried to identify these rulers (by taking their number as 9 taking Achyuta and Nandin as two names) with nine Nāga kings who ruled in Padmāvatī, Kāntipurī and Mathurā, according to the Purāṇas.[135] This seems to be far fetched as the number of the Nāga rulers did not remain restricted to nine; we know atleast 11 Nāga rulers from other sources[136] which may have further increased in course of time.

We agree that a few of the kings mentioned were Nāga rulers but certainly other rulers belonged to some other dynasties as inferred from their names.

Limits of Āryāvartta and the Identification of the Āryāvartta rulers :

As discussed above scholars have tried to place at least two rulers named Nāgadatta, Chandravarman in Bengal and one ruler Balavarman in Assam. As pointed out these identifications are not beyond dispute. Chattopadyaya is of the opinion that these nine kings including Chandravarman possibly ruled to the west of Allahabad.[137] What we want to state in the matter is that we have to wait for satisfactory identifications of some of these rulers. Till then we are unable to definitely refute the claims of those scholars who spread out the Āryāvartta rulers in Bengal and Assam on the basis of wider limits of Āryāvartta prevalent after the second century A.D. in accordance with the definition of Manu followed by later literary evidences which included the tract extending from Eastern to Western ocean (i.e. Bay of Bengal to Arabian Sea), and bounded on the North and South by the Himālaya and Vindhya respectively.[138]

Category

The Āṭavika Kingdoms

Just after the conquest of Āryāvartta rulers stated in the same line 21, mention is made about all the rulers of the forest kingdom being reduced to complete subjection, in the language of the inscription they were reduced to the status of servanthood (Parichārakīkṛita - Sarvvāṭavika - rājasya).[139] We know of a king Hastin as a ruler of Ḍabhāla (Jabalpur) along with eighteen forest kingdoms from two inscriptions dated in the Gupta Era 199 and 209 respectively from the Baghelkhand regions.[140] The Āṭavika-rāja, therefore, denoted Ḍabhāla and the wild region around, corresponding to the territory round present Jabalpur in Madhya Pradesh. Some scholars, however, have tried to include Alvaka (Ghazipur) along with Ḍabhāla or the Jabalpur territory.[141] But this area Alavaka (Ghazipur, Uttar Pradesh) already formed the part of his empire and in the present context only territory around Jabalpur region is intended.

Category : In the fourth category in line 22, are mentioned the kings of frontier states of Samataṭa, Ḍavāka, Kāmarūpa, Nepāla and the tribal states of the Mālavas, Ārjunayanas, Yaudheyas, Madrakas, Ābhīras, Prārjunas, Sanakānīkas, Kākas, Kharaparikas etc. who paid him all kinds of taxes, obeyed his orders and payed obeisance by way of personal attendence in his court. We shall identify them one by one.

The Frontier Kingdoms

All these kingdoms may not be necessarily Mleccha countries as interpreted by some scholars.[142]

Samataṭa

It has been identified with the part of East Bengal (South-East Bengal) and had probably its capital Karmmānta or Baḍ-Kāmtā in Commila district.[143]

Ḍavāka

Ḍavāka has been identified with modern Daboka in Nowgong district, Assam.[144]

Kāmarūpa

It consisted of the Western districts of the Brahmaputra valley which being the most powerful state and being the first to be approached from the Western side came to denote the whole valley with Prāgjyotish identified Kāmākhyā or Gauhati, as the capital of Kāmarūpa.[145]

Nepāla

In Nepāla, Jaydeva I, the Lichchhavi King of Nepal was a relation of Samudragupta, and his submission meant practically the submission of the Himalayan states on the Indian side. The Gupta Era was introduced in Nepal in the reign of Jayadeva-I.[146]

Kartṛipura

This kingdom most probably occupied the lower ranges of the Western Himalayas as seen from the context and included Kumaun, Almora, Garhwal and Kangra.[147] On a reconsideration we withdraw our earlier preference for Kahror or Karur between Multan and Loni assuggested by Dashratha Sharma.[147a]

Tribal States

Mālava

The same as Malloi of the Greeks, originally belonged to Jhang district in Punjab (now in Pakistan), from where they spread all over Punjab and by the time of Samudragupta had migrated to Rajasthan.[148] They occupied the territory round about Karkotanagar, near Tonk where their coins have been found in abundance, the dates of these coins range from the second century B.C. to the early fourth century A.D.[149] That they had migrated to Rajasthan from the Punjab is supported by the fact that the legend on some Mālava coins found in Rajasthan reads from right to left as in Kharoṣṭhī, which was prevalent in the Punjab and the North-West from very early times.[150]

Ārjunayanas

Ārjunayana coins have been found in the Mathurā region and scholars have assigned them the region lying West of Agra and Mathurā, within the triangle Delhi-Jaipur-Agra, roughly speaking comprising Bharatpur and Alwar states of the British period.[151]

Yaudheyas

During Samudragupta's time the Yaudheyas seem to have occupied northern Rajputana and South-East Punjab, and their territory extended upto the confines of the Bahawalpur State where their name survives in the name of the tract called Johiyawar.[152] They played a great role in driving an end to the Kushāṇa rule into the interior of India in the middle of the second century A.D. They were proud of their heroism and perhaps divided themselves into different regions to form a confederacy with some adjoining tribes like the Arjunayanas and the Kuṇindas.[153] According to P.L. Gupta in the third century A.D. they had settled between the upper courses of Sutlej and Beas with their headquarters at Sunet, near Ludhiana and formed the North-Western border of Samudragupta's empire.[154]

Madrakas

Originally from Sialkot (Madradeśa) now in Pakistan, they are known to have formed a corporation of warriors and enjoyed the title of *Rājan* as stated in the Arthaśāstra of Kauṭilya.[155] The legend of Sāvitrī and Satyavāna is connected with the Madra country for Sāvitrī was the daughter of Aśvapati king of Madra (deśa) whose brahmanical name was Śākala which formed in a tongue of land between two rivers known as Rechna Doab.[156] They were also divided into two branches called *Pūrva Madra* extending from the Ravi to Chenab and the *Apara Madra* from Chenab to Jhelum, as mentioned by Pāṇini.[157] According to P.L. Gupta possibly they occupied some region also in Rajasthan near Bikaner at the place named Bhadra at the time of conquest of Samudragupta. His view is based on the anology of words Madra and Bhadra being the same.[158] But this cannot be accepted as we have shown that they were two separate tribes.[159]

Ābhīras

We know of Abiravan between Herat and Khandhar which may have been the original home of the Ābhīras. They are taken as foreigners. From the inscription we know that they served as generals in the armies of Śaka Satraps of Western India. An Ābhīra king Īśwarasena ruled in the Maharashtra region about 248 A.D. During the time of Samudragupta, scholars have generally placed them in the province of Abhirwada between Jhansi and Bhilsa in Madhya Pradesh where they seem to have settled their kingdom.[160] P.L. Gupta, however, states that most likely they had not come to this area by that time and remained in the lower Indus Valley and the Western Punjab.[161] He assigns no reason for his view. "But as Ābhīras are often mentioned in the Kṣatrapa inscriptions of Saurashtra they may be located in Kathiawad and Gujarat,"[162] is the view of Diskalkar. The Purāṇas and the *Bṛihatsaṁhitā* place the Ābhīras in the Southern region during the period of Samudragupta. These tribes slowly spread out in different parts of India. They may have occupied Ahirwada in Madhya Pradesh at a later date. We know of their occupation of Rajasthan also at a later date as is evident from the Jodhpur inscription of Saṁvat 918, that the Ābhīra people of this area were a terror to their neighbours, because of their violent demeanour.[163]

Prārjunas

Following Smith, scholars have generally placed them in the Narasiṁhpur district of Central Provinces.[164] P.L. Gupta disagrees with the above identification and on the basis of Kauṭilya's mention of Prārjunakas along with the Gandharas, places them in the North-West.[165] But a more plausible location is Narasiṁhagarh in Central India.[166] On the basis of the contiguity of the other tribal states mentioned in the inscription i.e. Sanakānīkas, Kākas and Kharaparikas which fell more or less within the bounds of Central India we would like to place Prārjunas in Central Provinces.[167] Bhandarkar also places them near Narasiṁhagarh, not far from Bhilsa.[168]

Sanakānīkas

This tribe which consisted of aborginal people was gradually coming under the influence of Sanskrit culture. They seem to be ruled by hereditary chiefs which is borne out by the title *Mahārāja* being used by a Sanakānīka feudatory of Chandragupta-II, as well as his father and grand-father. They may be placed near Bhilsa.[169] They are mentioned in the Udayagiri Cave Inscription of Chandragupta-II and may be located in the Isagarh district, Gwalior near Bhilsa.[170]

Kākas

Kākas lived nearby evidently in Sanchi which was known as Kākanādabota. They have been mentioned in the *Mahābhārata* and are associated with the Vidarbhas, a well-known people occupying tracts of Madhya Pradesh.[171]

Kharaparikas

They are placed in Damoh district of Madhya Pradesh. Kharpara means a theif, rogue or a cheat. It is not certain to identify them with the five Karpatas of the *Mahābhārata*; Khara-Sāgara-Rāśis or Kharapatha of the Purāṇas. Generally scholars have identified them with the Kharparas mentioned in the Batihgarh inscription of the Damoh district of Madhya Pradesh.[172] Their identification with Mongols by Dashrath Sharma has been refuted by S.L. Katre.[172a]

Category: The rulers of foreign countries

Daivaputra and others

It is stated in the lines 23-24 of the Allahabad inscription that Daivaputras along with Śāhis, Śāhānuśahis, Śakas and Muruṇḍas and the people of Siṁhala and all other Islands are said to have acknowledged the suzerainty of Samudragupta by rendering to him all kind of service (sevā) such as coming to him personally (ātmanivedana), providing gifts of maidens and presents (Kanyopāyana - dāna)[173] and application (Yāchana) for charters bearing the imperial Gupta Garuḍa seal (Garu-tmadaṅka) so as not to be disturbed in the enjoyment (bhukti) and administration (śāsana) of their respective territories (Svaviṣaya).

1-3. Daivaputra - Śāhi - Śāhānuśahi

R.C. Majumdar states Daivaputra - Śāhi - Śāhānuśahi to be title borne by Kushāṇa kings and the whole expression may refer to one of them or according to some other scholars they denote not a single Kushāṇa king, but three of the smaller states into which the Kushāṇa empire was divided, the ruler of each of them appropriating one of the titles for himself.[174] We have already discussed it separately in an Appendix in our earlier work,[175] and have concluded that Daivaputra being a *taddhita* form from Devaputra refers to the descendants of Devaputra i.e. the Kushāṇa ruler Kaniṣka; Śāhi stands for the persons of a sub-branch of the Kushāṇas (if one give credence to the Jain legend *Kalakāchārya-Kathānaka*) and Śāhānuśahi stands for Sasanians.[176] According to R.K. Mookerjee, the Daivaputras were in possession of the Central Punjab, and the Śāhis and Shāhānuśahi were ruling beyond the Punjab and the frontiers in the region of corresponding to modern Afghanistan.[177]

We have also taken the Śakas and the Muruṇḍas as two separate entities,[178] and not as one word 'Śaka-Muruṇḍa' meaning Śaka lords as interpreted by some scholars.[179]

Śakas

During the time of Samudragupta the Śakas ruled over Malwa (Ujjain) and Kathiawar and were a power to be reckoned with. The Śaka homage to Samudragupta was not at all sincere and reliable since it was not tendered to his successor and Chandragupta-II who conquered them once again is remembered as "Śakāri" i.e. the enemy of the Śakas.[180]

Muruṇḍas

They were one of the foreign potentates who came of their own accord to offer allegiance to Samudragupta. We have ruled out earlier suggestions by scholars to confuse them with the Śakas or Kushāṇas,[181] as

Kushāṇas have been included in the Daivaputras and Śakas are mentioned separately as discussed above. It is clearly a non-Aryan word and can have no Aryan derivation. The old lexicons have identified them with Lampākas i.e. the inhabitants of modern Laghman in Afghanistan.[182] They dwelt in the North-West with other foreign tribes and are not to be compared with the aboriginal Munda tribes of Choṭā-Nāgpur region.[183]

Saiṁhalaka

Here we need a word of caution. As the inscription is in the form of a panegyric, the author poet Harisheṇa might have taken liberties to exaggerate the image of his patron by turning a normal diplomatic relationship of king of Ceylon with Samudragupta to that of subservience. We know from a Chinese text that Meghavarṇa, king of Ceylon, sent two monks to Bodh-Gaya to visit sacred spots who are stated to have experienced great inconvenience for want of suitable accommodation. To remove this difficulty Meghavarṇa sought to build a monastery and sent rich presents in the form of jewels for the purpose. Samudragupta sanctioned it and Hiuen Tsang was an eye-witness to this magnificient establishment to the North of the Bodhi tree.[184] The poet Harisheṇa seems to construe the Ceylonese king's prayer for permission to built a monastery into "application for charter confirming him in the enjoyment of his territories".[185]

The list does not exhaust here and seems to include all (other) dwellers in Islands (probably the islands of Southern Sea such as Jāvā, Sumātrā, etc.) Majumdar remarks in the context that without more positive evidence we are unbale to believe that these rulers in any way openly acknowledged the suzerainty of Samudragupta, or enjoyed their Kingdoms merely as fiefs on the basis of charters granted by him.[186] He maintained good foreign relations and diplomatic ties with the neighbouring countries and the buffer states. Thus his empire included the whole of India leaving out the countries of far South.

Significance of the conquests and the Aśvamedha Sacrifice

Soon after the narration of Samudragupta's conquests in line 24 itself Harisheṇa declares that thus he (Samudragupta) had conquered the whole earth and was having no rival on earth (Prithivyām-apratirathasya: unopposed by hostile chariots on earth).[187]

In ancient times people believed in the Divine right theory of Kingship which means that the king has divine right for rule and he is divine too in certain respects. Divinity was also consecrated on him by Brāhmaṇas by blessing the king after performance of certain sacrifices. In the next line, i.e. line 25 of the inscription Samudragupta has been compared with Parama Purusha (Achintya: Incomprehensible) who elevates the good and destroys the evil and has also been in line 26 compared with Dhanada (Kubera), Varuṇa, Indra and Antaka (Yama) in his qualities i.e. he possessed the attributes of these gods who are known for bestowal of wealth, sense of justice, administration - both mundane and spiritual (Indra is known as *Śatakratuḥ* i.e. performer of hundreds of sacrifices), and administration of justice respectively and who are known as governors of four regions (East, West, North and South) on earth.

Scholars have taken an illusion to his performance of Aśvamedha sacrifice in the expression *Suvarṇadāna* 'distribution' of gold, occurring in the Eran Stone Inscription or in the expression 'Aneka - go - hiraṇyakoṭipradasya' (the giver of many cows and crores of gold coins) occurring in Fleet's Inscription No.4.[188]

It seems they have forgotten to take proper notice of the expression'Aneka-go-Śatasahasra-Pradāyinaḥ' (who was the giver of many hundreds of thousands of cows) in line 25 of the Allahabad Pillar Inscription itself.[189] Not only that no scholar has paid proper attention to the expression "Kṛipaṇa-Dīnānāthātura-Janoddharaṇa-Sattradīkshābhyupagata-manasaḥ" "whose mind was engaged in providing relief to the low, the poor, the helpless and the afflicted and in the performance of the Sattra and Dīkshā ceremonies" in the beginning of line 26 of the Allahabad Pillar Inscription which has been misconstrued due to wrong translation of scholars.[190] It would imply that Samudragupta had performed many Soma sacrifices (Sattra : which involved many officiating Brāhmaṇas),[191] other religious ceremonies (Dīkshā),[192] and after their successful completion gave hundred and thousands of cows to the Brāhmaṇas in donation.

The mention of donation of only cows and not of gold which is significant by its mention along with the donation of cows in later inscriptions of the Guptas 'Aneka-go-hiraṇyakoṭipradasya' "i.e. the giver of numberous cows and crores of hiraṇya (gold coin)"[193] or taking the word 'koṭi' common to both and translating it as "giver of crores of cows along with gold" (the general custom is to donate the cow with her horns strung with gold provided one can afford to) since the word Aneka an adjunct of 'Go' (cow) is qualified by the number millions (Śata-Sahasra : hundred thousands). It is likely that Samudragupta donated much bigger number of cows and gold after the performance of Aśwamedha sacrifice which was much more significant and time consuming as compared to the above-mentioned Soma sacrifice. Only after that he may earned this complete appellation "Aneka-Go-hiraṇya-koṭi-pradaḥ". We may also compare the number 'millions' mentioned in the Allahabad Pillar Inscription with 'crores' in the later inscriptions of the Guptas.

Scholars are astounded not to find any mention of Aśvamedha sacrifice in the Allahabad Pillar Inscription which is proved beyond doubt by his coins and the specific mention of it by his successors in their records. Evidently Samudragupta performed this important sacrifice after the inscription was engraved.

We find the Aśvamedha type of Samudragupta's coins with the legend on the obverse; "Rajādhirājaḥ Pṛithi-vīma-vitvā divaṁ jayati aprativāryavīryaḥ"; "the king of kings, having conquered the earth, now conquers heaven with invincible valour" and on the reverse we have the legend "Aśvamedha Prākramaḥ".[194] The legends both on the obverse and reverse of the coins, taken together make it quite clear that Samudragupta had performed his horse-sacrifice after making all his conquestes which otherwise looks quite normal and could have created obstructions if performed earlier. The inscriptional evidence is very eloquent about his Aśvamedha sacrifice by the expression "Chirotsannāśvamedhāharttuḥ", "the reviver of the horse sacrifice which has for long been in abeyance" in the inscriptions of his successors[195] and curiously enough in the Nālandā and Gayā records (the so-called spurious records) of Samudragupta.[196] He has also been mentioned as 'anekāśvamedhayājin', 'the performer of many horse sacrifices' in the Poona Copper Plate Inscription of his grand-daughter Prabhāvatīguptā, Regnal year 13.[197]

However, some debate has been raised recently over the expression "Chirotsannāśvamedhāharttuḥ".[198] Fleet translates it as "who was the restorer of the Aśvamedha sacrifice, that had been long in abeyance".[199] This translation though accepted by most of scholars has been challenged on the ground that there have been many known instances of this ceremony extending over the whole period between Puṣyamitra and the rise of the Guptas.[200]

Gupta states that several kings before Samudragupta - Puṣyamitra of the Śuṅga dynasty, Khāravela of Kaliṅga, Śātakarṇī of the Sātavāhana dynasty, Pravarasena of the Vākāṭaka family and the Bhārasivas - had performed the Aśvamedha, and some of them not in the distant past.[201] He explains the word *Chirotsanna* as explained in the *Śatapatha Brāhmaṇa and the Taittirīya Saṁhitā* that Aśvamedha sacrifice had lost some of the details constituting the sacrifice as it was an extensive ceremony and Samudragupta performed it in a full-fledged manner.[202] In other words it would mean that "the Aśvamedha sacrifice which took long time to be performed[203] was enacted by Samudragupta".

The word 'Utsanna' also occurs in line 23 of the inscription where it means 'extirpated or suspended'. To place the whole expression "Aneka-Bhraṣṭha-rājyaotsanna-rājavaṁśa-pratiṣṭhāpana" which means that who had reinstated many ruling families previously dethroned and extirpated. Similarly we have to care for the word 'Āharttā' 'restorer' in the whole expression 'Chirotsannāśvamedhā-harttā' which would mean that is one *who restored the Aśvamedha sacrifice which was suspended for long*. We can support the above translation from the Monier Williams Dictionary which quotes from the *Śatapatha Brāhmaṇa* the word *"Utsanna Yajña"* to mean a suspended ceremony.[204] So we disagree with the interpretation of P.L. Gupta. Even if Aśvamedha sacrifice was not performed for one hundred years in the territories ruled by the Guptas, Samudragupta's successors are within their right to state that it was in abeyance for a long time and was restored by Samudragupta. So we can easily say without any hesitation that Samudragupta had a long career of conquests after which he also performed an Aśvamedha sacrifice.

Personality of Samudragupta

Samudragupta possessed all good qualities of head and heart. He was a great hero and a conqueror who had won this world by his conquests and the other world (heaven) by performance of Aśvamedha sacrifice. The poet states that as if this lofty pillar which was raised as an arm of earth to lead his good acts to heaven.[205] He has been shown as an ideal king in whose kingdom there was no unusual rivalry between the Goddess of learning (Satkāvya) and Goddess of Wealth (Śrī = Lakshmī) as depicted in the line 6 of the inscription "who put to shame the preceptor of the Lord of the Gods (i.e. Bṛihaspati) by his sharp and polished intellect and Tumburu and Nārada by lovely performances of music; established (his) title of 'King of Poets' by various worthy poetical compositions", has been stated in the line 27 of the inscription. He was soft-hearted, compassionate and was always eager to help the low, the poor, the destitute and the afflicted, has been shown in lines 25-26 of the inscription. He performed many sacrifices and is stated to have given many hundreds of thousands of cows (evidently as gifts to Brāhmaṇas) as mentioned in lines 25-26 of the inscription. He was known as "*Prākramānka*" for his various conquests (L.17) and maintained strict administration (L.23: Prachaṇḍa Śāsana).[206]

We can build up aspects of his personality on the basis of his coins. He issued many varieties of coins all in pure gold. The obverse legend on his standard type of coins reads : "Samaraśatavitatavijayo jitāripurajito divaṁ jayati" "The conqueror of unconquered fortresses of his enemies, whose victory was spread in hundreds of battles, conquers heaven....," and on reverse is the legend "Prākramaḥ".[207]

On Archer type of his coins, he stands fully dressed in one of his coins, with a bow on the left and an arrow on the right hand.[208] We have the legend on the obverse of these coins: "Apratiratha Vijitya Kshitim sucharitair (or avanipatir) divam Jayati". "Unopposed by hostile chariots, conquering the earth, he conquers heaven by his good deeds". On the reverse is the legend "Apratirathaḥ".[209]

On the Battle-axe type of coins, we find the legend on the obverse "Kṛitānta-parśur jayatyajitarajajetaji-taḥ, "Wielding the axe of Kṛitānta (the God of death), the unconquered Conqueror of unconquered kings is victorious" and on the reverse it carries the legend "Kṛitāntāparaśuḥ".[210]

In the Tiger-slayer type of coins the king is shown wearing Turban and Waist-cloth, is trampling on a tiger with the legend "Vāghraprākramaḥ". "Having the prowess of a tiger".[211]

In the lyrist type of his coins Samudragupta is shown wearing waist-cloth, is seated cross-legged on a couch, playing on the Viṇā (lute or lyre) which lies on his knees. The legend on this type of coins simply gives his name.[212]

On his Aśvamedha type of coins we have the legend on the obverse "Rājādhirājaḥ Pṛithivīm avitvā divaṁ jayatyaprativāryavīryaḥ". "The king of kings, having gained the earth, conquers heaven, with his irrestible heroism" and on the reverse is the legend "Aśvamedha-prākramaḥ".

On his coins Samudragupta is shown of tall stature and good physique, his body is marked by strong muscular arms and a fully developed chest.[213]

Lastly we may state that Chief characteristics of his personality are enumerated in some of the inscriptions of his successors. "Sarvva-rājochchhetuḥ Pṛithivyāmapratiratha-sya Chaturdadhisalilāsvāditayaśaso-Dhanada-Varuṇendrāntaka-Samasya Kṛitāntaparaśoḥ nyāyāgatanekago-hiraṇyakoṭipradasya chirotsannāsvamedhāharttuḥ" "Who was the exterminator of all kings, who had no antagonist to challenge him (Apratiratha) on earth whose fame was tasted by the water of the four oceans ('Chakravartin' who has control over all land and seas : by taking together *Pṛithivyāmapratiratha* with this epithet); who was equal to the gods Dhanada (Kubera), Varuṇa, Indra and Antaka (Yama), who was the very axe of the god Kṛitānta; who was the giver of may millions (crores) of lawfully acquired cows and gold; who was the restorer of the Aśvamedha-sacrifice that had been long in abeyance".[214] Surprisingly enough the same epithets are exactly used in the text of the Nālandā and Gayā records of Samudragupta of the year 5 and 9 respectively[215] which may prove that they are his genuine records but restored later by his successors with amendment in the first portion of the text in the light of his latest achievements.

The authenticity of all these epithets is proved beyond doubt by the evidence of his coinage and the Allahabad Pillar Inscription.

Samudragupta has often been compared with Napolean and called Indian Napolean but this description is odious and does him less than justice. While Napolean met his Waterloo in final years (final defeat at Waterloo got his glory eclipsed and he spent his last years in confinement in St.Helena) of his campaign,[216] Samudragupta had all his glorious career through the end which saw its climax in the performance of Aśvamedha sacrifice and earned for himself the title "Pṛithivyāmapratiratha" "Having no rival on earth".

Thus the personality of Samudragupta stands unrivalled in history along with his detailed Allahabad Pillar Inscription which finds no parallel among the inscribed records of all rulers in the world till now.

He had a long reign of about 20 years and ended his career around A.D. 373 after which his throne was seized by Rāmagupta of unsavoury reputation in history who is said to have surrendered his queen to the

Śaka ruler in order to safeguard his kingdom. He was finally killed by his brother Chandragupta-II who succeeded him as the Gupta Emperor around A.D. 375 as is evident by his Mathurā inscription of 380 A.D. (G.E. 61) which is mentioned as the fifth year of his reign.

Notes And References

1. R.K. Mookerjee, *GE.*, p.11; T.R. Sharma, *PGNGI.*, p.14.
2. See the previous chapter on Chronology and Genealogy.
3. Fleet, *CII.* III, pp.15-16; 28, 44, 51, 54.
4. Mookerji, *Op.cit.*, p.11.
5. Sadhu Ram, *SII.*, pp.114-117; D.C. Sircar, *SI.*, pp. 414-415
6. R.S. Tripathi, *HAI.*, p.238.
7. *Ibid.*, p.12.; R.C. Majumdar, *VGA.*, p.135.
8. Mookerji, *Op.cit.*, p.11.
9. *Ibid.*
10. Majumdar, *Op.cit.*, p.135.
11. *Ibid.*
12. T.R. Sharma, *PGNGI.*, p.21; P.L. Gupta, *IG.*, pp. 243-247.
13. Altekar, *GGCBH.*, p.xv.
14. *Ibid.*
15. *Ibid.*, *JNSI.*, xi, pp.151ff.
16. R.C. Majumdar, *VGA.*, p.129.
17. *Ibid.*, pp.128-129.
18. S. Chattopadhyaya, *EHNI.*, p.173.
19. *Ibid.*, p.174.
20. Allan, *CCGD, Introduction,* p. xix.
21. S. Chattopadhyaya, *EHNI.*, pp.173-174.
22. S.R. Goyal, *HIG.*, pp.120-121.
23. *Ibid.*, pp.115, 121.
24. *JNSI.*, xix, Pt. II, p.141.
25. *Ibid.*, p.153.
26. Goyal, *Op.cit.*, p.121.
27. *JNSI.*, V, p.41.
28. P.L. Gupta, *IG.*, p.247.
29. *Ibid.*
30. H.C. Raychaudhuri, *PHAI.*, pp.468-469.
31. Altekar, *JRASB*, 1937, Num. Supplement, XL.VIII., PP.105ff.
32. *VGA.*, p.127, Smith, *EHI*, 4th ed., pp.295-296.
33. Goyal, *HIG.*, p.115.

34. Gupta, *IG.*, p.81.
35. D.C. Sircar, *Studies in Indian Numismatics,* p.14.
36. Altekar, *GGCBH.*, Introduction, p.xLvi.
37. *Ibid.*, p.xLvii.
37a. S.K. Maity, *IGT.*, p.35.
38. Fleet, *CII.* III, No.2, pp.20-21;
    See also Chattopadhyaya, *EHNI.*, p.180.
39. *Ibid.*, p.20.
40. Fleet, *CII.* III, No.4.
41. *Ibid.*, p.27.
42. *Ibid.*, p.27-28.
43. *Ibid.* No.I, p.2.
44. *Ibid.*
45. *Ibid.*
46. *Ibid.* No.I, p.1; See the title of the Inscription.
47. *Ibid.*, pp.4-5.
48. *Ibid.*, p.4.
49. *Ibid.*, p.17, f.n.3.
50. P.L. Gupta, *IG.*, p.81.
51. *Ibid.*
52. *Ibid.*, p.82.
53. R.K. Mookerji, *GE.*, p.19.
54. Chattopadhyaya, *EHNI.*, p.180.
55. Vasudeva Upadhyaya, *GA.*, pp.126-127.
56. R.C. Majumdar, "The Gaya and Nalanda Plates of Samudragupta" in S.P. Gupta (Ed.) *RPHI.*, p.88;
    Chattopadhyaya, *Op.cit.*, p.180.
57. Upadhyaya, *Op.cit.*, p.126.
58. Fleet, *CII.* III. No.4.
59. Some scholars have given another interpretation of this epithet which we shall take up in the discussion on the personality of Samudragupta.
60. Fleet, *CII.* III, pp.11-12.
61. S.V. Sohoni, "Transfer of Power to Samudragupta", in the *Quarterly journal of the Mythic Society*, Bangalore, January-June, 1985, Vol. LXXVI, Nos.1-2, pp.23-28.
62. S. Chattopadhyaya, *EHNI.*, p.179.
63. *VGA.*, p.138 as quoted.
64. *Ibid.*, pp.137-139.
65. T.R. Sharma, *PGNGI.*, p.22.
66. B.G. Gokhale, *Samudragupta,* Bombay, 1962, p.40.
67. Altekar, *GGCBH.*, p.XVI.
68. *JRAS.*, 1893, p.81.

69. *ABORI.*, Vol.IV, p.83.
70. Altekar, *Op.cit.*, p.xvi.
71. R.K. Mookerji, *GE.*, p.18.
72. *Ibid..*, See Sadhu Ram, *SII,* pp.113-114; Rithpur Copper Plate Inscription of Prābhāvatī guptā of the time of Pravarasena-II Royal Year 19, Lines 2-3.
73. Goyal, *HIG.,* pp.125-127, 193-195.
74. P.L. Gupta, *IG.*, p.256.
75. Goyal, *Op.cit.*, p.125.
76. P.L. Gupta, *Op.cit,* p.256;
Goyal, *Op.cit.*, pp.194-195.
77. Goyal, *Op.cit.*, pp.126-127.
78. *Ibid.*
79. R.C. Majumdar, *VGA.,* pp.155-156
79a. For more details see Chapter VII of the present work: *The Place of Kāchagupta in the Gupta History.*
80. Raghuvaṁśa I verse 8 see U.N. Roy, *GRY.,* pp.111-112.
81. Fleet, *CII.* III, No.1 LL 9-10, p.12.
82. R.K. Mookerji, GE., p.19;
S. Chattopadhyaya, *EHNI.,* pp.181, 183;
P.L. Gupta, *IG.*, p.260.
83. D.C. Sircar, *SI.*, p.256; f.n.1.
84. R.C. Majumdar, *VGA.*, pp.139-140;
Goyal, *HIG.*, p.140.
85. *Harshacharita,* Thomas and Cowell, Chap.-VI, p.192.
86. *JRAS.*, 1898, p.449.
87. Goyal, *HIG.*, p.186.
88. Gokhale, *Samudragupta,* p.48.
89. Buddha Prakash, "*Samudragupta's Southern Expedition* : A New Interpretation", *Prof. K.A.N. Sastri Felicitation Volume,* Madras, 1971, pp.15-31., p.25.
90. M.G.S. Narayanan, "A Reappraisal of *Samudragupta's* Digvijaya", in *Dineśacandrikā* , Shri D.C. Sircar (Festschrift), Delhi, 1983, pp.15-31, pp.284-85.
91. S. Chattopadhyaya, *EHNI,* p.183.
92. R.K. Mookerji, *GE.*, p.23;
H.C. Raychaudhuri; *PHAI,* pp.477-79.
93. P.L. Gupta, *Op.cit.*, pp.266-67.
94. Fleet, *CII.* III, p.14.
95. *Ibid.*, L.24 cf. With this compare the legend on Samudragupta's Coins "Apratiratho vijitya kshitim Sucharitair divam jayati".
96. R.K. Mookerji, *GE.*, p.19.

97. *Ibid.*, pp.19, 23.
98. Goyal, *HIG.*, p.198.
99. *Ibid,*
100. Fleet, *CII.* III., p.9, L-21.
"Rudradeva - Matila - Nāgadatta - Chandravarma- Gaṇapatināga - Nāgasena - Achyuta - Nandi - Balavarmady aneka-Āryāvarttarāja". Mostly scholars have accepted Achyutanandin as one name and not two as Fleet has done.
101. Goyal, *Op.cit.*, p.198.
102. Fleet, *CII.* III, No.1, LL.20-24.
103. R.C. Majumdar, *VGA.*, pp.140-141;
Goyal, *Op.cit.*, p.198.;
Narayanan, *Dinesha-Chandrikā,* Op.cit., pp.283-284..
104. Majumdar, *VGA.*, p.140-141, 147.
105. Goyal, *Op.cit.*, p.198.
106. Diskalkar, *SSI.*, p.30.
107. Narayanan, *Dinesha-Chandrika, Op.cit.*, p.284.
108. *Ibid.*, pp.290-291, f.n.11.
109. Arthaśāstra (Ram Tej Sastri translation in Hindi), Kashi, 1964, pp.736-737.
Sarva-bandhana- mokshanamanugraham Dinānātha-vyādhitanāṁcha.
110. Narayanan, *Op.cit.*, pp.291, f.n.1.
111. cf. "Grihīta-pratimuktasya sa Dharmavijayī nṛipaḥ Śriyam Mahendranāthasya Jahāra, Na Tu medinīm. Raghuvaṁśa IV/43 i.e. He, the Dharmavijayī (Righteous) King by capturing Mahendranatha and liberating him snatched away his glory and not his kingdom" i.e. the defeated King was again reinstated in his kingdom by an act of anugraha (favour) by the Victor.
112. T.R. Sharma, *PGNGI,* p.223.
113. Fleet, *CII.* III, 12-13.
114. For details see T.R. Sharma, *PGNGI., Passim.*
115. See Appendix III in T.R. Sharma, *PGNGI.,* pp.316-17.
116. P.L. Gupta, *GE.*, p.264; Goyal, *HIG.,* p.208;
Chattopadhyaya, *EHNI.,* pp. 182-183.
117. T.R. Sharma. *PGNGI.,* pp.39-40.
118. P.L. Gupta, *IG.,* p.267.
119. T.R. Sharma, *Op.cit* ., p.47.
120. *IHQ,* Vol.I, p.254; P.L. Gupta, *Op.cit.*, p.268.
121. K.P. Jayaswal, *HI.*, pp.36, 142;
Fleet, *CII.* III, pp.282-83.
122. Proceedings of *IHC.,* VII. p.78;
D.C. Sircar, *SI.*, p.341. f.n.i.

123. Fleet,*Op.cit.*, p.283.
124. T.R. Sharma, *Op.cit.*, p.38.
125. D.C. Sircar, *SI.*, p.342.
"Pushkarnādhipateh Mahārāja Śrīsimhavarmaṇaḥ putraya Mahārāja śrī Chandravarmaṇaḥ Kritiḥ."
126. P.L. Gupta, *Op.cit.*, p.268.
127. T.R. Sharma, *PGNGI.*, p.261.
128. P.L. Gupta, *Op.cit.*, p.268.
129. R.G. Basak, *HNEI.*, pp. 15-16, 26-27.
130. *Ibid.*, p.16.
131. Fleet, *CII.* III, p.13; P.L. Gupta, *Op.cit.*, p.268.
132. R.G. Basak, *Op.cit.*, pp.25-26, f.n.2.
133. P.L. Gupta, *Op.cit.*, pp.268-269.
134. U.N. Roy. *GRY.*, p.129.
135. *Ibid.*, Pargiter, *DKA*, p.53.
136. A.S. Altekar, *VGA.*, p.37.
137. Chattopadhyaya, *EHNI.*, p.190.
138. T.R. Sharma, *PGNGI.*, pp.263-64;
U.N.Roy, *Op.cit*, pp.130-131.
139. Note the use of suffix Chvi in Sanskrit which is used to transform those who were earlier different i.e. from the status of independent rulers they were reduced to the status of his servants.
140. Fleet, *CII.* III, pp.112-116.
141. Raychaudhuri, *PHAI.*, pp.474-475.
142. T.R. Sharma, *PGNGI*, p.260.
143. P.L. Gupta, *Op.cit.*, p.270;
R.C. Majumdar,*VGA.*, p.142;
T.R. Sharma, *Op.cit.*, pp.260-261.
144. T.R. Sharma, *Op.cit.*, p.253.
145. *Ibid.*, p.254.
146. K.P. Jayaswal, *HI.*, p.144; *CII.* III, Preface, pp.135-136.
147. T.R. Sharma, *Op.cit.*, p.228.
147a. *Ibid.*, pp.228-229.
148. *Ibid.*, p.149.
149. P.L. Gupta, *Op.cit.*, pp.270-271.
150. T.R. Sharma, *Op.cit.*, p.147.
151. Ibid., p.128; P.L.Gupta, *Op.cit.*, p.271;
Chattopadhyaya, *EHNI.*, p.192.
152. T.R. Sharma, *Op.cit*, p.171.
153. *Ibid.*, pp.171, 174.
154. P.L. Gupta, *Op.cit.*, p.271.
155. T.R. Sharma, *Op.cit.*, p.145.

156. *Ibid.*, pp.145-146.
157. *Ibid.* ; P.L. Gupta, *Op.cit.*, p.272.
158. P.L. Gupta, *Op.cit.*, p.272.
159. T.R. Sharma, *Op.cit.*, pp.143-144.
160. Chattopadhyaya, *EHNI* ., p.193;
P.L. Gupta, *Op.cit.*, p.272.
161. P.L. Gupta, *Op.cit.*, p.272.
162. D.B. Diskalkar, *SSI.*, p.40.
163. T.R. Sharma, *Op.cit.*, p.126.
164. B.P. Panthari, *BKS.*, p.131;
• U.N. Roy, *GRY.*, p.151.
165. P.L. Gupta, IG., p.272.
166. S. Chattopadhyaya, *EHNI.*, p.193.
167. T.R. Sharma, *Op.cit.*, p.193.
168. *IHQ*. I, p.258.
169. T.R. Sharma, *Op.cit.*, p.166.
170. S. Chattopadhyaya, *EHNI.*, p.193;
D.C. Sircar, *SI.*, pp.271-272.
171. T.R. Sharma. *Op.cit.*, p.135.
172. *Ibid.*, pp.135-136.
172a. Sadhu Ram, "Allahabad Pillar Inscription of Samudragupta" in *AICL.*, p.152.
173. The expression *Kanyopāyana-dāna* has been variously interpreted by scholars:

i. "bringing presents of maidens, (giving) (Fleet, *CII.* III, p.14).
"gifts of maidens and presents (dāna) (R.K. Mookerji,*GE.*, p.26)
"offering of maidens, presents - Diskalkar, *SSI*, p.27.
presents of maiden - princess, payment of tribute of gifts (dāna) - (R.G. Basak, *HNSI.*, p.27).

ii. The word *Kanyā* to be added both with Upāyana and dāna i.e. *Kanyopayana* and Kanyādāna. The word *Kanyopāyana* meaning offering a bride.
Kanyādāna meaning to accept or seek princesses of the Guptas in marriage (A bestowal of a bride) - Sten Konow, *JBORS*, 1937, Vol.XXIII, Pt.IV, p.449., which looks to be far fetched and has been interpreted to suit the context of Rāmagupta.

iii. We do not agree with Sten Konow's suggestion made above and would like to interpret as follows: If the word "Kanyā" is to be taken as an adjunct both to "Upāyana" and "Dāna" Kanyopāyana (offering of maidens) and Kanyādāna (offering daughters in marriage).

174. R.C. Majumdar, *VGA*., p.147.;
P.L. Gupta, *Op.cit.*, p.275.
175. T.R. Sharma, *PGNGI*, Appendix IV, pp.318-20.
176. *Ibid*., pp.319-320.
177. R.K. Mookerji, *GE*., p.28.
178. T.R. Sharma, *PGNGI*., pp.319, 153, 162.
179. R.C. Majumdar, *VGA*., p.148;
T.R. Sharma, *Op.cit*., p.153.
180. T.R. Sharma, *Op.cit*., p.163.
181. T.R. Sharma, *Op.cit*., p.153; See the views of Sten Konow and R.K. Mookerji.
182. *Ibid*.
183. *Ibid*., p.154.
184. R.C. Majumdar, *VGA*., pp.149-150.
185. T.R. Sharma, *Op.cit*., p.159.
186. R.C. Majumdar, *VGA*., p.150.
187. The word Apratiratha literally translated "Unopposed by hostile chariots" (as translated by R.K. Mookerji, *GE*., p.33. See Archer type of coins) means "Unconquerable" i.e. whom no body dares to engage in warfare.
188. R.K. Mookerji, *GE*., p.30;
P.L. Gupta, *IG*., p.281.
V. Upadhyaya, *GA*, p.266.
189. Fleet, *CII*. III, p.14.
190. i. Fleet, *CII*. III, p.14
"whose mind busied itself with the support and the initiation of the miserable, the poor, the helpless and the afflicted".
ii. D.B. Diskalkar, *SSI*, p.27
"whose mind is occupied with the initiation of relieving the low, the poor, the helpless and the afflicted".
iii. U.N. Roy, *GRY*., p.685.
He wrongly translates the word "Sattra" as Mantra and the word "Diksha" as Saṁskāra meaning thereby that who has determined to provide relief to the miserable, the poor and afflicted.
191. *MW*., p.1138, Col.2, Sattra "Session", a great soma sacrifice (lasting according) to some from 13 to 100 days and performed by many officiating brāhmaṇas).
192. *Ibid*., p.480. Col.3 Dīkshā : Preparation or consecration for a religious ceremony, undertaking religious observances for a particular purpose and the observances themselves.
193. Fleet, *CII*. II No.4. Mathura Stone Inscription of Chandragupta-II, p.26, L.4; Ibid. No.10, Bilsad Stone Pillar Inscription of Kumāra-

gupta, p.43, L.2; V. Upadhyaya, *GA*., Nālandā Inscription of Samudragupta, p.126, L.2. Also see Fleet, Op.cit., No.60. Spurious Gayā Copper Plate Inscription of Samudragupta, p.256, L.3 which puts to doubt whether at least this part of the two inscriptions of Nalāndā and Gayā was written earlier even if they belong to later period.

194. R.K. Mookerji, *GE*., p.30.
195. Fleet,*CII*. III, No.4, Mathurā Stone Inscription of Chandragupta-II, pp.26-27, LL.4-5;
Ibid., No.10, Bilsad Stone Pillar Inscription of Kumāragupta, p.43, L.2.;
V. Upadhyaya,*GA*., Nālandā Inscription of Samudragupta, p. 126, LL. 2-3.
196. Fleet, *CII*. III, No.60, Spurious Gayā Copper Plate Inscription of Samudragupta, p.256. II.3-4.
197. Sadhu Ram, *SII*., p.117, L.3; *EI*, XV, p.41, ff.
198. P.L. Gupta, *IG*., pp.280-281.
199. Fleet, *CII*. III, p.28.
200. R.C. Majumdar, *VGA*., pp.153-154.
201. P.G. Gupta, *GE*., p.280.
202. *Ibid*., pp.280-81.
203. See for details, J.R. Basu, *India of the Age of the Brāhmaṇas* Calcutta, 1969, pp.94-98.
MW., p.115, Vol.3. The horse sacrifice, a celebrated ceremony, the antiquity of which reaches back to the Vedic period, in later times its efficacy was so exaggerated that a hundred such sacrifices entitled the sacrificer to displace Indra from the dominion of Swarga; kings who engaged it spent enormous sums in gifts to the Brāhmaṇas.
204. Utsanna:
MW., p.182, Col.I.
raised, elevated (opposed to ava-sanna); Vanished, abolished, decayed, destroyed, in ruins, disused, fallen into disuse, *Utsanna-Yajña* an interrupted or suspended sacrifice, S.Br.
205. See line 30 of the Inscription.
206. R.K. Mookerji, *GE*., p.30.
207. *Ibid*., p.31.
208. R.C. Majumdar, *VGA*., pp.156-157.
209. R.K. Mookerji, *Op.cit*., p.33.
210. *Ibid*., pp.33-34; R.C. Majumdar,*VGA*., p.157.
211. *Ibid*., p.134; Majumdar, *Op.cit*., p.157.
212. R.C. Majumdar, *Op.cit*., p.157.
213. *Ibid*., pp.157-158.

214. Fleet, *CII.* III, No.4, Bilsad Stone Inscription of Chandragupta-II, pp.26-28, LL.1-5; *Ibid.*, No.10.
Bilsad Stone Pillar Inscription of Kumāragupta, the year 96, pp.43-44, LL.1-2.
215. V. Upadhyaya, *GA*, p.126., Nālandā Inscription of Samudragupta, LL. 1-3.
Fleet, *CII.* III. No. 60, Gayā Copper Plate Inscription of Samudra-gupta, p.256, LL. 1-4.
216. B.G. Gokhale, *Samudragupta*, p.256.

6

Rāmagupta

For many years it was believed by scholars on the basis of epigraphic evidence that Chandragupta-II succeeded his father Samudragupta round A.D. 375 on the basis of his Mathurā Pillar Inscription of A.D. 380 (G.E.61) issued in the fifth year of his region[1] and the expression 'tatparigrahīta' (accepted by him) found in the Mathurā Stone Pillar Inscription of Chandragupta-II[2] as well as in the Bihar Pillar Inscription of Skandagupta[3] and Bhitari Stone Pillar Inscription of Skandagupta.[4] This belief continued till 1925 as pointed out by Altekar.[5] It was also known that Samudragupta had a long life and had many sons and grandsons (bahuputra-pautra) as mentioned in the Eran Stone Inscription of Samudragupta.[6] The same inscription also informed us that Airikiṇa (Eran) was his Bhoga-nagara (city of his own enjoyment : capital).[7] The inscription most probably belonged to this temple with a colossal figure of Viṣṇu as pointed out by General Cunningham.[8]

In the light of this background we shall consider the data which has been coming forth from time to time afterwards about an intermediate successor Rāmagupta who remains the most controversial figure in the Gupta history till now.

The story started unfolding when some fragments of a lost drama were noticed in the *Nāṭyadarpaṇa*, a work on the dramaturgy. Scholars found endorsement of this story in *Harshacharita, Sṛingāra-Prakāsha,* from some Rāshṭrakūṭa inscriptions, *Kāvya-Mīmāmsā* and the *Mujmalut-Tawarikh.* "The mention of a king Rāmagupta accepting to surrender his queen to the Śaka Chief who was killed by Chandragupta-II who disguised himself as queen and appropriation of the kingdom and queen to himself" in the above works was accepted as a literary tradition in the absence of concrete archaelogical evidence. Several articles were published on the identification of the Śaka Chief and the place of the event that took place. Even scholars like S. Levi who first brought Rāmagupta to light stated that dramas like *Devīchandraguptam* cannot be treated as reliable sources of history and doubted the historicity of Rāmagupta.[9]

It was argued that "the above traditions are late and have hardly any air of reality; and the absence of Rāmagupta's coins as well as the complete silence of the Gupta records about him lend further weight to this scepticism".[10]

But Rev. H. Heras stated that all those side names of the persons who have left no successor to the throne are actually omitted in the epigraphs and further expressed the hope for numismatic evidence some day. "The literary evidence in his favour is too great and too strong to be shaken by the lack of numismatic evidence... and that some day we shall happily hear of the appearance of a coin of Rāmagupta, so as to establish his historicity on numismatic evidence as well as it is on literary testimony".[11] According to Heras Kāchagupta occupies a definite place in Gupta history between Chandragupta-I and Samudragupta, we need to wait for numismatic evidence to come forth and assign a particular place to Rāmagupta.[12]

Fortunately enough, even when sufficient numismatic evidence was brought to light by K.D. Bajpai, it was also dismissed and Rāmagupta was merely accepted as a local ruler of Malwa, by scholars. The findings of three images of Jain Tīrthaṅkaras along with inscriptions, one of them fully available referring to the erection of the images by Mahārājādhirāja Rāmagupta, leave little room for doubting the historicity of Rāmagupta. But scholars have created a plethora of literature on Rāmagupta problem and added to it in some way or other by throwing their ingenious suggestions in an imposing manner or by their differing interpretations on the issues connected with it.[13]

Hence we need to discuss the whole issue again in the light of data and interpretations presently available with us. We divide our sources into two broad divisions.

(i) Literary Sources.
(ii) Archaeological Sources : which may be further sub-divided into:
   (a) Epigraphic Evidence
   (b) Numismatic Evidence

Literary Sources

We find the following literary sources at our disposal which throw some light on the episode connected with his personality. We may mention that when Sten Konow[14] wrote his article on Rāmagupta in 1937, king Rāmagupta was not known from any other contemporary source except the *Devīchandraguptam* and he had expressed his doubts about the historicity of Rāmagupa.[15] Now we know that excerpts from the *Devīchandraguptam* quoted in various works on Rhetorics and Dramaturgy, such as Bhoja's *Sṛiṅgāra-Prakāsha*, Rāmachandra and Guṇachandra's *Nāṭya Darpaṇa*, Sāgaranandin's *Nāṭaka-Lakshaṇa Kośa* and *Abhinava-Bhāratī* of Abhinavagupta, a commentary on Nāṭyaśāstra, had started taking notice of the scholars from 1923 onwards to 1936.[16]

Following are the literary sources available to us:

(i) *Devīchandraguptam* of Viśākhadatta (A.D. 400).[17]
(ii) *Mudrā-Rākṣasa* of Viśākhadatta (A.D. 400).[18]
(iii) *Harṣa-Charita* of Bāṇabhaṭṭa (A.D. 630).
(iv) Śaṅkarārya's *Commentary on Harṣacharita* (A.D. 1713).[19]
(v) *Kāvya-Mīmāṁsā* of Rājaśekhara (A.D. 940-1000).[20]
(vi) *Śṛiṅgāra-Prakāsha* of Bhoja (A.D. 1018-56).[21]
(vii) *Abhinava-Bhāratī* of *Abhinavagupta* (A commentary on Bharata's *Nāṭya Śāstra*: A.D.1050).[22]
(viii) *Nāṭaka-Lakshaṇa-Ratna-Kośa* of Sāgaranandin (A.D.1070-1100)[23]
(ix) *Nāṭya-Darpaṇa*, of Rāmachandra and Guṇachandra (A.D. 1200).[24]
(x) *Āyurveda-Dīpikā Ṭīkā* (A.D. 1150).[25]
(xi) *Majmal-ut Tawarikh*, of Abul Hasan (an Arabic work translated into Persian by Abdul Hasan : Hij. 417 A.D. 1026).[26]
(xii) *Rājāvalī* of Mṛituñjaya (A.D. 1808).[27]

Devīchandraguptam of Viśākhadatta

The Drama *Devīchandraguptam* written by Viśākhadatta who may have been a contemporary of Chandragupta-II[28] has not yet been traced out. Fortunately we are able to find it in fragments quoted in the works of Rhetorics like *Śṛiṅgāra-Prakāsha*,, *Nāṭya-Darpaṇa*, *Nāṭaka-Lakshaṇa Kośa*, *Abhináva-Bhāratī* and some other works which provide us enough material to build up the story. Goyal has tried to put them together at one place.[29] While P.L. Gupta has tried to arrange the original drama by collecting these fragments together along with an English translation which makes it an interesting reading.[30]

The story runs as follows: Rāmagupta was a weak and incapable king who agreed to surrender his wife Dhruvadevī to Śaka Chief when surrounded by enemies in his (enemy's) camp. The queen, after hearing this, looked like the moon just coming out from Rahu's mouth. Seeing the actions of her husband, like those of an impotent person, shyness, anger, sorrow and fear all took abode in her. *Kumāra* Chandragupta got distressed to know Rāmagupta in trouble by the Śaka Chief and was planning to go to enemy's camp to save his trouble and take vengeance upon the Śaka Chief, had decided to perform *Vetāla-Sādhana*, when he was told by Vidūshaka to go one step further to which the hero answered in affirmative. In the meanwhile he was supplied with a box full with garments and ornaments of Dhruvadevī by a maid servant of the harem. When Chandragupta got ready to go to enemy's camp dressed as a woman, he was addressed by Vidūshaka, 'O, will you move alone among the foes?" To which Chandragupta answered:

> "O'fool, do numbers count against solid matter? The lion, at the very smell of whom the herds of deer flee away, comes out of his den alone smiling when the elephants which are of strong back, mammoth body, full of strength and power and are strange-looking with their tusks, come into his sight. For a warrior number means nothing (*numbers do not count*)."

To reassure his subjects king Rāmagupta when he came to take along Dhruvadevī to hand her over to the Śaka (Chief), he encountered *Kumāra* Chandragupta who was ready to go in the guise of Dhruvadevī to kill the Śaka enemy.[31] He was discouraged for this act by Rāmagupta and proposed to discard the queen like a twig of grass which the queen heard behind the curtain.

Finally "Chandragupta is stated to have entered his enemy's camp at Alipura disguised as a woman to kill the lord of the Śakas".[32]

After killing the Śaka Chief Chandragupta when he came back he became very popular (shone like a moon in the sky who had destroyed the darkness like foes, with its white rays).[33] He pretended madness to save himself from the danger of his life at the hands of wicked Rāmagupta who might have been having apprehensions for the loss of his throne. He made some plan and entered the palace. Later developments are not known as the passages from the later portion of the drama are not available. But we can assure the same on the basis of the story reproduced in the *Majmal-ut-Tawarikh.* Entering the palace Chandragupta managed to kill King Rāmagupta and seized the throne for himself amid the plaudits of the people. He burnt the body of the king, married Dhruvadevī and restored order.[34]

We cannot accept the story in the drama *Devīchandraguptam* as a figment of imagination as suggested by some scholars.[35] It is based on some historical personages and thus reveals to us some important facts of the Gupta history.

Mudrā-Rākshasa of Viśākhadatta

It appears that Viśākhadatta was gravely concerned by the political disturbances in the empire. He wrote his plays when the country was in peril both internally and externally. During the period of the Nandas, on the one hand to which the play *Mudrā-Rākshasa* refers, there was fear of Greek invasions and on the other hand the fight ensued between Chandragupta and Mahāpadmananda. The same situation arose again after the death of Samudragupta when the country again felt the terror of the Śakas as well as the war of succession between the two brothers Chandragupta-II (who was nominated as his successor by Samudragupta) and Rāmagupta.[36]

Through the *Bharavākya* of the *Mudrā-Rākshasa* Viśākhadatta clearly serves his purpose of benediction for the welfare of his patron and his empire. The king in this case is Chandragupta Maurya, the Mlechchhas are the Greeks. While the *Bharatavākya* equally applies to Chandra-gupta-II, the Mlechchha was indiscriminately applied to all foreigners.[37] The *Bharatavākya* runs as follows:

> "Vārāhīmātmayone stanumavanavidh āvāsthītasyanurāpam, yasya Pragdantakoṭim Pralayaparigata Sisriye Bhūtadhātrī Mlechchhairudvijyamānā Bhujayugamadhunā Saṁśriye Rājamurtch, Sa Srimadbandhubhṛityaś chiramavatu Mahīm Parthivas' Chandraguptaḥ"

"May the King Chandragupta, *he the glorious along with his relatives and servants,* rule over the country for a long time. *The king on whose two arms rests the earth which has been shaken by the Mlechchhas;* the king, who, while being his ownself (sex) assumed the disguise of

Vārāhī (feminine form of the incarnation of Viṣṇu in which he delivered the earth from the deluge - and thus he is being compared with Viṣṇu himself), the form that was required in the performance of the duty of protection; and who rescued the sinking (pralaya-parigata) earth (Bhūta-dhātrīm) by the point of a dantakoṭi (dagger). [38]

The question of Chandragupta being loyal to his brother does not arise as he himself was the king. Moreover, no court-poet will mention his patron as a bhṛitya (servant) to his brother. Scholars have sometimes taken liberties by twisting the meanings of certain expressions as a they suit them as we have shown elsewhere, that Jagannath has taken "Srimadbhandhuh" to stand for the Chālukya overlord and the word 'Bhṛitya' 'servant' for Dantidurga while the whole expression simply means in the context: "He, the glorious along with his relatives and servants".[39] P.L. Gupta has taken the word 'Bhūtadhātrī which means 'earth' to stand for queen Dhruvadevī.

However, the importance of the play *Mudrā-Rākṣasa* is to praise Chandragupta for protecting the empire from a perilous political situation which it was in due to the invasions from the Śakas whom he is stated to have exterminated.

Harṣacharita of Bāṇabhaṭṭa

In the *Harṣacharita*, the same event is stated as follows :

> Aripure Cha Parakalatrakāmukam Kaminīveśaguptaś Chandraguptaḥ Śakapatimaśātayat".
>
> And in the enemy's City Chandragupta concealed in the guise of a woman butchered the Śaka lord who wanted another's wife".

This example is cited by Bāṇabhaṭṭa while quoting other examples of the kind from historical episodes. As pointed out by Levi, such a tradition was certainly known to Bāṇa, who wrote the above account in the Sixth *Uchchhavāsa* of the *Harṣacharita*. [40]

Śaṅkararya's commentary on Harṣacharita

Śaṅkararya, a commentator on the Harṣacharita wrote this as follows:

"Chandraguptaḥ Bhrātṛijāyām Dhruvadevi Prārthayamānaścha Chandraguptena Dhruvadevi veśadhāriṇā strīveśajanaparivṛittena Rahasi Vyapāditaḥ"
"That the Śaka Lord was asking for Dhruvadevī, the wife of Chandragupta's brother, and that Chandragupta disguised as Dhruvadevī and accompanied by followers dressed up as women, killed him".[41]

Kāvya-Mīmāṁsā of Rājaśekhara

In the tenth century Rājaśekhara quoted the same episode while giving an example of "Kathotha" (based on historical event) form of "Muktaka" (Free poetry) in the following verse:

Dattvā Ruddhagatih Khasādhipataye Devīm Dhruvasvāminīm,
Yasmātkhaḍitasāhaso nivavṛite Śrīśarmgupto nṛipaḥ
Tasminneva himālaye Gurugukākoṇakvanatkinnare,
Gīyante tava Kārttikeyanagarastrīṇām ganaiḥ Kīrttayaḥ".[42]

King Śarmagupta after being trapped and having lost courage came back therefrom after having committed to give his queen Dhruvasvāminī for the Khasa king, in the same caves and ravines of the Himalayas, O King! the ladies of Kārttikeyanagara are singing your glories to the tune of the Kinnaras".[43]

We cannot say definitely whether Rājaśekhara himself compiled this verse or has quoted from somewhere. We notice the changes in the verse that is instead of Rāmagupta, the name of the king is changed to Śarmagupta, and *Śakādhipati* is changed to *Khasādhipati*, Dhruvadevī is changed to Dhruvasvāminī. Śarmagupta is most likely a mistake for Rāmagupta, in some manuscripts we even get the reading Senagupta as well. Such changes by the passage of time are possible some time. The best example is that the name of the queen itself also is named as Dhurvadevī in most of the inscriptions of the Guptas[44] while in the Basāṛh Clay seal of Govindagupta she is known as Dhruvasvāminī as the wife of Chandragupta-II and the mother of Govindagupta.[45]

Śṛiṅgāra-Prakāsha of Bhoja

Some passages from the *Devīchandraguptam* are reproduced in the *Sṛiṅgāra-Prakāsha* which are as follows:

(a) According to Mr. Rangaswami Saraswati, the manuscript of Bhoja's *Śṛiṅgāra-Prakāsha* in the Government Oriental Manuscript Library, Madras, contains the notice : "Strīveśanihnutaḥ Chandraguptaḥ Śatroḥ Skandhāvāram Alipuram Śakapativadhāyāgamat". "Concealed in female attire Chandragupta went to the enemy's camp, to Alipura, in order to kill the Śaka-Lord".[46] Mr. R. Saraswati has pointed out that the place where the incident occurred is called Alipura which is corroborated by the passage in the *Harṣacharita* where the reading aripura is evidently a mistake for Alipura. If this view is not accepted, there would be tautology in the expression "Śatroḥ Skandhāvāram alipuram". Again, *Skandhāvāra* does not necessarily denote a camp, it means also a capital (Cf. Harṣcharita Niraṇaya Sāgara, ed.p.153). So the expression can be taken to mean "Alipura, the capital of the enemy," is the view of Mirashi.[47] In our view there is no need to complicate the meaning of expression which simply means "In enemy's camp at Alipura".

Mirashi locates this Alipura somewhere in or near the ancient country of Kulūta (Mr. K.P. Jayaswal also locates the place in the Doab of Jullundur -*JBORS*, XVIII).[48]

A manuscript of the *Harṣacharita* reads 'Nalinapura" instead of 'alipura' and Mrs. Rangaswami is of the opinion that Alipura is the right reading while Mirashi thinks that the original may have had Nalinipura which might be a synoym for Padmapura, according to Watters, a name of Huan-Tsang's Teng - Kuang "apparently a little to the West of the modern Jalalabad" and that Nalinapura in course of time may have been read as Alipura and then as Aripura.[49] We do not agree with this line of argument as pointed by Sten Konow.[50]

However, we prefer the reading 'alipura' in the context as it avoids the tautology in expression as 'Śatroḥ Skandhāvāra', means 'enemy's camp' which is also the meaning of 'Aripura' if it is not a placename. Therefore, the reading 'Aripura' in the *Harṣacharita* can be a mistake for 'Alipura' as quoted in the *Śṛiṅgāra-Prakāsha*

It will not be out of place taking together the evidence given in the verse of the *Kāvya-Mīmāṁsā* while commenting on this issue.

In the *Kāvya-Mīmāṁsā*, the enemy who reduced Rāmagupta to dire distress is called Khasa, while almost all other authorities name him Śaka. The author of this verse flourished in the 10th century, when the Khasas were ruling in Nepal. Mirashi agreeing with Altekar points out that the Khasas were not so powerful in the 4th century as to dictate terms to Rāmagupta and favours 'Śaka' as the correct reading.

Mirarshi identifies this Śaka enemy with the Kushāṇa king who ruled over Punjab and Kabul and is referred to as 'Devaputra Shāhi Shānushāhi' in the Allahabad Pillar Inscription of Samudragupta.[51]

Altekar identifies this Śaka ruler with Piro who succeeded his father Kidāra (ruler of Gandhara, Kashmir, Western and Central Punjab) in A.D. 375 who might have attempted to extend his power further eastward in the Punjab when the strong arm of Samudragupta was removed by death at about this time. To check this growing menace, Rāmagupta, the successor of Samudragupta, seems to "have undertaken an expedition against him in the Punjab, but was signally defeated and compelled to conclude as ignoble peace, requiring the surrender of the Gupta crowned queen."[52] A glance at the features of Piro shows that he was a cruel and lascivious ruler, who could well have imposed the above condition on his vanquished adversary.[53]

About the place of incident, Mirashi states that it occurred either near Jalandhara Doab or near Jalalabad. The King Kārttikeya to whom the verse "Dattvā Ruddhagatiḥ etc" is addressed in the *Kāvya-Mīmāṁsā* was Mahipāla-I of the Gurjara Pratihāra dynasty of Kanauja.[54]

(b) The other verse quoted from the *Devīchandraguptam* in the *Śṛingāra-Prakāsha* of Bhoja is addressed to Vidūṣaka by Chandragupta as follows:

Sadvaṁśān Pṛithuvarnyavikramabalan dṛiṣṭvādbhutan dantinaḥ
Hasasasyeva Guhāmukhādabhi-mukham niṣkramataḥ Parvatāt,
Ekasyāpi Vidhūtakesarajaṭābhārasya bhītāḥ mṛigāḥ,
Gandhādeva hareḥ dravanti bahavo vīrasya kim saṁkhyayā.[55]

" O fool, do numbers count against solid matter? The lion, at the very smell of whom the herds of deer flee away, comes out of his den alone, smiling, when the elephants, which are of a strong back, mammoth body, full of strength and power and are strange looking with their tusks, come into his sight. For a warrior, number means nothing (numbers do not count)."

(c) Another excerpt reproduced from the *Devīchandraguptam* is Madhava's statement addressed to Vasantasenā.

Ānandāśrujalam sitotpalruchorābadhnatā netrayoh,
Pratyaṅgeṣu varānane pulakiṣu svedam samatanvatā,
Kurvāṇena nitambayoru-pachayaṁ sampuraṇa-yayorapyasu,
Kenaspyaspriśataspayadhonivasana granthistavochchhvāsitaḥ.[56]

"Who has brought the tears of happiness in your eyes that are beautiful like the blue lotus? Why sweat has appeared in all parts of your body? Why are your well developed thighs so happy? Why are the hip-knots of your garment fluttering without being touched by anybody? "[57]

Abhinava-Bhārati of Abhinavagupta

The same verse from the *Devīchandraguptam* as in V(c) addressed to Madhavasenā by Chandragupta which was earlier addressed by Mādhava to Vasantasenā is quoted in the *Abhinava-Bhārati* .

"Ānanadāśrujalam...........Tavochchhavāsitaḥ"
This is the same in *Nāṭyadarpaṇa*,, as also in the *Sṛiṅgāra-Prakāsha*,[58]

Nāṭaka-Lakshaṇa-Ratna-kosha of of Sāgaranandin

In this work 'Ullekha' is defined as "Acting to do one's duty". The example is given from the *Devīchandraguptam* as follows:

"Yathā Unmattachandragupte - 'Loko Lochanandanasya Rataye Chandrodaye Sotusukaḥ Atra Kṛitakonmādam Chandraguptaḥ Parityajya Karttavyamāha -" Bhavatvena Jayaśabdena Rājakula-gamanam Sādhayāmi, "Iti". [59]

"For example, Chandragupta who feigned madness, "The world becomes intoxicated for love-making which is pleasing to the eyes as soon as the moon rises". Here Chandragupta declares open his duty leaving out his forged madness - "Okay, by uttering the word "Jaya" I comply with my determination to enter the palace".

Nāṭya-Darpaṇa, of Rāmachandra-Guṇachandra

This late twelfth century work became one of the first sources of information about the existence of the drama *Devīchandraguptam* whose some passages were brought to light by Silvain Levi some of which as shown are common to all the literary sources. According to Levi the hero of the drama must be Chandragupta-II, and it must have been composed by Viśākhadatta, the author of the *Mudrā-Rākṣasa* . We have build up our story of the *Devīchandraguptam* as described earlier on the basis of the passages from the *Nāṭya-Darpaṇa*. We quote some parts out of them here:

> "Devīchandragupte Divitīye aṅke
> Prakṛitīnām āśvāsnāya Śakasya
> Dhruvadevī-SaṁpradāneSbhyupagate
> Rājñā Rāmaguptena Arivadhārtham
> Yiyāsuḥ Pratipanna Dhruvadevī
> Nepathyaḥ Kumārachandraguptaḥ
> Vijñāpayan uchyate".[60]

"Rāmagupta who had come back to take Dhruvadevī to hand her over to the Śaka to reassure her subjects, speaks to Kumāra Chandragupta who is ready in the guise of Dhruvadevī to kill the Śaka. Dhruvadevī hears the talk from behind the curtain".

And also:

"Patyuḥ klīvajanochitena Charitenānena Puṁsassataḥ."[61]
"Seeing the actions of her husband like those of an impotent person".

Āyurveda-Dīpikā-Ṭīkā

In this renowned treatise on medical science an example is given which is based on the story of the Devichandraguptam......"Uttarakālaṁ Bhrātrādi Vadhena Phalena Jñāyate Yadayamunmatta Chhadmaprachārī Chandragupta Iti".... For instance, to obtain the murder of his brother and others at some later date Chandragupta fradulently declares himself mad".[62]

Majmal-ut Tawarikh

This work translated into Arabic from the Hindwani language (Sanskrit) was translated into Persian in 417 AH. (A.D. 1026) by Abul Hasan Ali.[63] The story runs thus:

After the death of Ayand, his son Rāsal became king. He reigned for sometime, until one rose up against him and expelled him from the kingdom. Rāsal (then) went southwards, and established himself there, he had two sons, one named Rawwāl, and the younger Barkamaris.

After the death of Rāsal, Rawwāl became the king. A certain king had a daughter of great intelligence. All the kings and princes of the Hindus sought her, but no one pleased her except Barkamaris who was very handsome. When Barkamaris brought her home his brother said, "As she pleased you so does she please me". Then he took the girl with her handmaids. Barkamaris said to himself: "The damsel chose me for my wisdom and there is nothing better than wisdom". So he gave himself to study, and associated with the learned and the Brahmanas, till he reached such perfection that he had no equal.

When the rebel who had expelled their father (Rāsal) heard the story of the damsel, he said, "Can they who do such things occupy such a position?" So he led an army and put Rawwāl to flight. Rawwāl with his brothers and all nobles went to the top of a mountain where a strong fortress had been built. Then they set guards on the summit and felt secure. But the enemy got possession of the mountain by strategem and besieged the fort, and was near upon taking it. Rawwāl then sent to sue for peace, and his enemy said - "send me the girl, and let every one of your chiefs send a girl. I will give these girls to my officers - then I will withdraw". Rawwāl was dejected, but he had a Wazir, blind of both eyes, named Safar, of whom he enquired what was to be done. He advised him to give up the women and save his life. He might then take measures against his enemy, but if he lost his life what would be the good of children and wife, and riches. They resolved upon this course, but just at this juncture, Barkamaris came in, and after making his salutation said: "I and the king are sons of the same father, if he will acquaint me with his opinion. It may be that I may be able to suggest something, - "Do not take my youth into consideration". So they informed him of the facts. He then said: "It seems proper that I should stake my life for the king: let an order be given for me to dress like a woman, and let all the officers dress their sons in like manner as damsels, and let each conceal a knife in our hair, and carry a trumpet also concealed, then send us to the king (Śaka Chief). When we are brought before the king they will tell him that I am the damsel, he will keep me for himself and give the others to his officers. When the king retires with me I will rip up his belly with the knife and sound the trumpet. When the other youths hear this they will know that I have done my work, and they must also do theirs. All the officers of the army will thus be slain. You must be prepared, and when you hear the trumpet, you must rally forth with your soldiers and we will exterminate the foe". Rawwāl was delighted and did as was proposed. It succeeded, not one of the enemy's horsemen escaped, all were slain and cast down from the mountain. Rawwāl's power increased. (The Wazir excites the King's suspicions against Barkamaris, who feigns madness).

One day in the hot season, Barkamaris was wandering barefoot about the city, and came to the gate of the King's palace. Meeting no hindrance he entered, and found his brother and the damsel sitting on a throne sucking sugar cane. When Rawwāl saw him he observed that there could be no porters at the gate, otherwise the poor mendicant would never have got in. Taking pity on him, he gave him a bit of sugar-cane. The mendicant took it, and picked up a piece of the shell of the cane to scrape and clean it with. When the King saw that he wanted to clean the cane, he told the damsel to give him a knife. She rose and gave the knife to Barkamaris, who cleaned the sugar-cane with it, and craftily watched until the King was off his guard. Then he sprang upon him and plunging the knife in his navel, ripped him up. After that he seized his feet and dragged him from the throne. He next called the Wazir and the people, and seated himself on the throne amid the plaudits of the people. He burnt the body of the King, took back the damsel and married her, and restored order.

Then he called the Wazir and said: " I know that it was you who counselled my brother in his dealings with me, but this was no fault nor is it blameable. It was God's will that I should be king, so continue to govern the kingdom as you did for my brother". Safar replied, "You have spoken the truth, all that I did was for the good and advantage of your brother, not out of enmity to you. But I have now resolved upon burning myself and cannot do as you desire. I was with your brother in life, and I will be with him in death". Barkamaris told him that he wanted him to write a book on the duties of kings, on Government and Justice. Safar consented and wrote the book, which is called "*Adabu-l Mulūk* " "Instruction of Kings" when it was finished he took it to Barkamaris and read it, and all the nobles admired and praised it. Then he burnt himself. The power of Barkamaris and his kingdom spread, until at length all India submitted to him. Such was Barkamaris.[64]

Thus it is evident that the story in the *Majmal-ut-Tawarikh* is based on the plot in the *Devīchandraguptam* of Viśākhadatta. Aparna Chattopadhyaya has shown some insconsistencies in the story of the *Majmal-ut-Tawarikh.* [65]

1. The first point which requires scrutiny is that according to this story Rawwāl (who is supposed to be Rāmagupta) and Barkamaris (who is identified with Chandragupta-II) are sons of a King Rāsal who went southwards and established a kingdom there, since he was dispossessed of his ancestral kingdom in the Hindustan proper. Can the timid and defeated Rāsal taking refuge in the South, that is beyond the Vindhya hills, out of fear for a rebel Chief, be, by any stretch of imagi-nation identified with Samudragupta, one of the greatest conquerors of the ancient world, who has been called an "Indian Napoleon" for his military genius?
2. Secondly, if we believe according to story that Rawwāl snatched away the married wife of his brother Barkamaris it looks strange that the same person of such a drastic nature succumbed before the threats of a rebel chief and decided to send that very lady to the latter?
3. Thirdly, Aparna Chattopadhyaya argues if we accept the historical value of the drama *Devīchandraguptam*, the scene of the drama is cer-tainly Pāṭaliputra because Pāṭaliputra was the capital of the Gupta Empire.[66] But in the *Majmal-ut-Tawarikh*, the secen is in Southern India where the family had fled due to the fear of a rebel and where the rebel appeared again demanding the queen, so this almighty rebel had first snatched away the northern and original kingdom, demanding the queen. Does such a picture apply at all to the Gupta Imperial family?
4. Fourthly, is it correct to identify Safar (Wazir) of the story with Śikharasvāmin, a Minister of Chandragupta-II? Śikharasvāmin not

only served himself as a counsellor of Chandragupta-II, but even his son Pṛithvīseṇa worked in the capacity of a Minister during the rule of Chandragupta-II's son and successor Kumāragupta.

But on the whole she reconciles with these variations. "The plot of the drama *Devīchandraguptam* by Viśākhadatta was in all probability carried to Arabia by sailors or traders. And this may be the reason that in *Majmal-ut-Tawarikh*, we find the tale of *Devīchandraguptam* narrated with certain modifications and changes tnat are inevitable in the case of tales carried by travellers, sailors and traders".[67]

Rājāvalī of Mṛituñjaya Paṇḍita

The last mention of Rāmagupta legend probably is reflected in this 19th century work *Rājāvalī* . Here the names of Mayūra dynasty of Indra-Prastha are given. A ruler named Rājapāla (Rakshapāla) was defeated by the Śaka-Lord (Śakāditya or Śakadatta), of Kumaon. Rājapāla is described as a cowardly and lascivious king. Afterwards Vikramāditya defeated the Śaka-Lord and got anointed as the king who had Samudragupta and Chandrapāla as his successors. Rājapāla may be compared with Rawwāl of the *Majmal-ut--Tawarikh*.. Vikramāditya may stand for Chandragupta Vikramāditya.[68] According to U.N. Roy it appears from the *Kāvya-Mīmāṁsā* of Rājaśekhara that Rāmagupta was defeated by the Śaka Chief in Kārttikeyanagara which Bhandarkar has identified with Baijanath in Almora District of the Kumaon Division, *Rājāvalī* also mentions the Śaka Chief to be the ruler of Kumaon.[69] Atkinson and Cunningham were the first to draw the attention of the scholars to this work.[70] It seems that some of the events of Rāmagupta episode such as Rāmagupta's defeat at the hands of the Śaka-Lord and then his being vanquished by Chandragupta Vikramāditya are in a way reflected in this work.

Archaeological Sources

Epigraphic Evidence

The Sanjan Copper Plates of Amoghavarṣa-I also give the evidence and also the story that Chandragupta-II killed his brother and married his wife.

In the Sāngli and Cambay Plates of Govinda-IV also we got some allusions to Chandragupta-II's marrying the wife of his elder brother after killing him.

(a) *Sanjan Copper-Plate Inscription (A.D. 871)*

This record has been published in the *Epigrahia Indica*, Volume XVIII. It states the following verse:

"Hatvā Bhrātarmeva Rājyamaharddevīmścha Dīnastatha,
Laksharn Koṭimalekhahyan kil kalau dātā saḥ Guptānvayaḥ,
Yenātyājitamusvarājyamasakṛidbahyārthakai kā kathā,
Hrīstasyonnati-rāṣṭratilako dāteti kīrtyāpi !!"

"The donar of Kali age who was one of the Gupta lineage, who had killed (his) brother, seized (his) Kingdom and wife and who wrote the gift of a lac as of a crore (in documents), was a despised donor. But his glory is ashamed before his (the Rāṣṭrakūṭa Amoghavarsha), who gave away more than once his own Kingdom. "What to speak of (donation) of material objects. Even after being exalted as a donor, he, the ornament of the Rāṣṭrakūṭa family, blushed at his praise".

The reference of killing one's brother and seizing his throne and wife to one ruler of the Gupta family is a pointer to Chandragupta-II and is based on the Rāmagupta episode of the *Devīchandraguptam.* Further the author of the inscription seems to have twisted the evidence of the Gupta records mentioning different figures of donations in favour of Samudragupta viz. (one lac of cows' as mentioned in the Allahabad Pillar Inscription (Go-śatasahasra-pradāyinaḥ)[71] with that of the menion of the 'donar of one crore of cows' (nyāyāgatakoṭihiraṇyapradasya)[72] in the Mathurā Inscription of Chandragupta-II, in favour of his patron.[73]

(b) *Cambay Copper-Plate Inscription* : (Śaka Era 852 = A.D. 930).

Mirashi[74] has drawn our attention to a verse in the Inscription which is written in reference to the Rāṣṭrakūṭa King Govinda-IV, high-lighting his achievements in comparison to *Sāhasānka* Chandragupta-II)

"Sāmarthye Sati ninditā pravihitā naivāgrajekrūrtā,
Bandhustrīgamanādibhih, Kucharitairāvirjjitam nāyaśaḥ !

Śauchāśauchaparāṅmukhaṁ na cha bhiya Paiśāchamaṅgīkṛitam
Tyāgenasamasāhasaiścha bhuvane yaḥ sāhasāṅkoSbhavat " !!

"He (Govinda-IV) became *Sāhasāṅka* only by virtue of his sacrifice and by his unparalled heroic deeds. Even having his might, he neither behaved with ignominious cruelty towards his elder brother nor earned ill-fame by his evil deeds of having intercouse with the wife of his brother; he did not resort to diabolical conduct out of fear by becoming indifferent to what is pure or impure".

Here the qualities of Govinda-IV who earned for himself the title of *Sāhasāṅka* who very much deserved it out of his virtuous conduct unlike the Sāhasāṅka (Chandragupta-II) who had shown cruelty towards his brother by killing him, bad conduct by marrying his brother's wife and who had resorted to diabolical conduct (Vetāla-sādhanā) without bothering for purity or impurity involved. (The black magic does not conform to the Aryan Code of Conduct according to Hindu Culture).

Clearly the verse narrated above projects on the career of Chandragupta-II and is based on the plot of the *Devīchandraguptam* in description of such events as killing his brother and seizing his kingdom and wife. Not only that the third line also receives a remarkable corroboration from a passage in the *Devīchandraguptam* where Chandragupta had due to fear resorted to diabolical conduct (niśivetāla-Sādhanāmadhyavasan Kumaraśchandraguputaḥ Ātrayena Vidūṣakena uktaḥ).[75] The association of a *Vetāla* with King Vikramāditya is supported by the stories recorded in the *Bṛihatkathā*, the *Kathāsaritsāgara* and the *Vetāla Pañchaviṁśati*. "The poet who composed the above stanza in the Sangli and Cambay charters of Govinda-IV has evidently drawn upon the *Devīchandraguptam* for his remarks about Chandragupta".[76]

(c) *Sāngli Copper Plate Inscription* : (Śaka Era 855 = A.D. 933)

In the Sāngli Copper Plate Inscription of Govinda-IV, the same verse "Sāmarthye sati.......SāhasāṅkoSbhavat", discussed in (b) above is quoted describing the killing of his brother. Snatching away his kingdom and wife and resorting to diabolical conduct by Chandragupta in comparison to Govinda-IV who did not do any such ignoble needs and was the real *Sāhasāṅka* by virtue of his sacrifice and unparalled valour (by performance of unequalled courageous deeds).

(d) *The Gupta Inscriptions*

The Gupta records stand in evidence to prove the historicity of the three characters mentioned in the *Devīchandraguptam* namely, Chandragupta-II, Dhruvadevī or Dhruva-svāminī and Rāmagupta. Several Gupta records throw light on the names Chandragupta-II [77] and Dhruvadevī [78] or Dhruvasvāminī.[79]

About King Rāmagupta whose historicity was earlier being doubted due to non-availability of any record, we have now three records found at a village named Durjanapura in modern Besnagar near ancient Vidiśā inscribed on the three Jain images where King Rāmagupta is mentioned with the title *Mahārājā-dhirajā* (available with the Vidiśā Museum) which leave no doubt that King Rāmagupta belonged to the Imperial Gupta family and should no longer be confused with a local ruler.

The three inscriptions published by Vasudeva Upadhyaya [80] are as follows:

Inscription on the first Image

1. BhagavatoSrhata. Chandrabhrabhasya Pratimeyam Kāritā.
2. Mahārājādhirāja Śrī Rāmaguptena Updeśāt Pānipā.
3. Trika Chandrakṣamāchārya Kṣamaṇa--Śirmana Paśiṣyāchārya.
4. Sarppa Sena Kṣamaṇa Śiṣyasya Gokulyantyāsa tputrasya Chelu Kṣamaṇa Syeti.

Inscription on the Second Image

1. BhagavatoSrhata. Punyadantasya Pratimeyam Kāritā
2. Mahārājādhirāja Śrī Rāmaguptena Upadeśāt Pāṇipatrika.
3. Chandrakṣama (ṇā) chāryya (Kṣamaṇa) Praśiṣyasya.
4. x x x

Inscription on the Third Image

1. BhagavotāSrhata (Chandraprabha) sma Pratimeyakāritā Mahārājādhirāja
2. Śrī (Rāmaguptena) Upadeśāt Pāṇipatri.
3. x x x
4. x x x

Even on palaeographical grounds scholars have placed these inscriptions in the Gupta period and the images on the basis of their style in between the Kushāṇa and Gupta Images of the fifth centuries A.D.[81]

It is not unlikely that Rāmagupta might have shown his inclination to jainism adopting a peaceful style of life probably after his humiliation by the Saka Chief which was more accentuated after the courageous deed of Chandragupta by saving the honour of the queen and kingdom at the risk of his own life. In the later period we find such examples in case of Budhagupta, Bālāditya and Vajra, the Gupta rulers of later date who embraced Buddhism.[82]

The three images described above show that the inscriptions include the name of *Mahārājādhirāja* Rāmagupta who seems to be responsible for installation of these images:

(a) The first image belongs to Jain Arhat Chandraprabha (Eighth Tīrthaṅkara). Even the word 'Bhagavān' is prefixed to his name. This image seems to have been accomplished by the said emperor at the instance of Āchārya Chelukṣamaṇa.[83]

(b) The second image belongs to Jain Āchārya Puṇyadanta (Puspadanta) - (the ninth Tīrthaṅkara) whose name is also prefixed by the word 'Bhagwan' and 'Arhata' [84]

(c) The third image may also probably belong to Jain Tīrthaṅkara Chandraprabha where the inscription is in a multilated condition.[85]

Numismatic Evidence

P.L. Gupta found in 1948 some coins which included six coins of Rāmagupta, discovered near Bhilsa (Besnagar) and published them in 1950.[86] They are ascribed by him to Rāmagupta.[87] Later in 1954 while commenting on these coins Altekar remarked:

> "It is well-known how Samudragupta had already occupied the territory upto Eran, within 40 miles of Besnagar, where the coins of Rāmagupta have been found. The coinage in ancient India was local and it is quite likely that a need for small copper currency, similar to the Nāga coinage current earlier there, may have been soon felt by the Gupta administration. Rāmagupta on his accession may have decided to meet it by issuing coins of the type which we have now found, which closely imitate the Nāga coinage, Or, it may be that even

> Samudragupta might have started his currency, and its specimens may be discovered, if a careful search is made for them in Malwa. Tiny coins of the type issued by Rāmagupta were not current in Bihar and U.P. and were, therefore, not issued in those provinces. It is not unlikely that gold coins of Rāmagupta may still be discovered in future".[88]

In 1961, K.D. Bajpai attracted the attention of scholars on some copper coins of Rāmagupta found at Eran, which were published by him in 1961, classifying them into two varieties (a) Siṁha variety; and (b) Garuḍa variety.[89] Immediately afterwards he seems to have found various copper-coins bearing the name of the King Rāmagupta in the Brāhmī script of the early Gupta period, both from Vidiśā and Eran and in one of his articles[90] published in 1962 classified them under the following four types:

> Lion type, Garuḍa type, Garuḍadhvaja type and border legend (Śrī Rāmagupta) type.

Having examined over two hundred coins of this ruler obtained from Vidiśā and Eran, bearing in pure Sanskrit the name Rāma, Rāmagupta or Śrī Rāmagupta in early Gupta Brāhmī script, Bajpai put forth his view that this king could be identified with Rāmagupta, son of the Gupta emperor Samudragupta, and the elder brother of Chandragupta-II. On the basis of several Śaka and Gupta stone inscriptions and their numerous coins known from the Eran—Vidiśā tract, he also suggested that the well-known episode of Dhruvadevī or Dhruva-svāminī probably took place in this very region.[91]

D.C. Sircar, however, tried to contradict the above opinion of Bajpai that the issuer of the coins bearing the name Rāmagupta can be the elder of Chandragupta-II.[92] He states that these coins "really show that, like some other rulers of his type (Harigupta and Indragupta), Rāmagupta was a Chief who issued coins in imitation of the Imperial Gupta money on the decline of the Guptas about the close of the fifth century A.D."[93] He further states that the similarity of twin coin types and names points to the fact that Rāmagupta, Harigupta and Indragupta enjoyed similar position, if not that, they belonged to the same region and also the same family. They appear to have flourished on the decline of the Imperial authority of the Guptas in the outlying provinces of the Gupta empire about the close of the fifth and the beginning of the sixth century A.D. The palaeography of the legend of coins in question does not appear to militate against this dating.[94]

Bajpai has refuted the above-mentioned view of D.C. Sircar in his article published in 1964.[95] His contention is that Rāmagupta cannot be grouped together with the rulers Harigupta and Indragupta mentioned by Sircar. No coins of Rāmagupta are so far known from Kauśāmbī, Ahichchhatra or Pāṭaliputra (the provenances of the Coins of Harigupta and Indragupta). Nor do we know of any coins of these two kings from Eastern Malwa, in which region several hundred of the monetary issues of Rāmagupta have been discovered. The palaeography of the legend on the coins of Rāmagupta is distinctly earlier than that obtained on the other two rulers' coins and this point does militate against the theory of Sircar in assigning a late date to Rāmagupta.

The only common feature in the coinage of Rāmagupta and Indragupta is the occurrence of the Garuḍa bird. But whereas we find the Garuḍadhvaja, the human form of Garuḍa (sometimes three-headed), the border legend in pure Sanskrit and the lion figure on many coins of Rāmagupta these features are conspicuous by their absence on the coins of Harigupta and Indragupta. There is a marked difference in the fabric of the coins also.

Harigupta and Indragupta are known to us from a couple of coins only. This presents a striking contrast when we notice that quite a large number of coins of Rāmagupta are known to us from Eastern Malwa.

Bajpai quotes Sircar to state that copper was a fairly costly metal in early India and states that this again proves that Rāmagupta must have enjoyed a much better status than that of Harigupta or Indragupta. Moreover, these two later rulers are nowhere mentioned as belonging to the royal house of Samudragupta, whereas there is ample testimony in favour of Rāmagupta referring to him as the elder brother of Chandragupta-II. It, therefore, does not seem to be at all plausible to bracket together Rāmagupta with the two rulers referred to above, who are not known to have been scions of the Imperial Gupta dynasty.

Bajpai has tried to show in his other articles[96] that Rāmagupta may be said to be first among the Gupta rulers to have issued Copper currency. His numerous coins, bearing certain characteristic features of the Imperial Gupta dynasty, should be borne in mind when we study the coinage of the Imperial Guptas, especially the copper and silver coins.

The monetary issues of Rāmagupta were coined in eastern Malwa especially in the mints of Eran and Vidiśā. Prior to the conquest of this area by Samudragupta, the mints at Eran and Vidiśā used to coin the money of the local Nāga rulers. We know from the Allahabad Praśasti of Samudragupta that he defeated Gaṇapatināga and some other Nāga Chiefs. Samudragupta, however, did not issue any copper coins for the Malwa region. It was left to his son Rāmagupta, who was evidently in charge of the Eastern Malwa, to issue such coins for circulation in that region. His coins were prepared in mints of the region previously occupied by the Nāgas which is borne out by numerous Nāga affinities found on the coins of Rāmagupta. The same procedure was followed by Chandragupta-II and Kumāragupta-I also, whose coins of Malwa type are too well-known.

It seems Allan and following him several other scholars have mistaken the figure of Garuḍa for an altar, a fire-altar or a Tulsi Vṛindāban. It may be pointed out in this context that different varieties of the bird Garuḍa found on the coins of Rāmagupta are met with on the silver and copper coins of the later Gupta rulers from Chandragupta-II onwards. On some coins of these kings, we also notice the figure of Garuḍa holding snakes, a feature, so common on several coins of Rāmagupta.

The depiction of coins of Rāmagupta reminds one of the multi-headed figures of Śiva, Kārttikeya etc. found on the Kushāṇa and certain tribal coins of India as also on some early sculptures.

This kind of three-headed Garuḍa can be seen on some silver coins of Kumāragupta-I, illustrated by Allan. Some coins of Chandragupta-II represent Garuḍa in the form noticed on Type IV of the coins of Rāmagupta. On the coins of Harigupta, Jayagupta etc., the imitation of the Garuḍa figure in seated posture is quite distinct, although in other details these coins differ from the coins of Rāmagupta.

Thus we are left with no doubt on the basis of numismatic evidence as also their palaeography that Rāmagupta flourished earlier to other rulers Harigupta, and Indragupta mentioned in the context.

The Place of Occurrence of the Incident of the Rāmagupta Episode

As discussed above the scholars have allocated different sites for the occurrence of the incident where Rāmagupta was surrounded by the Śaka Chief.

1. K.D. Bajpai puts it at Eran and Bhilsa region itself.
2. K.P.Jayaswal has identified it in the Jalandhar Doab, an old village named Alwal situated in the Himalayan region of Sabathu or a place adjacent to it.
3. D.R. Bhandarkar has placed it in the Kārttikeyapura, modern Baijanath in Almora district of Uttar Pradesh.
4. V.V. Mirashi has located it either near the Jalandhar Doab or to the West of modern Jalalabad.

Altekar has pointed out that probably Rāmagupta had gone to suppress the rising Śakas (Kushāṇa King) in the eastern Punjab, we would like to locate it in Alipur fort in Kangra district [97] which formed a part of Jalandhar division in early period, in case we fully trust the reading 'Alipura' as pointed out by Mr. R. Saraswati, instead of 'Aripura' in the Harṣacharita. Or we have to accept Bajpai's identification in case we give more credence to the later contemporary evidence in the time of Rāmagupta's successor Chandragupta-II who had come for a victorious campaign in the region which we shall discuss below in the identification of the Śaka-Lord, and also finalise our answer to the question of the place of incident as both these questions are interlinked.

Identification of the Saka-pati

Sten Konow [98] while commenting on the various kinds of homage paid to Samudragupta by *Daivaputraśāhiśāhanuśāhi* and the *Śaka-Muruṇḍas* interprets the word 'Kanyādānā' 'asking for the bestowal of a bride'. "What the 'Śaka-Muruṇḍa' asked for was if this analysis is right the hand of the Gupta princess and we would have a distinct indication of a state of things similar to that which is pre-supposed in the *Devīchandraguptam*".[99] Further Konow states that we cannot, of course, tell whether the Western Kṣatrapas asked for a Gupta princess more than once, or whether they were able to put pressure on the Guptas in this connection. But their downfall cannot have been far removed in time from the events mentioned in Hariṣeṇa's praśasti. According to Rapson, the latest date of Svāmī Rudrasiṁha-III, the last Mahākṣatrapa, is A.D. 388 or a little later and an Udayagiri inscription of Chandragupta-II is dated A.D. 401. There is no *a priori* objection to the indication contained in the *Devīchandraguptam* and confirmed by Bāṇa that the annihilation of the dynasty had some connection with the attempt to secure Gupta princess.[100]

Mirashi contradicts the above view stating that there is no evidence to suppose that war against the Satraps of Ujjayinī (Western Kṣatrapas) was the first campaign in which Chandragupta-II was engaged. We know that the Satraps continued to rule in Malwa till A.D. 388 at least, i.e. for more than 10 years after Chandragupta's accession. During this period he may have been occupied in other places, eg. in the Punjab and Kashmir, subjugating the Kushāṇas. It is noteworthy that the Minister Āmrakārdava, who made a gift to the monastery at Sāñchī , describes himself as 'Anekasamrāvāpatayaśaspatākaḥ' "who has acquired banners of victory and fame in many battles." [101] The many battles in which he had won renown were not evidently all of them fought in Malwa.[102]

It is more likely on the basis we reject the view of Bajpai on reconsideration of the whole matter and give more acceptance to our view of the scene of occurrence of the incident in the Kangra region of the Jalandhar Doaba. Accordingly we identify the Śaka ruler with the Kushāṇa ruler of the West, the word Śaka being used as a generic term for the foreign tribes even other than the Śakas, the view earlier expressed by R.D. Banerji, R.N. Dandekar and Dashratha Sharma.[103]

HISTORICITY OF RĀMAGUPTA

While reviewing the evidence in this context we shall put forth all the contradictory opinions doubting the historicity of Rāmagupta and reconsider them on their own merit before we conclude the issue.

Arguments against the historicity of Rāmagupta

Following objections have been raised by the scholars at one time or other to doubt the historicity of Rāmagupta.

1. The story in itself, even apart from such supernatural elements as *Vetāla-sādhanā* (Goblin worship) is not only unusual, almost bordering on the romantic and incredible, but is also *prima facie* so much opposed to our knowledge of facts and belief in practices of the period, that nothing but the strongest evidence should induce us to place any credence in it.[104]

2. In the epigraphs Chandragupta-II is described as '*tatparigṛahīta*' implying that he was selected by his father for the throne, and, therefore, the question of Rāmagupta becoming king after Samudragupta cannot arise. Even if it be assumed that Rāmagupta seized the throne by force, in that case, there would have been a civil war before Chandragupta would have meekly submitted to the act of usurpation, and in the case of such an event, certainly Rāmagupta who was so weak and imbecile as to surrender his own wife to an invader, would have been defeated.[105]

3. Bhandarkar has identified Rāmagupta of the *Devīchandraguptam* with Kāchagupta stating the name Rāmagupta is a misreading in the drama for Kāchagupta and even attributed the Kācha Coins to Rāmagupta.[106] While Jayaswal has concluded that Kācha and Rāma are two different names of the same king.[107]

4. The details of the incident, as given in the *Harṣacharita*, the *Kāvya-Mīmāṁsā* of Rājaśekhara, and the Rāshṭrakūṭa records differ from one another regarding the place of the incident, the identification of the "other's wife" and also the victim of Chandragupta's strategem. In fact, the earliet record on the subject is the *Harṣacharita* which gives us two facts, that the Śaka King was possessed of lust for another's wife, and that Chandragupta killed him in the guise of the female. Later on the following details were added to the story: (a) that this 'otherman' was Rāma or 'Sārma or Sen Gupta', (b) that Dhruvadevī who is described in the epigraphic records as the wife of Chandragupta-II, was originally the wife of this Rāmagupta, and, (c) that Chandragupta killed Rāmagupta and married his widowed queen.[108]

5. While the murder of a brother for the sake of kingdom is by no means unusual, the marriage of his brother's widow by the rebel and the regicide clashes with our cherished notions about morality and social custom prevalent in those times.[109] It is difficult for instance to believe that the inheritor of the mighty empire of Samudragupta could be so decisively defeated by a Śaka King that he had no means of saving his army or kingdom save by consenting to an act, which would be regarded as the most ignominious by any king in any age, or country, not to speak of

the mighty emperor of the golden age of India, who had the blood of Samudragupta running in his veins. A story preserved in the *Mujmul-ut-Tawarikh* has been seized upon as a suitable explanation of this enigma. It has been suggested that the king and his retinue were beseiged in a hill-fort, and his army having been defeated by the Śaka King he was at the complete mercy of the latter. But even such a situation can hardly condone the utter infamy and disgrace involved in the proposed means of escape from it. It should be remembered that we cannot explain it merely as a caprice a monarch, who might be imbecile or insane, for we are asked to believe that his action had the full approval of the people, even if it was not instigated by them. The code of honour in the golden age of India must be assumed to have been very different from the later and more degenerate days, when, in similar peril, women, preferring death to dishonour, are known to have thrown them-selves in blazing fire, and men rushed out and fought till they avenged the insult with the last drop of their blood.[110]

6. No gold coins of Ramāgupta have been found in the hoards of Gupta coins discovered so far.

7. There is no mention of a ruler Rāmagupta in the genealogy of the Gupta rulers.

8. There is no inscription referring to Rāmagupta of the Gupta dynasty.[111]

9. The copper coins discovered from Eastern Malwa bearing the name Rāmagupta belonged to a local ruler of Malwa region like some other rulers of region namely Harigupta and Indragupta who issued coins in imitation of the Imperial Gupta money.

Historicity of Rāmagupta

Even the scholars who doubted the historicity of Rāmagupta agreed that the story connected with Rāmagupta cannot be dismissed off-hand, as altogether a figment of imagination, we must not rush to the other extreme of accepting, *in toto*, plots of drama and popular tales as reliable facts.[112] To which we may add that we should also avoid indulgence in imposing certain alterations or to modifications in the facts of the story from its sources until and unless we find some solid evidence to the contrary to tilt the balance in favour of our theory.

We have already discussed the evidence in favour of establishing the identity of King Rāmagupta who no more remains legendary figure. For the sake of convenience, we shall review our evidence in relation to the arguments raised above by many scholars doubting the historicity of Rāmagupta.

The arguments of the scholars collated together against the historicity of Rāmagupta, generally deal with the following issues:

(a) Changes and variations in the contents of the story or even the contents of the story;
(b) Lack of epigraphic evidence;
(c) Lack of numismatic evidence.

1. The inclusion of the super-natural element such as *Vetāla-Sādhanā* in the story is not queer as far as the name Chandragupta Vikramāditya is concerned. His name is traditionally associated with *Vetāla* carried on even to-day in the Television serials of "Vikrama and Vetāla" being reproduced now-a-days for which we have an earlier tradition stated in the *Kathāsaritsāgara* and the *Vetāla-Pañchaviṁśati.*

2. The mention of the expression 'Tatparigṛihīta' (accepted by him i.e. Samudragupta or nominated as a successor to the throne by him) does not come on the way of the historicity of Rāmagupta. It is well-known that Samudragupta had many sons as stated in his Eran Inscription. It may not be doubted that Chandragupta had no elder brother. Moreover, in the Gupta period law of primogeniture was not strictly followed. So there does not seem to be any law of succession among the Guptas. There is a probability of war of succession after the death of Samudra-gupta in view of Rāmagupta episode as we find in the case of Govinda-gupta, the seals from Vaiśāli mention the name of Govindagupta and not of Kumāragupta, in case of Skandagupta (with Purugupta?) and in the 6th century we observe a competition among Narasiṁhagupta Bālāditya, Bhānu-gupta and Vainyagupta for gaining the empire. Such

rival claims find an allusion in verse 4 of the Allahabad Pillar Inscrip-tion of Samudragupta itself. Moreover, the expression *Tatparigṛihīta* in respect of Chandragupta-II only indicates that his succession was approved by his father. The poet can write it under the influence of his patron to show his claim to the throne.

3. That D.R. Bhandarkar gave his view that the ruler who intervened between Samudragupta and Chandragupta-II was Kācha and doubted the very basis of Rāmagupta, deeming 'Rāma o' to be a copyist's error in place of 'Kācha o' in the *Devīchandraguptam* and that Kācha type of coins prove the existence of one Kāchagupta. Jayaswal gave his view that Kācha and Rāma are the different names of the same person. Both these theories are not agreeable to us.

   R.S. Tripathi has stated that the attribution of the Kācha coins to Rāmagupta by Bhandarkar is not at all convincing.[113] Moreover the Kācha coins bear on them the legend *Sarvarājochchhetā* "an extermi-nator of all kings", an epithet unthinkable for a weakling Rāmagupta; even hyperbole has its limits.[114] Jayaswal's theory that Kāchagupta and Rāmagupta are one and the same person cannot be relied upon as now on the basis of copper coins in case of Rāmagupta and gold coins of Kāchagupta it is proved beyond doubt that the two are seperate entities. How will Rāmagupta not like to inscribe his name on the gold coins if he were the author and why will he change his name to Kāchagupta on them? We propose to discuss the whole issue in a separate chapter.

4. Chattopadhyaya while stating the *Harṣacharita* to be the earliest evidence for Rāmagupta has forgotten the evidence of the *Devī-chandraguptam* certainly on earlier work which formed the original basis of the episode and must not have remained out of reach to the scholars for a few centuries at least where from they seem to have quoted in their works on Rhetorics and the same travelled in the form of a Sanskrit work or in the form of story carried by the sail-ors and traders to Arabia as discussed by us earlier. These changes are the result of the original plot of the *Devīchandraguptam* which formed the basis for future writers. About the certain discrepancies pointed out about the names, we may state they do some time crop in and may be unavoidable in certain case with the passage of time

in various manuscripts and editions which are the order of the day even now.

5. Regarding the notions of morality in the Gupta period about widow marriage, Majumdar himself seems to have reconciled with the issue by stating that it has been pointed out that neither Śāstric injunctions nor social practices prohibited a marriage between a widow and the younger brother of her husband.[115] But it seems the posterity did not approve of this action of Chandragupta as is clear in the verses of the Rāshṭra- kūṭa plates. He, however, has raised his objection more seriously and vociferously about the explanation provided in the *Mujmal-ut-Tawarikh* of Rāmagupta's agreeing to surrender his queen to settle peace with the victor Saka foe. To this we may answer that due to dictates of time and circumstances anything is possible in human affairs to which exceptions are very few. We may quote from P.L. Gupta:

> "If Rāmagupta agreed to send his wife to Śaka King, he was neither imbecile nor insane, rather he was acting in accordance to the code of polity and there was nothing for the people to disapprove. The Rajputs and their women are well-known for their *jauhara*; yet we find amongst the very Rajputs, Ratnasena agreeing to send his wife Padminī to Allauddin Khilji in very much similar circumstances as those of Rāmagupta".[116]

6. So far, for the absence of gold coins of Rāmagupta in the Gupta hoards, we may state that he had a very short reign and for political reasons he could not have been in the capital Pāṭaliputra to issue such coins. It is even more likely that the throne of Pāṭaliputra, might have been occupied by the usurper Kācha, his uncle or brother and Rāmagupta's rule might have remained confined to Eastern Malwa of which he was probably the Viceroy before the death of Samudragupta.

7. About the absence of Rāmagupta's name in genealogies of the Gupta records, we may mention that the official records give the genealogy and not the succession of kings, "the omission of the name of Rāmagupta need casue no surprise".[117]

8. Recently found three incriptions on the Jain images discovered near Vidiśā where the name of Rāmagupta is inscribed as "Mahārājādhirāja Śrī Rāmagupta" disapprove the theory of his being a local chief or subordinate ruler like Harigupta and Indragupta. The bearing of the title 'Mahārājādhirāja' leaves us with no doubt that Rāmagupta was a paramount sovereign belonging to the Imperial Gupta family if we do not discard the literary evidence altogether.

9. We have already elaborately discussed this point shown on the basis of palaeography and numismatic evidence that Rāmagupta should not be bracketed with the other two rulers of Harigupta and Indragupta who may be local chiefs and belonged to a later period.

Thus we may conclude that Rāmagupta was a historical figure connected with the family of the Imperial Guptas. Due to some upheavals in the empire after the death of Samudragupta his authority may have remained confined to Eastern Malwa of which he seems to have been the Viceroy during the life-time of his father Samudragupta.

However, some scholars have tried to give a new turn to the story of Rāmagupta as described in the drama *Devīchandraguptam* of Viśākhadatta by way of their conjectural thinking stating that Viśākhadatta wrote this drama at the instance of his pattern Chandragupta-II to whitewash his misdeeds of marrying his brother's wife which could be otherwise sinful by showing Rāmagupta as a weak, wicked and cowardly king who snatched away his brother's married wife and later agreed to ignominious peace by consenting to surrender his queen to the Śaka foe and that "it was the misdeeds of the latter that forced Chandragupta to caputre power in his own hands".[118]

Further another conjecture that Rāmagupta was killed in the course of war [119] which disparages with the plots of the *Devīchandraguptam* and the *Mujmal-ut-Tawarikh*. We find no basis for such changes or modification until and unless we get any documentary evidence to the contrary.

Notes And References

1. R.K. Mookerji, *GE.*, p.44.

2. Fleet,*CII*. III, No.4; Mathurā Stone Inscription of Chandragupta-II, p.27, LL. 8-10.
.....Mahārājādhirāja - Śrī Samudraguptasya putrena tat-parigṛhī-tena "by the son of *Mahārājādhirāja* Samudragupta - accepted by him..." See also *Ibid*, No.1, p.12 fn.1 which means "accepted (as his favourite son and chosen successor) by him (Samudragupta)".
3. *Ibid*.., No.12, p.50, L.19.
4. *Ibid*.., No.13, p.53, L.4.
5. A.S. Altekar, *GGCBH*., p.xxii.
6. Fleet, *CII*. III, No.2 Eran Stone Pillar Inscription of Samudragupta, p.20, L.19.
7. *Ibid*., L.25.
8. *Ibid*., p.20
9. J.A. Tome, *CCI*. II, 193, pp.193-218. as quoted in Panthari *BKSY*., p.159.
10. R.S. Tripathi, *HAI*., p.249.
11. H. Heras, "Kācha Gupta and Rāma Gupta", *JBRS*, XXXIV, 1945, pp.19.ff.
12. *Ibid*.
13. Some important articles and works deserve mention here: S.Levi (*JA.CCI*.II, pp.201 ff); R.Saraswati, (*IA*., L.11, pp.181 ff); A.S. Altekar (*JBORS*., XIV, 223 ff); R.D. Banerji, (*AIG*., pp.26 ff); D.R. Bhandarkar, (Malaviya Commemoration Volume, pp.189 ff); K.P. Jayaswal, (*JBORS*., XVIII, 17ff); Winternitz (Aiyangar Commemoration Volume, pp.359 ff); Sten Konow, (*JBORS*., 1937, Vol. XXIII, Pt.IV, 444 ff); V.V. Mirashi, (*Studies in Indology*, Vol.I, 1960, *IHQ*., X. 48; *IA*., LXII, pp. 201 ff; N. Dasgupta *IHQ*. XX, pp.351 ff; *IC*. IV, 216); V.Raghavan, (Journal of Benaras Hindu University II, pp.23-54, 307); H.C. Raychaudhuri (*PHAI*., 7th Ed. p.488 note 2); S.Bhattacharya, (*JIH*., XXXXV, pp. 727 ff), Rev. H. Heras, (*JBRS*., XXXIV, 1948. pp.19ff; *ABORI*., XII, pp.455 ff); K.C. Ojha, (*JBRS*., XXXVII, 1951); A.S. Altekar, (*GGCBH*., 1954, pp. XXII-XXVI; *JBORS*., XIV, pp.223ff; *JBORS*., XV, pp.141 ff.; K.D. Bajpai, (*IHQ*., XXXVIII, No.1, March 1962; Also see *JNSI*., XVIII, VIII, XII, IX, XIII, *JIH*., XLII. 1964, pp.389 ff); R.C. Majumdar, (*VGA*., pp.161-165); J. Agarwal, (*VIJ*., Vol. IV Pt.I, March 1966. pp.58 ff); D.C. Sircar (*JIH*., Vol. XL, 1962); A. Chattopadhyaya (*JOI*., June 1969, XVIII); S. Chattopadhyaya,(*EHNI*., 1968, ED. pp.199 ff); B.S. Sitholi, (*JNSI*., XII, pp. 38ff); S.R. Goyal, (*HIG*.,

pp.223-237); Goyal, PBRI, Part III, 1969, pp.158-169); H.C. Raychaudhuri, (*IHQ.*, XXXV, pp.333 ff); A.H. Dani, (*JNSI.*, XXVI, pp.11.ff); A.K. Narain, (*JNSI.*, XII, pp.107ff); S.K. Maity, (*IGT.*, pp.34-37); R.S. Tripathi (*HAI.*, pp.248-49); P.L. Gupta (*IG.*, pp.135-141, 148, 149, 150, 290-296;*NSI.* XII, pp.103 ff); B.P. Panthari (*BKS.*, pp.151-170); U.N. Roy (*GRY.*, pp.175-209); V. Upadhyaya (*GA.*, pp.267-274); T.R. Sharma (*JOI.*, XXXI, No.1 Sept. 1981, pp.1-6).

14. Sten Konow,"Devīchandraguptam and its Author", *JBORS* XXIII, pp.444 ff.
15. K.D. Bajpai, *JNSI.*, XXIII, pp.340ff.
16. P.L. Gupta, *IG.*, p.135.
17. T.R. Sharma, *JOI.*, XXXI, No.1, Septemeber 1981, "Date of Viśākhadatta", pp.1-6.
18. *Ibid.*
19. U.N. Roy, *GRY.*, p.183: R.N. Dandekar assigns the date A.D. 1713 for Sankarārya.
20. R.P. Sharma, *Rājaśekhara Aura Usakā Yuga,* Patna, 1977, p.50.
21. Baldeva Upadhyaya, *Sanskrit Sāhitya Kā Itihāsa,* Kashi, 1958, p.642.
22. *Ibid.*, p.640; Kane puts it between A.D. 980-1020 as quoted by U.N. Roy, *Op.cit.*, p.184.
23. Nagendra (Ed.) *Kāvya Prakāsha* (Commentary of Acharya Vishweshwara), Varanasi, 1960, *See Introduction*, pp.76-77.
24. U.N. Roy, *GRY.* p.176; Rāmachandra Guṇachandra were both disciples of Hemachandra who was contemporary of Chālukya King Kumārapāla (A.D. 1143-1163).
25. P.L. Gupta, *IG.*, p.150.
26. M.L. Sharma, Bhārata Kā Itihāsa (Hindi Translation of *History of India*, Vol.I by Elliot and Dowson,) Agra, 1973, p.69.
27. U.N. Roy, *Op.cit.*, p.189.
28. T.R. Sharma, *JOI.* XXXI, pp.1-6.
29. S.R. Goyal, *PBRI.*, pp.165-168.
30. P.L. Gupta, *IG.*, pp.135-140.
31. *Ibid.*, p.137 Italics ours.
32. *Ibid.*, p.138. "Strī Vesha nihunataḥ Chandraguptaḥ Śatroḥ Skandāvaram Alipuram Śakapativadhāyāgamat".
33. *Ibid.*, p.139.
34. *Ibid.*, p.156.

35. *PHAI.*, pp.488-89 fn.2.
36. T.R. Sharma, *JOI.*, XXXI, p.6 cf; P.L. Gupta, *IG.*, p.141.
37. *Ibid.*
38. P.L. Gupta, *Op.cit.*, p.141. Italics ours. We differ with P.L. Gupta in rendering the translation on these points.
39. T.R. Sharma, *JOI.*,XXXI, p.3.
40. Sten Konow, *JBORS.* XXIII, Pt.IV, pp.444-445.
41. *Ibid.*, p.445.
42. G.S. Rai, *Kāvya-Mīmāṁsā*, Varanasi, 1964, p.124.
43. P.L. Gupta, *IG.*, p.149. We have made some amendment in the translation given by P.L. Gupta and have translated the word "Dattvā" as "Having committed to give". If already given then question of Chandragupta in the guise of Dhruvadevī (Dhruvsvāminī) in the enemy's camp will not arise.
44. T.R. Sharma, *PGNGI*, p.30.
45. *Ibid.*
46. Sten Konow, JBORS., XXIII, p.445.
47. V.V. Mirashi, *Studies in Indology,* Vol.I, 1960.
48. *Ibid.*
49. *Ibid.*
50. Sten Konow, *JBORS.*, p.445.
51. V.V. Mirashi, *Studies in Indology,* 1960, Vol.I.
52. A.S. Altekar, *VGA.*, pp.22-23.
53. *Ibid.*, p.23. See the plates in Num.Suppl.XLVII, Plates I and II, Nos. 15-21.
54. V.V. Mirashi, *Studies in Indology,* 1960, Vol.I.
55. U.N. Roy, *GRY.*, p.182 cf.; P.L. Gupta, *Op.cit.*, p.136.
56. *Ibid.*, cf. P.L. Gupta, Op.cit., pp.138-139.
57. P.L. Gupta, *Op.cit.*, pp.138-139.
58. U.N. Roy, *Op.cit.*, p. 184.
59. *Ibid.*
60. S.R. Goyal, *PBRI.*, p.167.
61. *Ibid.*, p.168; A part of the verse "Ramyām ....Tāmyati" is quoted above.
62. P.L. Gupta, *Op.cit.*, p.150.
63. H.M. Elliot and John Dowson, *History of India* (as told by its own Historians), Allahabad, Vol.I, p.100.
64. *Ibid.*, pp.110-112.

65. Aparna Chattopadhyaya, "*Mujmal-ut-Tawarikh* and the Rāmagupta problem of Gupta History", *JOI*., in June 1969, Vol.XVIII.
66. We have our reservations on this point as the incident occurred at a different place viz. Kulūta, Jalalabad or elsewhere.
67. Aparna Chattopadhyaya, *JOI.,* 1969, XVIII.
68. U.N. Roy, *GRY*., p.189.
69. *Ibid.*
70. Edwin T. Atkinson, *The Himalayan Districts of North Western Provinces of India,* pp.411-412; Cunningham, *ASRI.,* p.139.
71. Fleet, *CII*. III, No.I, p.8, L.25.
72. *Ibid.,* No.4, p.26, L.4.
73. The increase in figures, otherwise may not be far from truth as pointed out by us earlier in view of the fact that the earlier figures are before the performance of Aśvamedha sacrifice (Sacrifices).
74. *IHQ*., Vol.X, 1934.
75. P.L. Gupta, *IG*., pp.135-136.
76. V.V. Mirashi, *Studies in Indology*, Vol.I, 1960.
77. T.R. Sharma; *PGNGI*, p.22.
78. *Ibid*., p.30 3(a).
79. *Ibid*., p.30 3(b).
80. V. Upadhyaya, *GA*., p.128.
81. P.L. Gupta, *IG*., p.295.
82. Cf. U.N. Roy, *GRY*., pp.206-208.
83. Cf. *Ibid*., p.206.
84. *Ibid.*
85. *Ibid.*
86. *JNSI*., Vol.X, 1950.
87. *JNSI*., Vol.XII, p.106.
88. A.S. Altekar, *GGCBH*., p.xxv.
89. *JNSI*., Vol. XII, p.107.
90. *IHQ*., Vol.XXXVIII, No.I, March 1962.
91. K.D. Bajpai, *IHQ*., Vol. XXXVIII, No.I, March 1962.
92. D.C. Sircar, *JIH*., Vol.XL, 1962.
93. *Ibid.*
94. *Ibid.*
95. K.D. Bajpai, "Identification of Rāmagupta" in *JIH*., Vol. XLII. 1964, pp.389 ff.
96. K.D. Bajpai, *JNSI*. XXIII, XVIII, VIII, XII, XI, XIII.
97. IA., 1923, p.183.

98. Sten Konow, *JBORS*, 1937, Vol.XXIII, p.449.
99. *RC.*, pp.CXLIX ff.
100. Sten Konow, *Op.cit.*, p.449.
101. Fleet. *CII*. III., No.5, p.32.
102. V.V. Mirashi, "Further light on Rāmagupta" in *Studies in Indology.*, Vol.I, 1960.
103. U.N. Roy, *GRY.*, p.196.
104. R.C. Majumdar, *VGA.*, p.163.
105. S. Chattopadhyaya, *EHNI.*, pp.200-201.
106. D.R. Bhandarkar, Malviyaji Commemoration Volume, 1932, pp. 204 206; R.C. Majumdar, *VGA.*, p.163.
107. S. Chattopadhyaya, *Op.cit.*, p.201; Jayaswal, *JBORS.*, XVIII, pp.17ff
108. S. Chattopadhyaya, *Op.cit.*, p.201; R.C. Majumdar, *VGA.*, p.163.
109. R.C. Majumdar, *VGA.*, p.163.
110. Ibid., p.164.
111. S. Chattopadhyaya, *EHNI.*, p.201.
112. R.C. Majumdar, *VGA.*, p.164.
113. R.S. Tripathi, *HAI.*, p.249 f.n.3.
114. S. Chattopadhyaya, *EHNI.*, p.201.
115. R.C. Majumdar, *VGA.*, p.163.
116. P.L. Gupta, *IG.*, p.292.
117. R.C. Majumdar, *VGA.*, p.163.
118. S.R. Goyal, *HIG.*, p.231.
119. *Ibid.*, p.237.

7

The Place of Kāchagupta in the Gupta History

Earlier while discussing the problem of Samudragupta's accession to throne we have pointed out that Kāchagupta had a different identity than Samudragupta but did not fix his place in the Gupta history. He may be an elder brother of Samudragupta who revolted against his succession but failed to achieve his objectives as indicated by us.[1] He probably succeeded in his venture second time when after the death of Samudragupta he seized the throne of Pāṭaliputra while Rāmagupta who was a viceroy at Eran during the life-time of Samudragupta (Eran being his Bhoga Nagara must be a place of strategic importance), might have declared himself a king of Eastern Malwa. Chandragupta-II who has been described as a prince (*Kumāra* Chandragupta) in the *Devīchandraguptam* even though a 'tatparigrihīta' (accepted by his father) as described in the inscription during the period of this turmoil had to accept the service of his elder brother Rāmagupta. He seized the opportunity and became the King of Eastern Malwa after killing his cowardly brother Rāmagupta who had accepted ignominious peace at the hands of the Śaka foe and also married the queen Dhruvadevī whose honour he had saved by killing the Śaka Chief. Later on it seems that Kāchagupta who might have been age-old either died to clear his way for throne of Magadha or Chandragupta took his opportunity to capture the throne of Magadha where he seems to have shifted his capital to Pāṭaliputra and appointed some Sanakanika as his chief which is evident by Udayagiri Cave Inscription of the year G.E. 82 (A.D. 401).

Our contention is that it is second time during his rule after the death of Samudragupta that Kāchagupta seems to have issued his gold coins 16 of them found in Bayana hoard which resemble the Aśvamedha type, (known to be the latest of Samudragupta's coins issued after the inscription at Allahabad Pillar) and the tiger-slayer types of Samudragupta.[2]

It may be remarked that Kāchagupta has been such a confusing personality in the Gupta history that Smith first advocated the identity of Kācha and Samudragupta in 1889, later in 1893 he abandoned this view and accepted the opinion of Rapson that the two were different rulers, but again he reverted back to his first opinion in 1902.[3] Not only that even Altekar himself got confused to believe that Kachā might also have been a king of another dynasty, who ousted Samudragupta for a while; about which opinion he was not firm.[4] Later he believed Kācha to be identical with Rāmagupta, the eldest son of Samudragupta, according to the literary tradition.[5] Again he changed his opinion after the discovery of the copper coins of Rāmagupta on the logic that one and the same ruler could not have issued the coins with the name of Rāmagupta on copper coins and Kācha on gold coins, even if he had these two names.[6] Again he believed him to be a local ruler of Malwa and not the Gupta emperor about which also at the same time he expressed his doubts.[7]

We may sum up the main opinions of scholars about Kāchagupta as follows:

(1) Smith,[8] Fleet,[9] Allan [10] and Sircar [11] believe Kācha to be identical with Samudragupta.
(2) Some scholars believed Kācha as an abbreviation of Ghaṭotkacha and identified the king with the father of Chandragupta-I.
(3) Bhandarkar believed Kācha to be identical with Rāmagupta.[12]
(4) According to Banerji, Kācha coins represent the memorial coins or medals issued in his memory by Samudragupta for his martyrdom in a war against the Kushāṇas during the time of Chandragupta-I. He further stated: "If the coins bearing the name of Kācha are real coins and not medals then Kācha was most probably the elder brother of Samudragupta whose rule was very short."[13]
(5) Mookerji [14] and B.S. Shitholey believed that Kācha, the issuer of the coins, was an usurper who did not belong to the Gupta dynasty.[14a]
(6) Heras,[15] Dandekar,[16] Goyal [17] and Gupta [18] believed him to be a rival prince most probably his brother who led a rebellion against Samu-

dragupta in which he succeeded and ruled for some time in which he struck his gold coins till he was overthrown by Samudragupta.

Goyal tried to interpret the whole problem in context of verse 4 of the Allahabad Pillar Inscription that Samudragupta was the leader of the orthodox Brahmanical faction of the courtiers while Kācha became the leader of the liberal faction, the view which we have already rejected during our discussion on the problem of Samudragupta's accession. In the present context, however, the concluding remarks of P.L. Gupta merit attention in view of further discussion on the problem, "Nothing is known about this intruder or the rival brother of Samudragupta. Quite possibly, he was attracted by Samudragupta's extraordinary deeds and submitted to him. It is also not likely that he kept on fighting against Samudragupta and was killed".[19] We shall now take up all the points raised above for scrutiny and review them in the light of evidence at our disposal before we conclude.

(1) This theory no longer finds support from the scholars. We cannot impose without any strong evidence, the name Kācha on Samudragupta. There is no instance of the Gupta kings having used their different names on the coins, even if they had the tradition of having more than one name. Instead, of name, they do sometime make use of their different 'biruds' signifying them.

(2) The second point has been refuted both by Altekar [20] and Fleet [21] on linguistic grounds as well. The name of the king is Kācha and not Kacha, and Kācha can hardly be regarded as an abbreviation of Ghaṭotkacha. Moreover, Ghaṭotkacha was a mere feudatory and could hardly have issued any coins.

(3) This point we have already refuted as Altekar also withdrew his endorsement of the theory of Bhandarkar that Kācha and Rāmagupta are one and the same person. After noticing the finds of copper coins of Kācha it is impossible to believe that he will change his name to Rāmagupta at the height of his power while issuing the gold coins. We subscribe to the same view.

(4) There is no evidence to show that Samudragupta had any brother who was killed in any war against the Kushāṇas. Moreover, commemorative coins or medals are unknown to Hindu tradition.[22]

(5) We need not postulate any such hypothesis of an outside usurper occupying the throne in view of the definite evidence of literary tradi-

tion of the usurpation of the throne by one of his brothers.[23] Moreover, the official records give the genealogy and not the succession of kings; hence the omission of his name causes no surprise.

(6) We agree with the theory that Kācha was a rival brother of Samudragupta who might have contested for the throne even if Samudragupta had been named as his successor by Chandragupta-I openly in the full durbar and who might have at last submitted to Samudragupta. But we cannot rule out that he died and did not raise his head again when Samudragupta died around A.D. 370 or so. Rather there is more possibility of this act in view of his coins resembling the late variety of Aśvamedha type of coins.

Maity [24] has classified the coins of early period taking the pure gold content into consideration and divided them into following five categories:

(1) Vasudeva coins - 118 grains.
(2) The Chandragupta-Kumāradevī coins - 109 grains.
(3) The Archer, Tiger and Lyrist coins of Samudragupta - 104-05 grains.
(4) The standard, Battle-Axe, Aśvamedha coins of Samudragupta, Kācha and Archer coins of Chandragupta-II - 99-99 grains.
(5) The Archer coins of Kumāragupta-I - 92 grains.[25]

Further he states that they should be normally taken to be issued in the same chronological order.

Thus Kācha coins come in the same bracket of the standard, Battle-Axe and Aśvamedha coins of Samudragupta which are also closer to Archer coins of Chandragupta-II.

Altekar[26] has classified the 16 Kācha coins found in the Bayana hoard into two categories. Fifteen of these belong to variety A, but the sixteenth constitutes a new variety called variety-B. While the former variety is already well-known to us from several specimens, the latter is disclosed to us by the present hoard for the first time.

The obverse of the coins of variety-A shows the king standing to left in a posture remarkably similar to that of Samudragupta on his standard type. Like Samudragupta Kācha also is shown offering oblations on altar, but in variety-A the Garuḍa standard is missing in his front and he is seen holding a *Chakradhvaja* (wheel standard) in his left hand. Under the left arm is engraved the issuer's name Kācha. The circular legend reads as follows:-

"Kācho Gāmavajitya Karmabhiruttamairdivam Jayati". "Having conquered the earth, Kācha (now) wins the heaven by his excellent works".

The reverse of the coins of Kācha closely resembles that of the Aśvamedha and the Tiger-slayer types of Samudragupta. The reverse legend is "Sarvarājochchhettā". "The exterminator of all kings".

Variety-B, altogether a new variety, bears general resemblance to Variety-A, but it shows slightly greater Indianisation. The king on the obverse wears a *dhoti* instead of the Kushāṇa trousers. The most striving feature of the variety is the presence of *Garuḍadhvaja* on the obverse, which is absent in Variety-A.

Then if Kācha was neither the name of Samudragupta nor of Rāmagupta, it settles the issue to the extent that he was a separate entity most probably a brother of Samudragupta. But it raises the question where to place him and on what grounds? An indepth study of the numismatic evidence can only help the matter.

Sarvarājochchettā, the legend on the reverse of the coins of Kācha, no doubt appears as the exclusive epithet of Samudragupta in the later Gupta official records; [27] however, we need not merely on that account ascribe Kācha coins to Samudragupta, because Samudragupta's own records do not show the use of that epithet for that emperor. The obverse legend on the coins of Kācha is no doubt a close paraphrase of the obverse legend on the coins of Samudragupta. This circumstance, however, should not be taken to mean that they were identical. It may perhaps show that the two kings were not far removed from each other. This has earlier been shown in classification given on the basis of the gold content of the coins as well.

So it is not unlikely that Kācha outlived his brother Samudragupta. He rose in revolt again after his death and most probably captured the throne of Magadha keeping the authority of eldest son Rāmagupta confined to Eastern Malwa.

It was later when Chandragupta-II took the reigns of administration after killing his elder brother Rāmagupta that he might have successfully attempted to seize the throne of Magadha.

However, we may point out that the tradition of having surname 'Gupta' after the first name which started with Chandragupta-I continued in all the names of the Gupta family. In case of Kācha we find it missing. The only evidence of connecting him to the Gupta family we have at our disposal is that of the *Mañju-Śrī-Mūla-Kalpa* which referring to the history of Samudragupta and his times, mentions that he had a younger brother named Bhasma who ruled for a short period and that Kācha is synonymous with Bhasma according to lexicographers.[28]

It is not improbable that Kācha ruled for a shorter duration. Either he died as an age-old person by the time Chandragupta-II became king of Eastern Malwa and paved the way for him for the throne of Pāṭaliputra as well or was killed by him in an encounter to seize the throne of Magadha.

Notes And References

1. See *Infra* Chapter V(b).
2. A.S. Altekar, *GGCBH* ., p.LXXIII.
3. *Ibid.*, p. LXXIV.
4. *Ibid.*, p. XVI, f.n. 1.
5. JNSI, IX, p.131.
6. A.S. Altekar, *Op.cit.*, p.XVI
7. *Ibid.*
8. See, f.n. 3 above.
9. Fleet, *CII*. III., p.27, f.n.4. "Kacha must, in that case, be a personal and less formal name of Samudragupta".
10. Allan, *CCGD* ., Introduction, p.xxxiv.
11. D.C. Sircar, *SI* ., p.268, f.n.2.
12. D.R. Bhandarkar, *Malviyaji Commemoration Volume,* pp.189 ff.
13. R.D. Banerji, *AIG* ., p.9.
14. R.K. Mookerji, *GE* ., p.18.
15. H. Heras, *ABORI* ., IX, p.83.
16. R.N. Dandekar, *The History of Guptas,* pp.40-43
17. S.R. Goyal, *HIG*., pp.124-128.
18. P.L. Gupta, *IG*., p.256.
19. *Ibid.*
20. Fleet, *CII* . III., p.27 f.n. 4.
21. A.S. Altekar, *GGCBH*., p.LXXV.
22. P.L. Gupta, *IG*., pp.255-256.

23. *Ibid.*, p.256.
24. S.K. Maity, *IGT*., pp.34-35.
25. *Ibid.*, p.35.
26. A.S. Altekar, *GGCBH*., p.LXXIII.
27. R.K. Mookerji, *GE*., p.39.
28. P.L. Gupta, *IG*., p.256.

8

Chandragupta-II

Sources

We need to divide the sources for the study of Chandragupta-II into two groups:

(A) His early career and sources;

(B) His later achievements and sources.

HIS EARLY CAREER AND SOURCES

Eventhough Chandragupta-II was accepted or nominated by his father Samudragupta (Tatparigṛihīta)[1] as shown in his records and the records of his successors, it seems that he was intervened by his elder brother Rāmagupta who was a Governor of Malwa in his father's life-time [2] and who most probably asserted his authority in the Malwa region while Kācha seized the throne at Pāṭaliputra, as discussed earlier in the light of the evidence of the *Devīchandraguptam*, the *Harṣacharita*, the Rāshṭra-kūṭa records and the Arab account *Mujamal-ut-Tawarikh* etc., and the numismatic evidence. He became king after killing his ignominious brother Rāmagupta and it might have taken him some time to seize the throne of Magadha for himself.

HIS LATER ACHIEVEMENTS AND SOURCES

We know about his various names, date of accession, his queens and achievements from the inscriptions and coins. We may mention that in the Bayana hoard of the Gupta coins discovered in 1946, out of 1821 coins, while we find 183 coins of Samudragupta, we find 983 coins of Chandragupta-II.[3]

His Date of Accession and the Inscriptions

(1) Among the inscriptions, we may mention that his first inscription is the Mathurā Pillar Inscription of the G.E. 61 =A.D. 480 which mentions also his regnal year 5 (Pañchama).[4] His reign, therefore, commenced in A.D. 375. Other inscriptions are:
(2) Mathurā Pillar Inscription - undated;
(3) Udayagiri Cave Inscription dated G.E. 82;
(4) Udayagiri Cave Inscription - undated;
(5) Gadhwa (village Gadhwa in Tehsil Karchhana in Allahabad District) Stone Inscription dated G.E. 88;
(6) Sāñchī Stone Inscription dated G.E. 93;
(7) Mehrauli Iron Pillar Inscription-undated.

Thus mostly these inscriptions cover the area from Allahabad to Malwa region. Date-wise Sāñchī Stone Inscription seems to be his last inscription in A.D. 412 (G.E. 93).

His Name and Family.

In the inscriptions he has been mentioned as a son of *Mahārājadhirāja* Śrī Samudragupta born on his Chief Queen Dattadevī. He has been mentioned as Chandragupta and Devagupta in the Vākāṭaka records. He had a third name Devaśrī. He had also Devarāja as his favourite name. In the Mandasor record of year 524 he has been described as "who is like a moon in the galaxy of Gupta kings with the famous name Chandragupta", which very much resembles line 5 of the Mehrauli Iron Pillar Inscription. "His name was Chandra and he was holding the glory of a full moon on his face". He has been mentioned as "Apratiratha" (having no equal or antagonist), "Paramabhāgavata" (the devout devotee of Viṣṇu) and "Mahārājadhirāja" or sometimes even as "Bhaṭṭāraka-Mahārājadhirāja". In the Supia Stone Pillar Inscription of the G.E. 141 of the time of Skandagupta his title *Vikramāditya* is also mentioned.[5]

He had two queens named Dhruvadevī or Dhruvasvāminī [6] and Kuberanāgā. Most probably the prime place was accorded to Dhruva-devī as her son Kumāragupta-I was named as his successor by Chandra-gupta-II while either he had no son from the other queen Kuberanāgā who was the first queen in status who could be named as the successor and her daughter Prabhāvatīguptā was married to the Vākāṭaka prince Rudrasena-II around A.D. 380 which shows that Chandragupta must have married her mother Kuberanāgā by A.D. 360 or so which puts his date of birth around A.D. 340. This point has been discussed by us earlier in our discussion on the Gupta chronology.

His matrimonial alliances

His matrimonial alliances may be deemed as master-strokes of his diplomacy.

Matrimonial alliance with the Vākāṭakas

As we have noted that Chandragupta had already good relations with the Nāgas by marrying their daughter Kubernāgā.

He married his daughter Prabhāvatīguptā with the Vākāṭaka Prince Rudrasena-II around A.D. 380. This Vākāṭaka matrimonial alliance brought to the Gupta family several off-shoots and extended political influence.[7] As a result, the Vākāṭaka politics came under the influence of the Gupta Empire. Though Pṛithvīsheṇa Vākāṭaka had a long reign, his son Rudrasena-II, the son-in-law of Chandragupta-II died soon and had a short reign followed by the regency of Prabhāvatīguptā.

It has been inferred on the basis of the Talgunda Pillar Inscription that the Kadamba King Kākusthavarman married his daughter to Kumāragupta (or to his son).[8] Several grants of the Western Gangas indicate that Kākusthavarman is to be assinged to A.D. 435 - 475.[9] Certain chiefs of the Kanarese Districts who claimed descent from Chandragupta (Vikramāditya), referred to their great ancestor as *Ujjayinī-Puravarādhīśvara*, 'Lord of Ujjain', the 'best of the Cities', as well as *Pāṭaliputravarādhīśvera*, 'Lord of Pāṭaliputra, the best of the Cities".

Objectives behind the Matrimonial alliances

According to Smith, "the Vākāṭaka *Mahārāja* occupied a geographical position in which he could be of much service or disservice to the northern invader of the dominions of the Śaka Satraps of Gujarat and Saurashtra. Chandragupta adopted a prudent pre-caution in giving his daughter to the Vākāṭaka prince and so securing his subordinate alliance".

Some scholars contest any influence of Chandragupta over Gujarat and Saurashtra due to lack of any inscription belonging to his or his son's (Kumāragupta's) reign or any silver coins.[12] Gupta even goes so far to say on the basis of numismatic evidence (the latest coins available) that long before the accession of Chandragupta-II to the throne i.e. A.D. 351 (Śaka era 273), the Western Kṣhatrapas had ceased to exist in Rajasthan and Malwa.[13] And that scholars relying on one-sided evidence had suggested that "during this long-military campaign Chandragupta-II defeated Svāmī Rudrasena-III and brought to amend the almost unbroken rule of the Western Kshatrapas in Malwa, Gujarat and Saurashtra, which had lasted for more than three hundred years.[14] Goyal has gone even to preclude any such possibility in the matter as Vākāṭaka King Pṛithvīsheṇa-I was not an ambitious person and secondly these two events (the event of the marriage of Vākāṭaka prince with Chandragupta-II's daughter and the Śaka conquest) are separated from each other by about 20 years.[15] But on the other hand Goyal himself admits that the Gupta conquest of Saurashtra and Gujarat took place between the interval of A.D. 388 to A.D. 409, [16] which shows the beginning of the campaign against the Śaka is not far removed from the event of marriage. So we cannot agree with Goyal that the marriage alliance with the Vākāṭakas was without any political motives. It is not unlikely that the kings thought of some side-advantages while concluding their acts of matrimony.

Whatever may be the case in the matter discussed above, Rudrasena-II, ascended the throne after the death of his father Pṛithvī-Sheṇa in A.D. 385. Just after 5 years he suddenly died which gave the reigns of administration in the hands of Chandragupta-II's daughter Prabhāvatīguptā, as the regent queen of her minor sons, and carried it on for a period of about 20 years with the valuable assistance of the officers that had been deputed by her father from Pāṭaliputra.[17] This certainly gave enough occasion for Chandragupta-II to extend his direct influence in the South and West. The Poona plates of Prabhāvatī-Guptā which use the Eastern Gupta script and beign with Gupta genealogy, instead of the Vākāṭaka one, make it fairly clear that they were drafted by a Gupta Officer, imported from Pāṭaliputra. Not only that Chandragupta-II seems to have played some important role in the training of his young grandsons. It is a known literary tradition that the poem

Setubandha composed by King Pravara-sena was revised by Kālidāsa who might have been engaged as a royal tutor to educate the Vākāṭaka Princes.[18]

Chandragupta-II had extended his influence in South also. From the Kāvya work *Kuntaleśvara Dautyam,* we may infer that Chnadragupta-II exerted his influence over the Kuntala King Śrīkṛishṇavarman through his emissary Kālidāsa. He mollified the hard feelings of Śrīkṛishṇa-varman towards the Vākāṭaka ruler Pṛithvīsheṇa; he had subdued the father of Śrīkṛishṇavarman who was now a constant danger to the Vākāṭakas. Chandragupta-II overcame this danger by his diplomacy and influence.[19]

His Conquests

Nothing is very clearly known about Chandragupta-II's conquests. However, we can get some allusions to the world conquest from his inscriptions and the Śaka conquest on the basis of his coins.

From the Udayagiri cave inscription of Chandragupta-II we know that a cave-temple was dedicated to Lord Śiva on Udayagiri hill, about 2 miles to the North-West of Bhilsa by one Vīrasena, 'a minister of peace and war' of Chandragupta-II who was a resident of Pāṭaliputra. He had accompanied his royal master to Udayagiri while the latter was 'seeking to conquer the whole world' (Kṛitsna-pṛithvī-jayārthena)[20] which is also endorsed by his last inscription at Sāñchī dated G.E. 93 (A.D. 412) which tells us about a military officer Āmrakārddava "who carried the flag of glory of many victories (over his head)" "Anekasamarāvāpta-Yaśaspatākaḥ".[21] It seems Chandragupta-II had already his feudatory ruler in the Udayagiri region which is proved by Udayagiri Cave Inscription G.E. 82 (A.D. 401) which makes mention of one Sanakānīka Mahārāja who was a feudatory ruler (Śrī Chandragupta-pādānudhyātasya) of Bhaṭṭāraka Mahārājadhirāja Śrī Chandragupta who got a Viṣṇu image carved out on the wall of the Udayagiri cave. He might have proved helpful to Chandragupta-II in his victory campaign. It may be remarked that Eastern Malwa which was atleast an important principality if not a capital being the Bhoganagara "City of his enjoyment" as stated in the Eran Inscription of Samudragupta was now being governed through feudatory rulers.

Some scholars have taken a clue of these statements in the above-stated inscriptions to stand for a protracted struggle in the Malwa region itself, [22] while Mirashi seems to have rightly pointed out that 'evidently all battles were not fought in Malwa'.[23]

The Śaka Conquest

Mirashi has also stated that there is no evidence to suppose that war against the Satrapa of Ujjayinī was the first campaign in which he was engaged. We know that the Satraps continued to rule in Malwa till A.D. 388 atleast i.e. for more than 10 years after Chandragupta's accession. During this period he may have been occupied in other places eg. in the Punjab and Kashmir, subjugating the Kushāṇas.[24]

Chandragupta-II's victory over the Western Satraps is adduced by the absence of the Śaka coins in the region after A.D. 388, as the coins of Rudra Simha-III are of the Śaka year 310 = A.D. 388, the latest.[25] Further Chandragupta-II issued his rare silver coins which are more or less direct imitations of those of the latest Western Kṣhatrapas. We may quote from Rapson's catalogue, "like their proto-types, the coins of the latest mahākṣatrapas, which they closely resemble in style and fabric, they have on the *obverse* the date accompanied by some equivalent of the word *varṣe*, behind the king's head, and retain some traces of the old inscription in Greek characters, while on the *reverse* they substitute the Gupta type (a peacock) for the Caitya, with crescent and Star".[26] The earliest date of the Kṣatrapa coins as restruck by Chandragupta-II is 90 + X (Gupta Era) = A.D. 409 (410). [27] In this connection Chattopadhyaya remarks that thus it was during this interval (between A.D. 388 and 410) that the Gupta conquest of Saurashtra and Gujarat took place. It is possible to limit the period further, for the Udayagiri inscription of the year A.D. 401 shows the occupation of Eastern Malwa by the Guptas and it is improbable that the Western Kṣhatrapas were able to resist for long the victorious progress of Chandragupta-II. [28]

Other Conquests

We are left with no doubt that Chandragupta-II was to inherit a vast empire from his father Samudragupta. This might have been intercepted by the possibility of the events of the Saka encounter with Rāmagupta which was eventually successfully countered by Chandragupta-II himself and who might have struggled for the throne of

Magadha against his uncle Kāchagupta. Taking advantage out of this turmoil, the kings in the North-West and Bengal which were the border kingdoms, might have raised their heads for independence so much so that the kings in Bengal seem to have formed a confederation against him and challenged him who were routed by him while the people in the North-West most probably the Kushāṇa rulers were also defeated by him and in this outrageous campaign he uprooted them in their breeding ground in Bactria. The above-mentioned conquests of Chandragupta-II are reflected in the mention of his victories in the *Meharauli Iron Pillar Inscription of Chandra* which is ascribed to him. Scholars expressed different opinions about the posthumous character of the inscription and also with regard to the identification and also with regard to the identification of King Chandra of the Meharauli Iron Pillar Inscription. It is pertinent to discuss them first taking into consideration the text of the inscription before we proceed further.

Identification of King Chandra of the Meharauli Iron Pillar Inscription (and the posthumous character of the Inscription?)

The Iron Pillar standing near the well-known Kutub Minar on which this inscription is inscribed was not originally located at its present site, the village called Meharauli (Mihirapuri), about 9 miles south of Delhi. It was originally located on a hill near the Beas and was brought to this place by a ruler of Delhi like the two pillars of Asoka from Topra (near Ambala) and Meerut which were brought over to Delhi by Feroze Shah Tughlaq.[29]

TEXT

1. Yasy = Odvarttayataḥ Pratīpamurasā Śattrun = Sametyāgatān = Vaṅgeshvāhava - varttinoSbhilikhitā Khaḍgena Kīrttir-bhuje.
2. Tīrtvā sapta mukhāni yena samare sindhor - jjitā Vāhlikā Yasyādya-pyadhivāsyate jalanidhir = vvīryyānilair - ddakshiṇaḥ //
3. Khinnasy-eva visṛijya gām narapater - ggāmāśritasyaetrām mūr(t)yā karmma-jitāvanim gatavataḥ kīrt (t) yā sthitasya kshitau.
4. Śāntasyeva mahāvane hutabhujo yasya pratāpo mahānnadhayā-pyutsṛijati praṇāśista-ripor =Yyatnasya śesaḥkshitim //

5. Prāptena sva-bhuj-ārjitañ-cha suchirañ = ch = aikādhirājayam
kshitau chandrāhvena samagra - chandra-sadṛiśīm vaktra-śriyam
bibhratā.
6. Tenāyam praṇidhāya bhūmipatinā bhāvena[30] vishno (shnau) matim
prāṅśurviṣṇupade girau bhagavato Vishnordhidhvajaḥ Sthāpitaḥ //

Thus we find that the inscription is composed of three verses which are as follows:-

V-1. He, on whose arm fame was inscribed by the sword, when in the battle in the Vaṅga countries, he kneaded (and turned) back with (his) breast the enemies who, uniting together, came against (him), he, by whom, having crossed in warfare the seven mouths of the (river) Sindhu (Indus), the Vāhlikas were conquered - he, by the breezes of whose prowess, the Southern Ocean is even still perfumed.

V-2. He, the remnants of his efforts of destroying his enemies do not even now leave the earth (people remember them) as if the glory of a king who being wearied even bodily attains the land earned out of his actions, does not leave the earth like the remnants of a fire in the great forest even after it is extinct (has burnt out).[31]

V-3. And by him who attained sole supreme sovereignty on earth (which) he had earned by his own arm (and) who, having the name of Chandra, possessed a beauty of countenance like (the beauty of) the full-moon, having with devotion fixed his mind upon (the God) Vishṇu, this lofty standard of the divine, was set up on the hill (called) Viṣṇupada.

The Meharauli Iron Pillar Inscription is not posthumous

The debate on the posthumous character of the inscription is based on the incorrect translation of verse 2 by Fleet [32] in the *Corpus Inscriptionum Indicarum* Volume III where he has captioned the inscription as posthumous.[33] D.R. Bhandarkar drew the attention to the faulty translation by Fleet and held the view that the inscription was not posthumous. Dasharatha Sharma gave his own translation after questioning Fleet's translation which has been discussed by S.S. Rana where he has made his own suggestions which we have not fully accepted and have brought out our own translation trying to make it perfect.[34] However, Rana has made two valid points to prove his contention that the inscription is not posthumous "Had it been done at the instance of a successor of Chandra there would have been a mention of it in the inscription. Moreover, a successor of Chandra could not have been so discourteous as to refer to him by mere Chandra without an honorific". [35] The engraving was done at the same time as soon as the pillar was raised and we do not have a tradition of posthumous inscriptions belonging to the Gupta kings. It seems that king himself commanded the inscription to be inscribed. Even the palaeography of the inscription goes in favour of this view as the dates assigned by scholars range between the second half of the 4th century A.D. and A.D. 410.[36]

We may state that such expressions that "who having conquered the earth, the invincible one wins heaven by his good deeds" are quite comon in the engravings on the coins of the Gupta Emperors viz. the standard type, Archer type and Aśvamedha type of coins of Samudragupta and the coins of Chandragupta-II [37] which can compare with the expression "the land earned by his (good) actions" in the Meharauli Pillar Inscription.

Hence we are of the view that the inscription is not posthumous.

Identification of King Chandra

Once we are sure that the inscription is not posthumous, we can concentrate our mind on the identification of the ruler named Chandra. We notice that the inscription credits King Chandra with the following achievements:

(1) Conquest of the Vaṅga countries (Vaṅgeshu) by his battling alone against a confederacy of enemies united against him (śatrūn-sametyāgatān);
(2) Conquest of Vāhlikas in a running fight across the seven mouths of the river Sindhu;
(3) Spread of his fame as a conqueror upto Southern seas;
(4) Achievement of sole supreme sovereignty in the world (aikādhirājyam) by the prowess of his arms. The inscription then relates

> the king celebrated his conquests by setting up this pillar in honour of Lord Vishṇu on the hill known as Vishṇupada.

A big controversy has arisen among the scholars by way of varied identifications of King Chandra by various scholars starting with Chandragupta Maurya and ending with Chandragupta-II which include Chandravarman of Pushkarana, the Nāga kings Chandrāṁśa and Sadāchandra, Devarakshita of the Purānas, Chandragupta-I, Samudragupta as well.[38]

We need not discuss all the theories mentioned above as most of them except his identification with Chandragupta-II are no longer valid. We have discussed the weaknesses of Goyal's theory of identification with Samudragupta in our earlier work.[39]

Sharma was inspired to write his article on Chandra of the Meharauli Pillar Inscription in 1945 [40] after Majumdar tried to identify him with Kanishka (though afterwards he changed his view in favour of Chandragupta-II).[41] However, it will be appropriate to review the arguments in favour of the identification of King Chandra of the inscription with Chandragupta-II.

It was believed earlier that Chandragupta-II had inherited a big empire from his father Samudragupta and did not engage himself much in warfare. His only military achievement was the destruction of the Śaka power in Central and Western India. He was supposed to have a peaceful succession and a still more powerful reign. How can we identify him with one who had above all a military career, who carried his victorious arms in Bengal and across the Indus, who acquired sole supreme sovereignty on this earth earned by his own arm? This picture changed afterwards with the discovery of passages from the *Devīchandraguptam* of Viśākhadatta in many works and the researches of scholars brought to light the new facts that Chandragupta-II was not a direct successor for his father, emerged on the scene when everything seemed lost under a major military defeat by a Śaka-Chief, and as a result the Gupta Empire was tottering.[42] He killed the Śaka Chief and had to kill his ill-fated brother Rāmagupta who had earlier tried to sue for peace by agreeing to ignominious terms of the Victor of surrendering his queen. This became necessary as according to the *Devīchandraguptam* the Minister has filled the mind of the King Rāmagupta with suspicion and anger about Chandragupta-II. After killing his brother he married his queen who had earlier wed him in a

Svayamvara as stated in the Arab work *Majumal-ut- Tawarikh*. As Rāmagupta's empire was confined to the Malwa region it might have taken some more time to gain his hold on Pāṭaliputra which was perhaps seized by his uncle Kācha after the death of Samudragutpa. So we can say that Chandragupta had earned his sole supreme sovereignty by the prowess of his own arm and which he enjoyed for a long period of about 40 years.

It is no wonder that he might have resolved to destroy the Śaka and Kushāṇa chiefs in their breeding ground Balkh.[43] We do not agree with the view of scholars that this Balkh should be located in the Punjab as we have discussed this point in sufficient detail in our earlier work that the Bahlika country was bordered on the Oxus, the ancient Bactria.[44] His father had completely destroyed all possible enemies in the Āryyāvartta, and had thus left no thorn in Uttarāpatha. That Chandragupta-II started on such a campaign is fully evident by his own inscription. The Udayagiri inscription of 82 records his camapign for the conquest of all earth. Numismatic evidence of his conquest of Kṣatrapa dominions is complete.[45]

At a time when the Gupta empire lay at the feet of the invaders from the West, it is not impossible that the neighbouring kings of Samataṭa, Ḍavāka etc. might have decided to fish in troubled waters. That the war in Bengal was not of Chandra's own seeking, but was rather forced on him by his enemies in Bengal is clear from the Meharauli Inscription.[46] We know that only a part of Bengal i.e. Samataṭa was conquered by Samudragupta, Ḍavāka and Kāmarūpa being in Assam, their subjugation does not mean the occupation of the whole of the Vaṅgas. Gupta inscriptions are recovered from Puṇḍravardhana, Damodarpur and Rajshahi districts of Bengal only after the reign of Chandragupta-II. Moreover, it seems that the people of Samataṭa had revolted and were joined by other neighbouring kings and King Chandra suppressed the revolt with his force; eventually the whole of Vaṅga may have come under his suzerainty.[47]

Raychaudhuri[48] has drawn our attention to a large hoard of Gupta coins (1100,[49] majority being the Garuḍa type of Kumāragupta-I) found in the Satara district.[50] Such a large hoard cannot be explained by the existence of comercial inter-course alone. The association of Kuntala with the Guptas does not, however, begin with Kumāragupta-I, it goes back to Chandragupta-II. Gupta rulers of Banavasi trace their origin to Chandragupta Vikramāditya. "If the Halsi grant is dated in the Gupta era, Kākusthavarman must have been a contemporary of Chandragupta-II and the Gupta King to whom he gave his daughter in marriage according to Tālaguṇḍa inscription must be Kumāragupta-I".[51]

Thus the mention of *Gupta-pārthivakāla* in the Tālaguṇḍa inscription, the large find at Satara (or Ahmedabad) and the records of the Rāshtrakūṭas and the Gupta rulers of Banavasi leave no room for doubt about Chandragupta-II's association with the South, so much so that the kings of the South make it a point to compare their own achievements with the great Gupta monarch. And it may be truly said of Chandragupta-II "by the breezes of whose prowess the Southern ocean is even still perfumed" (yasyādyāpi adhivāsyate jalanidhir - vvī)ryyānilair-ddakshinaḥ).[52]

To consider the last point in the contents of the inscription that King Chandra "bowing down in devotion caused this lofty pillar to be established on the hill called Viṣṇupada".[53] Chandragupta is called Paramabhāgavata (indicating him thereby as - devout devotee of Lord Vishṇu) invariably in all later family records and in some of his own also. His devotion to Vishṇu was so great that he influenced considerably the religious leanings of the later members of his family which is evident from the inscriptional records viz. Gadhwa, Bihar, and Bhitari inscriptions.[54]

Not only the contents of the inscription but also on the palaeographical grounds the scholars are more than satisfied to put the inscription in the time of Chandragupta-II.[55]

Sharma has also refuted the objections raised by Allan against this identification, the most serious objection being the use of abbreviated name Chandra while defending the same in case of his coins (on the obverse of Flower Vase type), this being due to lack of space. G.R. Sharma states in this regard, "But if the mint-master could violate the traditions, which are necessarily more conservative in numismatics, there is no reason why the writer of a Praśasti in verse should not use the abbreviation Chandra if the exigencies of metre demanded it." [56]

Similarly the absence of genealogy cannot be valid ground against it, we do not find any mention of genealogy even in the Udayagiri cave inscription of the year 82 which even omits the imperial titles of the king.

On the basis of stylistics also we would like to put the inscription in the time of Chandragupta-II. The verse forms in Indian inscriptions start only with Samudragupta and come into vogue after Chandragupta-

II. According to K. Chattopadhyaya, the writer of the Meharauli Pillar Inscription may be identical with Vīrasena alias Śāva of the Udayagiri Inscription (mentioned above) which also is in verse.[57]

On the basis of the above discussion we are of the opinion that King Chandra of the Meharauli Pillar Inscription should be identified with Chandragupta-II.

Altekar while subscribing to the same view states that we may presume that the first ten years of his reign (C. 375 - 385 A.D.) were spent in his encounter against the Kidāra Kushāṇas and the revolts in Bengal.[58]

His characteristics as known from the Numismatic Evidence

In history Chandragupta-II is credited with the title of Vikramāditya which is borne out by the biruds "Vikramāditya", "Vikrama and Vikramānka" on his coins, which show his valour. Altekar has pointed out that out of 983 [60] coins in Bayana hoard, 798 belong to the Archer type which shows he was fond of fighting and seems to have fought and won many battles as discussed above. The Tiger-slayer type of his father are replaced by Lion-slayer type of Chandragupta-II which also may signify his conquest over the Gujarat region. While it is astonishing that why *Chakra-Vikrama* type of coins are rare which show the emperor as receiving a gift from God Vishṇu, manifesting himself before his devotee for that purpose, Chandragupta being known himself as a devout Vaishṇava.[61] It is a unique coin of its type so far not known from any second specimen,[62] which P.L. Gupta thinks to have been issued on the occasion of a sacrifice (Yajña) like the *Aśvamedha* coins of Samudragupta. On the basis of a scanty and unreliable evidence of a short record found on a stone horse at Varanasi read as *Chandragu*, some scholars have tried to adduce the performance of Aśvamedha by Chandragupta-II which is not acceptable to Gupta[63] and equally to us. But the mention of Chandragupta-II as "donor of numerous cows and gold in thousands of crores" in the Poona Copper-Plate Inscription of Prabhāvatīguptā, (Regnal year 13) [64] gives an illusion to his having performed an Aśvamedha sacrifice.

It will be illuminating to quote a few legends on his coins:

(i) On some of the Archer type of coins we find the legend on the obverse;

"Devaśrīmahārājadhirāja Śrī-Chandra-guptaḥ"

"His Majesty, glorious Chandragupta, King of Kings".[65]

(ii) In the Lotus Reverse type, on some of the varieties we find "Chandra" written vertically under the left arm of the standing King.[66]

(iii) On each type of coins we have the obverse legend "Deva-śrī-Mahārājadhirāja Śrī-Chandraguptaḥ" and on reverse the legend "Śrī Vikramaḥ". One specimen of Indian Museum contians the additional word "Vikramādityasya" and beneath couch the word "Rūpākṛiti".[67]

(iv) On the horseman type, the obverse legend is "Paramabhāgavata-Mahārājadhirāja-Śrī-Chandraguptaḥ"

"Chandragupa, the overlord of kings and a devout worshipper of Vishṇu".[68]

(v) On Chhatra type of his coins we find the prose legend in Class I:

"Mahārājādhirāja Śrī Chandraguptaḥ"[69] while in Class II we have the metrical legend (a hemistich)

"Kshitimavajitya Sucharitairdivam-jayati Vikramāditya"

"Having conquered the earth, Vikramāditya wins the heaven (also) by his meritorious actions".[70]

(vi) We have his coins of Lion-Silver type divided into various classes such as Class I: The Lion-Combatant type, has the legend on all the coins of this variety as :

"Narendrachandraḥ prathitarano raṇe jayatyajeyo bhuvi Siṁha-vikramaḥ".

"The moon among the kings, who is famous for his warfare, who is valorous like a lion and who in invincible, is victorious on the battle-field".

The other two classes are known as Lion-Trampler type and the Lion-Retreating type. On some of the Lion-Trampler type of coins we have the legend on obverse "Devaśrī-mahārājādhirāja-Śrī-Chandraguptaḥ" and the legend on the reverse is "Siṁha-vikramaḥ".[71]

(vii) After the conquest of his Western Kṣhatrapa kingdom Chandra-gupta-II had to keep up its silver coinage stamping on it some Gupta features. We find the Śaka era replaced by Gupta era. The reverse shows a completely Gupta design, the figure of

Vishṇu's bird Garuḍa, standing facing with outspread wings and the corresponding legend, the King as a devotee of Vishṇu in Variety-A.

"Paramabhāgavata-Mahārājādhirāja-Śrī Chandragupta-Vikramādityaḥ".

While in Variety-B, the legend is:

"Śrī Guptakulasya Mahārājādhirāja - Śrī Chandragupta-Vikramāṅkasya".[72]

(viii) Chandragupta-II was also the first to issue copper coins. While on some coins we have the legend on the obverse "Śrī Chandra" completed by the legend "Guptaḥ" on the reverse or on some coins we have simply the name "Chandra" only without the suffix "Gupta".[73]

Thus Chandragupta-II's numismatic innovations comprise the figures of Couch, Chhatra, Lion, Horse, and Garuḍa and of Goddess Lakshmī on Lotus in place of the throned Goddess (Ardochso), and also silver and copper coinage.

His coins give Chandragupta-II the following titles: Rupākṛiti, Vikramāditya, Vikramāṅka, Siṁhavikrama, Narendrachandra and Paramabhāgavata which is also mentioned in his inscriptions.[74]

The Bilsad inscription of his son Kumāragupta applies to Chandragupta-II the epithet - "Svayamapratiratha" "Having no rival or antagonist" apart from the title "Paramabhāgavata" preceeding it.[75] The title "Sarvarājochchhetā" "Exterminator of all Kings" is added by his daughter Prabhāvatīguptā in her Poona Copper Place Inscription of regnal year 13.[76]

Fa-Hien's evidence about his reign (399-414 A.D.)

During the reign of Chandragupta-II the celebrated pilgrim, Fa-Hien came overland from China to India. After crossing the Gobi desert and the mountains tracts of Khotan, the Pamirs, Swat and Gandhara, he reached Peshawar where from he entered Punjab, passed on to places like Mathurā, Samkasya, Kannauj, Śrāvastī, Kapilavastu, Kuśīnagara, Vaiśālī, Pāṭaliputra, Kāśī etc. He then proceeded to Tāmraliptī (Tamluk, Midnapore Dsitrict), where he embarked for Ceylon and Java on his voyage homeland.[77]

Fa-Hien, a Chinese Buddhist Monk is stated to have lived in India during six years of Chandragupta-II's reign to obtain authentic copies of the scriptures. But he seems to have remained so much absorbed in his pursuit of Buddhist studies that he even forgot to record the name of the Emperor. The account of his travels records the peacefulness of India, the rarity of serious crime and the lack of severity of punishment but gives more information about the state of temples and monasteries, and repeats Buddhist legends and only a few passing phrases mention social conditions.[78] He found Buddhism flourishing still, but theistic Hinduism very wide spread. In his remarks on social customs he noted that all respectable people were now vegetarians, meat-eating being confined to low-castes and untouchables.[79] Writing about the people of Madhyadesa, he says that they had "no shambles or wine-shops in their market places". They do not keep pigs and fowls, nor do they eat onions and garlic, nor drank wine.[80] We may mention that all such statements cannot be trusted fully.

He further mentions that here Brahmanism was predominant and the King was himself a devout Vaiṣṇava. The relations between the "Brahman heretics" and the Buddhists were generally cordial, and nowhere is there any hint of prosecution of any religion. Indeed we learn from inscriptions that some of the high officers of Chandragupta-II, like Śāba-Vīrasena and Āmrakardava were Śaiva and Buddhist as mentioned in his Sāñchī Stone Inscription, the year 93 and Udayagiri Cave Inscriptions.[81]

The wealth and prosperity of Magadha deeply impressed the pilgrim who stated with admiration that the people "vied with one another in the practice of benevolence and righteousness" "The heads of the Vaiśya families establish houses for dispensing charity and medicines". There was an excellent hospital, endowed by nobles and house-holders in the capital where the poor and the destitude patients were supplied food and medicines free according to their needs. Besides rest-houses exisisted in large towns as well as on the highways for the comforts of travellers.[82]

Thus we may conclude that the Chandragupta-II consolidated the empire which was built up with great efforts by his father, while saving it at the initial stages from dilapidated conditions created firstly due to perhaps turmoil in the family after Samudragupta's death and the humiliating defeat of his elder brother at the hands of Śaka enemy and his submission to the ignoble conditions of peace. Later he suppressed the revolts in Bengal and stopped the Śaka inroads into India for all times to come.

From his Sāñchī record of the year 93, the last known record of Chandragupta-II, we may assume that his reign ended around A.D. 412-413 as the Bilsad Inscription of the year 96 = A.D. 415 mentions Kumāragupta-I as the *Mahārājādhirāja*. Thus Chandragupta-II's reign ended after a reign of long rule of about 38 years, sometime between A.D. 412 and 415.

Notes And References

1. Fleet, *CII*. III No.4 Mathurā Stone Inscription of Chandragupta-II, p.27, LL.9-10.
   Ibid., No. 12, Bihar Stone Pillar Inscription of Samudragupta, LL.19-20.
   Ibid., No. 13, Bhitari Pillar Inscription of Skandagupta, L.4.
2. It is likely that the Guptas followed the Mauryan pattern of adminstration in appointing princes of Royal blood as Governors of Provinces. See S.K. Maity. *IGT*., p.44.
3. A.S. Altekar, *GGCBH*., p.iii.
4. V. Upadhyaya, *GA*., p.130, Mathura Pillar Inscription, LL. 3-4.
5. T.R. Sharma, *PGNGI*, pp.22-23.
6. *Ibid*., p.30.
7. R.K. Mookerji, *GE*., pp.45-46.
8. *Ibid*., p.46.
9. *Ibid*.
10. H. C. Raychaudhuri, *PHAI*., p.491.
11. *Ibid*., p.490; V.A. Smith, *JRAS*., 1914, p.324.
12. P.L. Gupta, *IG*., p.303.
13. *Ibid*., p.302.
14. *Ibid*., pp. 301-302; *ASI. AR*, 1913-14, p.245.
15. S.R. Goyal, *HIG*., pp.243-44.
16. Ibid., pp.246-247.
17. A.S. Altekar, VGA., pp.110-112.
18. *Ibid*., p.112.
19. P.L. Gupta, *IG*., p.303; Supra, p.143.
20. V. Upadhyaya, *GA*., p.130 - Udayagiri Cave Inscription, L.5.

21. *Ibid.* Sāñchī Inscription, G.E. 93, L.4.
22. S.R. Goyal, *HIG.*, p.248.
23. V.V. Mirashi, ***Studies in Indology***, Vol.I, 1960 "Further light on Rāmagupta".
24. *Ibid.*
25. R.K. Mookerji, *GE.*, p.44.
26. Rapson, Catalogue, p.cli; S. Chattopadhyaya, *EHNI.*, p.204.
27. R.K. Mookerji, *Op.cit.*, p.44.
28. S. Chattopadhyaya, *EHNI.*, pp.204-205 .
29. R.K. Mookerji, *GE.*, p.66;
Fleet, *CII.* III, pp.139-141
T.R. Sharma, *PGNGI*, p.309.
30. We have taken the reading "Bhāvena" instead of "Dhāvena" by Fleet, *CII.* III, p.141, L.6. Also see T.R. Sharma, *Op.cit*, p.309.
31. We have given our own translation leaving out Fleet's and Diskalkar's (who copies Fleet) due to some inaccuracies. Attention to this aspect was drawn by D.R. Bhandarkar and Dasarath Sharma and following them by Rana who while correcting the imperfections of Sharma's translation seems to have over-indulged in making some suggestions uncalled for in the context viz. the King (i.e. Samudragupta?) about which he is not certain himself, or even in construing the stanza in Sanskrit, they are apparent in the first two lines. See S.S. Rana, "The Meharauli Iron Pillar Inscription is not Posthumous" in *AICL.*, pp.127-128.
32. Fleet, *CII.* III, p.141.
33. *Ibid.*, pp.139-140.
34. See the translation with f.n. 31.
35. S.S. Rana, *Op.cit.*, p.129.
36. *Ibid.*, pp.129-130.
37. P.L. Gupta, *IG.*, pp.81-83.
38. S.R. Goyal, *HIG.*, pp.201-202.
G.R. Sharma, "Chandra of the Meharauli Pillar Inscription" in *IHQ.*, Vol.XXI. No.4, December 1945, p.202.
39. T.R. Sharma, *PGNGI.*, pp.309-313.
40. G.R. Sharma, *IHQ.* XXI, pp.202 ff.
41. R.C. Majumdar, *Ancient India,* 1952, p.246. for his earlier views of identification with Kanishka, see *RPHI*, pp.89 ff. which are refuted by G.R. Sharma in his article referred to above.
42. G.R. Sharma, *Op.cit*, pp.209-210.

43. G.R. Sharma, *Op.cit*, p.210 cf., U.N. Roy, *GRY*., pp.229-230.
44. T.R. Sharma, *Op.cit*, pp.170-171.
45. G.R. Sharma, *Op.cit*., pp.210.
46. Fleet, *CII*. III, p.141, L.1. ".....Śatrūn Sametyāgatān Vaṅgeshvāhava-varttino...."
47. T.R. Sharma, *Op.cit*., p.310. cf., G.R. Sharma, *Op.cit*. p.210.
48. G.R. Sharma, Op.cit., p.210 as quoted by Sharma.
49. Altekar gives this number of silver coins as 1395. See A.S. Altekar, *GM*., p.151, P.L. Gupta, *Gupta Sāmrājya*, p.92 states the total number to be 1395 out of which 1100 are of Kumāragupta.
50. P.L. Gupta, *Gupta Sāmrājya*, p.92 has pointed out that these coins were discovered in 1861 in Sanauda District Ahmedabad, which Smith by mistake mentioned in Satara District. The same mistake has been carried over by Allan and Altekar. See *Ibid*. f.n. 9, B.P. Sinha has also mentioned this point. See B.P. Sinha, *DHM*., p.1. f.n.1.
51. G.R. Sharma, *Op.cit*., p.211.
52. Fleet, *CII*. III., p.141, Meharauli Iron Pillar Inscription of King Chandra, L.2.
53. *Ibid*., L.6.
54. G.R. Sharma, *Op.cit*., p.211.
55. *Ibid*.
56. *Ibid*., p.212.
57. *Ibid*.
58. A.S. Altekar, *GGCBH*., XXVI-XXVII.
59. P.L. Gupta, *Op.cit*, p.304.
60. Altekar, *GM*., p.62. The number given here is 973 which should be corrected to 983 as in *GGCBH*., pp.iii and lxxvii.
61. Altekar, *GGCBH*, p.LXXVII.
62. *Ibid*., p.LXXVI.
63. P.L. Gupta, *IG*., pp.303-304.
64. Sadhu Ram, *SII*., p.117; *EI*., XV, p.41 ff. Line 6 in the inscription.
65. A.S. Altekar, *GGCBH*., p.LXXIX.
66. *Ibid*., p.LXXX.
67. *Ibid*., p.XCI; R.K. Mookerji, GE., p.53.
68. *Ibid*, p.LXXXII.
69. *Ibid*.
70. *Ibid*., p.LXXXIV.
71. *Ibid*., p.LXXXVII.
72. *Ibid*., p.LXXXIX.

72a. A.S. Altekar, *GM*., p.105.
73. R.K. Mookerji, *GE*., p.56.
74. *Ibid*.
75. Fleet, *CII*. III., p.43, Bilsad Stone Pillar Inscription of Kumāragupta, L.5.
76. Sadhu Ram, *SII*., p.117. Line 5 of the inscription.
77. R.S. Tripathi, *HAI*., p.253.
78. A.L. Basham, *WTWI*., p.65.
79. *Ibid*., pp.65-66.
80. R.S. Tripathi, *Op.cit*., pp.254-55.
81. Fleet, *CII*. III., pp.29 and 34;
R.S. Tripathi, *Op.cit*., p.255.
82. R.S. Tripathi, *Op.cit*., p.254.

9

Govindagupta

Earlier it was believed that Kumāragupta-I ascended the throne of his father Chandragupta-II soon after his death. This being not a confirmed view now in the light of the doubts of scholars who place Govindagupta between A.D. 412 and A.D. 415 i.e. the short interval of three years between the last known inscription at Sāñchī dated year 93 = A.D. 412-13 and the first known record, the Bilsad inscription, year 96 = A.D. 415, of Kumāragupta-I. Whether it is true or not in the light of the debate being carried on by the scholars on the problem, it will be pertinent to discuss this issue before we take up the discussion on Kumāragupta-I.

Identity of Govindagupta became known from the Barsah clay seal where he is known as *Mahārāja* Govindagupta son of Mahādevī Dhruvasvāminī (Dhruvadevi in other inscriptions) who is wife of *Mahārājādhirāja* Chandragupta-II. It runs thus: "Mahārajādhirāja - Śrī Chandragupta-Patnī Mahārāja Śrī Govindagupta mātā Mahādevī Śrī Dhruvasvāminī".[1]

On this basis the scholars concluded that Govindagupta was his father's Governor in the province of Tira or Tirabhukti during the time of his father Chandragupta-II.[2] Some scholars have gone so far to interpret that it shows that Govindagupta-I being the younger brother and that no special significance should be attached to the expression "tat-pādānudhyāta" ' meditating on or favoured by the feet of ' in records which is just a conventional expression showing cordial relations between father and son and should not be taken as a token of nomination or immediate succession.[3] Accordingly, P.L. Gupta believed that Govindagupta who was the crownprince during the time of his father, came to the throne immediately after him.[4] But he had only a short rule of not more than two years.[5] Probably he was overthrown by his younger brother Kumāragupta-I and met violent end.[6]

B.P. Sinha sees the possibility of transfer of Govindagupta by his father Chandragupta-II in later years of his reign, from the governorship of Tirabhukti to that of the Western Malwa region.[7] We know about Govindagupta from the Mandasor Stone Inscription of the time of Prabhākara, Mālava (Vikrama) year 524 (A.D. 467) where it is stated "the Lord of the earth i.e. King Chandragupta, produced a son whose exalted name was Govindagupta, who as famous as Govinda (Vishṇu) for the glory of his virtues and who resembled the sons of Diti and Aditi, i.e. the demons and Gods".[8] The poet meant that Govindagupta resembled demons in physical strength and valour and gods in spiritual virtues.

Some scholars have tried to carry too much of interpretation on a verse of the Tumain inscription of the year 116 (A.D. 435) stating that Kumāragupta-I was protecting the earth like a good wife, whom he has seized by force.[9] The whole interpretation has gone wrong on the basis of the wrong translation of the word "Upaguhya" which means "embracing" rather than 'seizing by force', the same word is found in verse 4 of the Allahabad inscription of Samudragupta used to mean "embracing" "Āryyoh-ity=upaguhya......."[10] Thus the whole expression given in the Tumain inscription should be translated as "Kumāragupta son of Chandragupta who is like Mahendra who protected the entire earth like one protects one's wedded virtuous wife by means of embraces full of force and bonds of love".[11]

However, we cannot ignore the evidence of the Mandasor inscription of Kumāragupta-I and Bandhuvarman of the Mālava year 493 = A.D. 435 which has to be read with the Mandasor record of the Mālava year 524 = A.D. 467 referred to above, to arrive at some conclusion for the present.

The first record of A.D. 435 mentions that at this time Kumāragupta was reigning over the whole earth while the noble Bandhuvarman, the best of kings, son of King Viśvavarman governed the city of Daśapura (Mandasor).[12] The use of the Mālava era [13] instead of the Gupta era shows that Bandhuvarman wielded some degree of autonomy in his reign, eventhough he might have been a feudatory of Kumāragupta-I. This is again corroborated by the evidence of the dating in Mālava era in the Mandasor record of Prabhākara of the Mālava era 524 (A.D. 467).

Chattopadhyaya takes the record to be a posthumous one and mentions one Govindagupta and Indra is represented as being suspicious of Govinda's power.[14] He is undoubtedly identical with Govindagupta of the Vaiśālī seal who may have been transferred to govern Western Malwa during the life time of his father as pointed out by Sinha.[15]

It may lead us to infer that Govindagupta, brother of Kumāragupta revolted and became independent in the Mandasor region.[16] We may agree with Chattopadhyaya that as the Mandasor record of A.D. 435 (436) referred to above, describes Kumāragupta-I as ruling the earth, it may be presumed that this revolt possibly occurred after this date. This supposition is strengthened by the fact that in the second part of this Mandasor record which bears the date 529 (A.D. 473),[17] there is no reference to any Gupta king. Possibly after this revolt Mandasor region became an independent unit.[18]

We may not necessarily believe that this event of revolt occurred after A.D. 467 as pointed out by Goyal who seeks bleak possibility of such revolt on the part of Govindagupta in A.D. 467 at the time of death of Skandagupta, being himself about the age of 90 years.[19] We have already pointed out that the inscription of A.D. 467 (Malwa 524) is posthumous. The revolt could have taken place after A.D. 436 even during the life time of Kumāragupta or after his death when the family became torn as expressed in the Bhitari Inscription of Skandagupta, the uncle might have availed of the opportunity to assert his independence in Western Malwa.

We may conclude by stating that at the present state of facts to our knowledge we have no definite evidence to say whether Govindagupta ruled between the short interval of A.D. 412 and 415 or not? Whether he was Kumāragupta-I's younger or elder brother! What we can say is that he could have ruled for a very short period, that so far we have found no coins belonging to him. Being a collateral, he does not appear in the genealogical table in the inscriptions of Kumāragupta and his successors.

It is also likely that Kumāragupta defeated or ousted Govindagupta and seized the throne, and after his accession avoided all references to his elder brother.[20] It is also likely that Govindagupta might have been transferred as a Governor to Western Malwa from Tirabhukti (North Bihar) during the life-time of his father Chandragupta-II and continued to rule there even after his death in the same capacity during the reign of

Kumāragupta [21] and asserted his independence during the final years of his reign when the country faced trouble within due to family strife for succession to the throne. This is quite likely in the light of contemporary events taking place after the death of Kumāragupta-I. Ghaṭotkachagupta who earlier served under the crown-prince Govindagupta at Vaiśālī had been transferred as Governor of East Malwa by his father Kumāragupta-I. It appears from the Mandasor record of the Malwa year 524 = A.D. 467 issued by Dattabhaṭa, son of Govindagupta's general Vāyurakshita, that he was also the Viceroy in Malwa under his brother Kumāragupta-I.[22] Ghaṭotkachagupta on the death of Kumāragupta and consequent turmoil in the empire assumed sovereignty, issued gold coins and assumed āditya title of *Kramāditya* like Gupta emperors.[23]

V.S. Agrawala has attributed the construction of the Devagarh Temple to Govindagupta on the basis of an inscription discovered on a pillar in the courtyard of the Devagarh temple which is as follows:

> "Keshavapurasvāmipādāya Bhāgavata
> Govindasya dānam".

The Gupta rulers were devoted Vaishaṇavas and called themselves as Bhāgavatas.[24]

About this we may state that had it been so he might have addressed himself as the Parama-Bhāgavata as has been done by his father Chandragupta-II and brother Kumāragupta-I in their inscriptions. It might be an act of a private individual who might have given the donation to instal merely the pillar in the courtyard in line with the tradition which started during the Gupta period or even earlier.[25]

Notes And References

1. *ASR. AR.*, 1903-04, p.107.
2. S. Chattopadhyaya, *EHNI.*, p.210.
3. P.L. Gupta, *IG.*, p. 313.
4. *Ibid.*, pp.311-312.
5. Fleet, *CII.* III, p.43 Bilsad Inscription of Kumāragupta-I, LL.5-6. "........Mahārājādhirāja Śrī Chandragupta puttrasya mahādevayām Dhruvadevyām = Utpannasya Mahārājādhirāja Śrī Kumāara-guptasya....."

6. V. Upadhyaya, *GA* ., p.12.
7. B.P. Sinha, *DHM* ., pp.7-8.
8. EI., *XXVII* ., p.15, LL. 3-4.
   "Govindavat-khyāta-gunaprabhāvo,
   Govindagupto rjjita-ṅ āmadheyam
   Vasundhreshastanayam prajajñe
   sa dityditylostanayassarupam".
9. P.L. Gupta, IG., p. 312.
10. Fleet, *CII* .III., No.I Allahabad Pillar Inscription of Samudragupta, L.7. See translation, p.11.
11. V.Upadhyaya, *GA* ., p.139: Tumain Inscription (G.E. 116) L.2:
    "Śrīchandraguptasya Mahendrakalpaḥ
    Kumāraguptaḥ tanays-samagrām Raraksha
    Sadhvīmiva Dharmmapatnim Vīryyāgrahai
    tairupaguhya bhūmim".
12. Fleet, *CII* .III, p.86.
13. *Ibid* ., Line 19 of the Inscription.
14. S. Chattopadhyaya, *EHNI* ., p.214.
15. See f.n. 7 above.
16. S. Chattopadhyaya, *Op.cit* ., p.214.
17. Fleet, *CII* .III., p.83, Line 21 of the Inscription.
18. S. Chattopadhyaya, *Op.cit.*, p.214.
19. S.R. Goyal, PBRI., p.185
20. T.R. Sharma, *PGNGI* ., p.23.
21. U.N. Roy, *GRY.*, p.615. Roy, however, assigns his transfer to Western Malwa during the time of Kumāragupta.
22. R.K. Mookerji, *GE.*, p.73; B.P. Sinha, DHM., p.7.
23. B.P. Sinha, *DHM.*, p.8.
24. V.S. Agrawala, *Studies in Indian Art* , p.224.
25. P.L. Gupta, *IG.*, p.312; U.N. Roy, *GRY.*, p.613.

10

Kumāragupta-I

We have the minimum number of inscriptions belonging to Kumāragupta's reign in comparison to other Gupta rulers. But all of them belong to the private individuals and we do not have even a single royal panegyric or royal record at our disposal. Fortunately enough most of these records although dealing with the construction of some temple, installation of some image for worship or donation of land for sacred purpose provide us with the details of genealogies, the name of contemporary ruler, dates, location and even the names of officers which include sometimes the names of residents, members of the village council or village elders. Kumāragupta-I is the first among the Gupta rulers whose inscriptions, five in number, have been found in Bengal. We know about 18 inscriptions of his reign out of which one Mandasor inscription even-though inscribed in Mālava Samvat 529 = (A.D. 472) about 17 years after his reign, gives the details of prosperity during Kumāragupta's time.

The list of 18 inscriptions in chronological order is as follows:[1]

(1) Bilsad (in Etah District of Uttar Pradesh, a village in Aliganja Tahsil) Stone Pillar Inscription G.E. 96 (A.D. 415);

(2) Gadhwa (in Allahabad District, Uttar Pradesh.) Stone Inscription, G.E. 98 (A.D. 417);

(3) Gadhwa Stone Inscription (discovered by Raja Shiva Prasada);

(4) Gadhwa Stone Inscription (discovered by Cunningham);

(5) Udayagiri (Udayagiri hills in Madhya Pradesh, East Malwa) Cave Inscriptions, G.E. 106 (A.D. 425);

(6) Mathurā (in Uttar Pradesh) Jaina Image Inscription, G.E. 113 A.D. 432);

(7) Dhanaidaha (Village Dhanaidaha in the Natore sub-division of the Rajshahi District, East Bengal, now in Bangladesh) Copper Plate Inscription G.E. 113 (A.D. 432);

(8) Tumain (in Esagarh District in erstwhile Gwalior State, now in Guna District, Madhya Pradesh, East Malwa) Stone Inscription, G.E. 116 = (A.D. 436);

(9) Karamdanda (a Village in Faizabad District, Uttar Pradesh) Stone Linga Inscription, G.E. 117 (A.D. 436);

(10) Kalaikuri- Sultanpur (Village Kalaikuri, 8 miles from Nowgong in Sultanpur in Bogra District, East Bengal, now in Bangladesh) Copper Plate Inscription, G.E. 120 (A.D. 439);

(11) Damodarpur (village Damodarpur about 8 miles West of Phulbari in the District of Dinajpur, East Bengal, now in Bangladesh) Copper Plate Inscription No.1, G.E. 124 (A.D. 443);

(12) Mathurā, Broken Image Inscription, G.E. 125 (A.D. 444);

(13) Damodarpur Copper-Plate Inscription No.2, G.E. 128 (A.D. 447);

(14) Baigram (in Bogra District, modern Bangladesh) Copper Plate Inscription, G.E. 128 (A.D. 447);

(15) Mankuwar (on the bank of Yamuna in Karchana Tahsil of Allahabad District, Uttar Pradesh) Buddha Image Inscription, G.E. 129 (A.D. 448);

(16) Sānchī (in Madhya Pradesh, East Malwa), Stone Inscription, G.E. 131 (A.D. 450);

(17) Mathurā Buddha Image Inscription, G.E. 135 (A.D. 454);

(18) Mandasor (in Madhya Pradesh, West Malwa) Stone Inscription, Malwa Samvat 529 (A.D. 472).

Thus out of 18 inscriptions of Kumāragupta, 9 are from Uttar Pradesh, 4 from Madhya Pradesh, and 5 from Bengal if we classify them according to their geographical location.

We have his first Bilsad Inscription of G.E. 96 (A.D. 415) and may assume that after Chandragupta-II he assumed the reigns of administration with his accession to throne either in A.D. 415, or a little earlier but after A.D. 412 the last known record of his father Chandragupta-II. There is no evidence at our disposal to put the short interval for Govindagupta's reign and scholars should avoid imposing their own assumptions without any valid evidence in the matter.

Apart from a good number of inscriptions, we have large number of coins in gold, silver and copper to know about his reign.

Period of Reign

He had a long period of reign as known from his inscriptions and coins. The first known inscription is of A.D. 415 and the first known date of his son and successor Skandagupta is G.E. 136 = (A.D. 455) which coincides with his last coins of the ṣame date. This shows that he ruled for 40 years or more, beginning ạtleast from A.D. 415 to A.D. 455. Kumāragupta was able to maintain the great empire inherited from his father, is evident from his inscriptions found all over Uttar Pradesh, Madhya Pradesh, Bengal and his son's (Skandagupta's) inscription having been found from the Junagarh region.

His family

From the inscriptions we know of the titles of Kumāragupta as 'Mahārājādhirāja', Paramabhaṭṭāraka' 'Paramadaivata' and 'Mahendrāditya' which is borne out by the *Ārya-Manju-Śrī-Mūlakalpa* as his name *Mahendra.*[2] He had his queen *Mahādevī* Anantadevī, the mother of Purugupta.[3] He had another son who is said to have immediately succeeded him, by name Skandagupta as stated in the Bihar and Bhitari Stone Pillar Inscriptions[4] where his mother's name is not mentioned which has been found against the tradition of the Gupta records. Dandekar states that Anantadevī, mother of Purugupta was a Kadamba princess.[5] This may find support from the Tālaguṇḍa Pillar Inscription of Kadamba King Kākusthavarman which refers to his matrimonial connection with the Guptas. Scholars, on the basis of the contemporary Western Ganga records have shown Kākusthavarman and Kumāragupta as contemporaries.[6] We also get information of another marriage probably with the sister of one of his ministers as mentioned in the Bihar Stone Pillar Inscription of Skandagupta.[7]

On the basis of the evidence of Yuan Chwang, R.K. Mookerji has pointed out that Budhagupta was another son of Kumāragupta.[8] Ghaṭotkachagupta was another son of Kumāragupta as found out from an inscription on a Vaiśālī seal and the Tumain fragmentary inscription of G.Ẹ. 116 (A.D. 435) found near Eran.[9]

His conquests

Generally it is believed that Kumāragupta-I had a peaceful reign and made no conquests at all. It was towards the end of his reign that he faced disturbances which were taken care of by his son Skandagupta. In this respect R.K. Mookerji has remarked: "Kumāragupta-I may be credited not merely with the negative and static work of maintaining *in toto* his imperial inheritance but also with some positive and bold exploits in adding to the extent of that inheritance by some new conquests and records".[10]

In fact, his first inscription, Bilsad Stone Pillar Inscription itself states: "Mahārājādhirāja Śrī Kumāraguptasya abhivarddhamāna - Vijayarājya - Saṁvatsare" [11] "in the augmenting victorious reign of the Mahārājādhirāja, the glorious Kumāragupta". It seems he became worshipper of the war-god Kārttikeya (Skanda) from the beginning of his career so much so that he named one of his sons as Skandagupta and this is evident from his coin-types with the figure of Kārttikeya seated on a peacock.[12]

Incidentally his own name was also based on Kumāra (Kārttikeya).[13] His rhinoceros type of coins also indicate that he was resolute in putting down those who posed to be trouble-makers with their heads high with pride. He was also a 'Khaḍaga-Trātā' 'protector with sword'.[14] It has been suggested that his gold-coins of Rhinoceros - Slayer type may suggest that he had succeeded against his contemporary Kings of Kāmarūpa as the particular animal is rather peculiar to Assam.[15] Raychaudhuri has stated that the assumption of the title"Vyāghra-balaprākrama" "displaying the strength and prowess of a tiger", on coins of the *tiger-slayer* type, by Kumāragupta may possibly indicate that he attempted to repeat the Southern venture of his grandfather and penetrated into the tiger infested forest territory beyond the Nerbudda (Narbada).[16] All these suggestions cannot be simply dismissed as an act of fancy as suggested by Goyal.[17] The find-spots of his silver coinage with the figure of Garuḍa stamped on the reverse indicate that he was able to retain intact his father's conquests in Western India. The other class of silver coins with the Peacock Reverse points by its provenance to his hold on the central parts of the empire, the valley of the Ganges.[17a]

Chattopadhyaya states that Kumaragupta-I performed an Aśvamedha sacrifice. Whether is was the outcome of any new conquests by the

King is, however, uncertain.[18] The same is the view of Majumdar.[19] We may quote from Tripathi in this regard:

> "Certain gold coins of Kumāragupta-I prove that he performed the Aśvamedha sacrifice. Unhappily, inscriptions do not throw any light on his conquests, but it may be safely said that he could not have indulged in this imperial celebration without having won some successes in war".[20]

Mookerji also remarks that the celebration of horse-sacrifice is a sure proof of some considerable conquests achieved by the King.[21] Goyal, however, suggests that these coins were issued by him to celebrate his southern adventure.[22]

We know from an inscription from Mandasor of the Mālava year 529 (A.D. 472) which falls within the reign of later King Purugupta, that in A.D. 436 it was Kumāragupta-I, who was ruling the Gupta empire.[23] Thus the first date refers itself to the time of Kumāragupta-I, and of his feudatories, the second date is later, and belongs to other Kings.[24]

It has been suggested that during Kumāragupta-I's reign the Gupta empire took a southwardly direction which is indicated by the discovery of a big hoard of 1395 silver coins of Kumāragupta at Samand in the Satara District,[25] and a small find of his 13 coins from Ellichpur in Berar.[26] It has also been suggested that he seems to have captured the Traikūṭakas in the South Gujarat area which is proved by the resemblance of Class-III of his silver coins with those of the Traikūṭakas.[27]

Some scholars have suggested some reverses on the part of Kumāragupta around A.D. 448 on the basis of the mention of his title as 'Mahārāja - Śrī ' instead of 'Mahārājādhirāja' in the Mankuwar (Allahabad District Stone Image Inscription of the year 129 (A.D. 448).[28] Fleet wrote in this connection as follows:

> "The use of the subordinate title may be perhaps due to carelessness or ignorance on the part of the drafter of the inscription. Or possibly it may indicate an actual historical fact, the reduction of Kumāragupta, towards the end of his life, to feudal rank by the Pushyamitras and the Hūṇas, whose attacks on the

Gupta power are so pointedly alluded to in the Bhitari inscription of Skandagupta.".[29]

Thus Fleet did not give his firm opinion about Kumāragupta's reduction to the rank of a feudatory as alleged by Gupta.[30] We may not agree with the examples quoted by Mookerji also supported by P.L. Gupta of the Damodarpur Copper Plate Inscriptions one of the year 443, and the other of the year 448 along with the Baigram Plate Inscription of 448 to state that Kumāragupta was a paramount sovereign, as these inscriptions fall prior to the date of Mankuwar. Inscription by one year which is enough period for anybody's decline in status.[31] However, we are of the view that too much importance should not be attached to this small inscription of two lines which is a private record and may not necessarily reflect the decline of status of Kumāragupta in the region.

Altekar has pointed out that the numismatic activity continued unabated in the reign of Kumāragupta-I. In the Bayana hoard itself we have 628 of his coins in number as compared to 983 of his father. They, however, show very much greater activity; for we have got 14 types of this ruler represented in the hoard as against 6 of his father Chandragupta-II.[32] Some of the legends on these coins merit our attention.

On Archer type of coins - class I on the obverse we have the circular legend 'Mahārājādhirāja Śrī Kumāraguptaḥ' and on the reverse 'Śrī Mahendraḥ'.

On Classes II and III which maintain the heavier standard of 128 grains, in variety A, we get the legend on the obverse:

"Vijitāvaniravanipatiḥ Kumāragupto divam jayati"
"King Kumāragupta who has conquered the earth,wins heaven"

and in variety B, we get the legend:

"Jayati mahītalamekaḥ Śrī-Kumāraguptaḥ Sudhanvī" "The excellent archer Kumāragupta conquers the earth".[33]

2. The legend on the obverse of Swordsman type is "Gāmavajitya Sucharitai Kumāragupto divam jayati" is adopted from the Chhatra

type of Chandragupta-II, only the word "kshitim" of the original is substituted by its synonym 'gām'.[34]

3. On Horseman the reverse legend is "Ajita - Mahendraḥ" throughout.

In variety A, Class I of these Coins, the legend is "Pṛithvītalāmbaraśaśī Kumāragupta Jayatyajitaḥ", "Kumāragupta who is a veritable moon in the firmament in the form of earth's surface, and who is invincible, conquers (the enemies)" and in variety B which is represented by a single coin in this hoard, the legend is "Jayati nṛiporibhirajitah", "Victorious is the King, never conquered by the enemies", Variety C has the legend "Kshitipatirajito vijayī Kumāra-gupta divam jayati", "The unconquered and victorious King Kumāra-gupta wins the heaven". On Class II in variety A the legend is "Guptakula-vyomaśaśī Jayatyajeyo jitamahendraḥ", "The invincible Mahendra who is unbeaten, a veritable moon in the firmament of the sky in the form of the Gupta family is successful". In variety B, the legend is "Guptakulāmala Chandro Mahendrakarmajito Jayati" "The spotless moon in the firmament of the Gupta family, the invincible (hero) who is as valorous as Mahendra, conquers (the enemy)".[35]

4. On the Aśvamedha coins of Kumāragupta, the observe legend reads: "Devo jitaśatruḥ Kumāraguptodhirājaḥ" "King Kumāragupta, who has conquered the enemies, is King of Kings". The reverse legend is "Śrī-aśvamedha-Mahendraḥ".[36]
5. The obverse legend on the Elephant Rider type is "Kshataripurājatrātā Kumāragupto jayati ripūn", "Kumāragupta, whose enemies have been destroyed, and who protects feudatory Kings, conquers the enemies", and the reverse legend is "Śrī - Mahendragajaḥ".[37]

If these coin-legends are an indication to deduce historical information we may state that he had a big victorious career which was unspotted by any reverses which made him shine like the Moon in the firmament of the Gupta dynasty. But this Moon got eclipsed by invading enemies such as Puṣyamitras and Hūṇas in the later years of his life which is evident from the contents of Bhitari Stone Pillar Inscription of Skandagupta[38] and it was left to his son Skandagupta to restore the fallen fortunes of his family which he could do after his father had expired.[39]

From the list of inscriptions referred to earlier it is clear that various religious sects such as Vaiṣṇavism, Shaivism, Buddhism, Jainism, etc. flourished during his reign and were helped by way of religious and charitable endowments. We also know about the sale of land for purposes of donation along with the details of procedure from his inscriptions in Bengal. These inscriptions also give us insights into the Gupta administration during Kumāragupta's reign which is made effective by the emperor with the aid of his ministers, provincial governors (Uparikas) who were called 'Uparika-Mahārāja' incharge of a Bhukti (province) who with the approval of King appointed the Vishayapatis (District Magistrates) who worked in their district office (Vishayādhikaraṇa) and the district headquarters (Adhishṭhānas). The District Magistrate was associated with an Advisory Council consisting of:

1) *Nagara-Śresṣṭin* "President of the Town Corporation (Mayor),
2) *Sārthavāha,* representative of the Guild of Merchants,
3) *Prathama-Kulika,* Chief of the guild of Artisans,
4) *Prathama Kāyastha,* Chief of the Union of Writers or Scribes, Chief Secretary.[40]

Another event of significance during Kumāragupta-I's reign was that he probably sought to establish good diplomatic relations with the Sung Emperor of China which is evident from the visits of delegations from China as well as an exchange of an Indian envoy during his period.[41]

We cannot say with certainty that after Kumāragupta's death in A.D. 454 - 455 if there was any war of succession among the rival Princes Skandagupta and Purugupta though the joint evidence of inscriptions and coins indicates that there was a struggle for succession among the sons of Kumāragupta-I and Skandagupta emerged successful out of the conflict.[42]

Notes And References

1. We have compared the list with P.L. Gupta's list in IG., p.23 who gives a list of 14 inscriptions while U.N. Roy gives a list of 18 Inscriptions. See U.N. Roy, *GRY*., p.273; the following 4 are added by U.N. Roy :
   (i) Gadhwa Inscription (discovered by Cunningham);

(ii) Mathurā Broken-image inscription, G.E. 125 (A.D. 444);
(iii) Sāñchī Inscription, G.E. 131 (A.D.450);
(iv) Mathura Buddha Image Inscription, G.E.135 (A.D. 454).

2. T.R. Sharma, *PGNGI.*, p.24.
3. *Ibid.*,p.30.
4. Fleet, *CII* .III. Nos. 12 and 13.
5. R.N. Dandekar, *HG.*, p.102.
6. R.K. Mookerji, *GE.*, p.72.
7. Fleet, *CII.* III. p.50, see Transaltion line 3.
8. R.K. Mookerji, *Op.Cit.*, pp.72-73.
9. *Ibid.*, p.73.
10. *Ibid.*, p.74.
11. Fleet,*CII.* III. p.43, Bilsad Stone Pillar Inscription of Kumāragupta, the year 96, L.6.
12. P.L. Gupta, *IG.*, p.319.
13. T.R. Sharma, *Op.Cit.*, p.24.
14. P.L. Gupta, *Op.Cit.*, p.320; A.S. Altekar, *GM.*, p.138.
15. *IHQ.*, XXXI, pp.175 ff; See S.R. Goyal, *HIG.*, p.257, f.n.2.
16. H.C. Raychaudhuri, *PHAI.*, p.503.
17. S.R. Goyal, *HIG.*, p.257, f.n.2.
17a. R.K. Mookerji, *GE.*, pp.73-74.
18. S. Chattopadhyaya, *EHNI.*, p.214.
19. R.C. Majumdar, *VGA.*, p.174.
20. R.S. Tripathi, *HAI.*, p.74.
21. R.K. Mookerji, *GE.*, p.74.
22. S.R. Goyal, *Op.Cit.*, p.257.
23. Fleet, *CII.* III. No.18, LL.19, 21.
24. R.K. Mookerji, *Op.Cit.*, pp.70-71.
25. B.P. Sinha suggests that it is Sahad in Ahmedabad District and not in Satara District (B.P. Sinha, *DHM.*, p.19 f.n.I).
26. S.R. Goyal, *HIG.*, pp.256-257.
27. *Ibid.*, p.257; R.K. Mookerji, *Op.Cit.*, p.74.
28. Fleet, *CII.* III. No.11, p.47. L.2.
29. *Ibid.*, p.46.
30. P.L. Gupta, *IG.*, p.318.
31. R.K. Mookerji, *Op.Cit.*, p-71; P.L. Gupta, Op.Cit., p.318.
32. A.S. Alterkar, *GGCBH.*, p.XCIII.
33. *Ibid.*, pp.XCIV-XCVI.
34. *Ibid.*, p.XCVI.
35. *Ibid.*, pp.XCVII-XCIX.
36. *Ibid.*, pp.XVIII-CIX.
37. *Ibid.*, p.CIX.

38. Fleet, *CII*. III. p.54, Lines 11 and 15 of the inscription.
39. *Ibid.*, L.12 of the inscription.
40. R.K. Mookerji, *G.E.*, pp.76-84.
41. P.L. Gupta., *IG.*, pp.318-319. P.C. Bagchi, *India and China.*, pp.72-73.
42. R.C. Majumdar, "The Successors of Kumāragupta I" in S.P. Gupta (ed) *RPHI.*, p.100.

11

Skandagupta

The debate with regard to Skandagupta's accession to throne after his father's death or its intervention by other sons of Kumāragupta mainly Ghaṭotkachagupta and/or Purugupta is still being carried on among scholars. The problem has become more complicated with the statement of P.L. Gupta ending the reign of Kumāragupta around A.D. 450 [1] while earlier on the authority of a date on one of the silver coins this event has been placed in 136 G.E. (455-56 A.D.) which tallied with the date 136 G.E. given in the Junagarh Inscription of Skandagupta. So far his reign period had been accepted from A.D. 455 to A.D. 467. [2] Gupta has doubted any reading of date on the above-mentioned coin as well as the dates 134 and 135 read by V.A. Smith on the silver coins of Kumāragupta-I and that he could not have continued to rule long after 450 A.D. has also been accepted by Sinha.[3] Sinha even has ascribed the Bihar Stone Inscription of Skandagupta [4] to Purugupta [5] which he quotes Sircar to have tentatively accepted. [6]

The problem arises in case of death of Kumāragupta in A.D. 449-450 as to who ruled between A.D. 450-455. The other issues usually raised are the war of succession among the sons of Kumāragupta after his death and the issue of legitimacy of Skandagupta to throne on the basis of the absence of his mother's name in his Bhitari Pillar Inscription.

Before we arrive at the issues let us have a look into the sources for the study of Skandagupta.

INSCRIPTIONS

1. Junagadh (in Kathiawad) Rock Inscription of Skandagupta the years 136, 137 and 138 (A.D. 455, 456 and 457);

2. Kahaum (in Deoria District, Uttar Pradesh) Stone Pillar Inscription year 141 (A.D. 460);
3. Supia (Rewa District, Madhya Pradesh) Pillar Inscription of the time of Skandagupta, year 141 (A.D. 460);
4. Indore (Bulandshahr District, Uttar Pradesh) Copper Plate Inscription, the year 146 (A.D. 465);
5. Bhitari (Ghazipur District, Uttar Pradesh) Stone Pillar Inscription of Skandagupta, undated;
6. Bihar (Bihar Sharif in the Patna District, Bihar) Stone Pillar Inscription of Skandagupta, undated (No. 12 of Fleet, *CII*. III which has been now relegated to Purugupta as referred to above);
7. Gadhwa (Allahabad District) Sand Stone Inscription the year 148 (A.D. 467) - supposed to be belonging to the reign of Skandagupta. [7]

COINS

Skandagupta's gold coins are not marked by variety of types issued by his predecessors. His gold coinage shows only two or probably three types as described below.[8]

1. Archer Type
2. King and Lakshmi Type
3. Horseman Type

Silver Coinage

Altekar states that Gupta mints were fairly active during the reign of Skandagupta. This silver coinage is as copious as that of his father, he supplanted some of its types by introducing new ones like the Bull and the Altar types. But his gold coinage is much less copious and shows only three certain types referred to above.[9]

Skandagupta's silver coins were issued both in the Western and Central Provinces of the empire and thus may be categorised as the Western issues and Central Provinces issues. The Western issues have three Reverse types representing Garuḍa, Bull (Nandi) and Altar on the reverse of the coin while the obverse has the King's bust in common. The Garuḍa type do not show variety of fabric as in the case of Kumāragupta, the Bull types were circulated in Kathiawar while the Altar were found in Cutch. The Central Provinces issues do not show any alteration from those of Kumāragupta.[10]

LITERARY SOURCES

The literary sources include the *Ārya-Mañju-Śrī-Mūlakalpa*, the account of Yuan Chwang, the *Chandragarbha-Pariprichchhā* and the *Kathā-saritsāgara*.

His Names, Epithets and Family

From a verse in the *Ārya-Mañju-Śrī-Mūlakalpa* we know that among the names of the successive Gupta kings, viz. Samudra, Vikrama, Mahendra and Devarāja corresponding to Samudragupta, Chandragupta-II Vikramāditya, Kumāragupta Mahendrāditya and Skandagupta; Skandagupta has been described one among Sakarādya i.e. one whose name begins with 'sa',

> "Samudrākhya Nṛipatis'chaiva, Vikramaśchaiva Kīrttitaḥ!
> Mahendra Nṛipavaro mukhyaḥ, Sakarādyamataḥ Param //646//
> Devarājākhya Nāmasau (bhaviṣyati) Yugādhame!
> Vividhākhyo nṛipaḥ śreṣṭhaḥ, budhimān Dharmavatsalaḥ //647//[11]

Thus we know a new name 'Devarāja' given to him which is not mentioned in the inscriptions. However, he has been stated as the equal of Indra (Śakropama) in his Kahaum Stone Inscription, year 141 (A.D. 460).[12] We know that the name 'Devarāja' was assumed by his grandfather Chandragupta-II and in line 26 of the Allahabad Pillar Inscription Samudragupta has been mentioned as equal to Indra and whose other name Mahendra is assumed by his father Kumāragupta-I. Not only that Skandagupta also assumed the title of *Vikramāditya* earlier assumed by his grandfather. The title we find on the reverse of the Altar type of his coins.

> "Paramabhāgavata Śrī Vikramādityaḥ Skandaguptaḥ"
> In the Garuḍa and Altar types he is also called "Kramāditya"

About his family we do not have any information from the Gupta inscriptions. P.L. Gupta, however, has adduced some information about his marriage from a coin-type which was earlier taken as King and Lakshmī type, the Lakshmī making a choice for him as her husband as described in verse 5 of the Junagarh Rock Inscription which Gupta thinks to be a queen, and purports to his marriage.[13]

Probably he had a son named Kumāragupta-II who succeeded him after his death. [14]

His Reign Period

So far the scholars had considered his twelve year reign period from A.D. 455-A.D. 467 [15] on the basis of Junagarh inscription beginning with the year 136 (A.D. 455) as well as on one of his silver coins and terminating with G.E. 148 (467-68 A.D.), the last known date read on one of his silver coins. [16] B.P. Sinha states that the period of his reign cannot be fixed with certainty ".......there is no compulsion to place the beginning of his reign latest in 136 or 455-56 A.D. He may have even come to the throne in the year 135 (454-55 A.D.), the year of Mathurā Stone-Image Inscription which does not mention the King's name probably because of raging war of succession". [17] And if we believe the theory of P.L. Gupta that Kumāragupta-I died in A.D. 450 as referred to earlier, then his reign period may even go earlier.

Date of Accession

P.L. Gupta has pointed out that there is no date after 130 (A.D. 449) which can be read with certainty on Kumāragupta's coins. [18] The Sāñchī Stone Inscription dated in the year 131 (450-51) concerning the grant of money to a Buddhist Vihāra of Kākanādaboṭa does not refer to any king. We have earlier seen that Mathurā Stone Image Inscription of the year 135 (454-55 A.D.) does not mention the king's name. On these grounds Sinha states that this might be due to raging war of succession and probably the years between 131 and 135 were years of turmoil when the rival sons of the kings were struggling for the throne and in this confused state of affairs the name of the king emperor could not be used by the scribes of the inscriptions. [19] Further he states, "however, Skandagupta was in complete saddle by 137 or 138 (= 457-458 A.D.)". [20]

On the other hand we have the views of scholars like Majumdar, Raychaudhuri, Basak, Mookerji and others on this issue.

Majumdar in one of his articles read at a meeting of the *Asiatic Society of Bengal* [21] had advocated the theory of fratricidal war in which Skandagupta came off victorious after defeating his brothers. This theory has been refuted by Raychaudhuri stating that there is nothing to show that "the struggle at the end of Kumāra's reign, referred to in the Bhitari Pillar Inscription, was a fratricidal conflict". That the hostile powers were outsiders not belonging to the Gupta line i.e. the Pushymitras and Hūṇas. There is not the slightest reference to a fratricidal war. There is nothing to show that Skanda shed his brothers' blood and that the epithets "amalātmā" "pure-souled", and "parahitakārī" "the benefactor of others", applied to him in the Bhitari inscription and coin-legends, were unjustified. [22]

Further he states that the view that Skandagupta was the immediate successor of Kumāragupta-I seems to be confirmed by a verse in the *Ārya-Mañju-Śrī-Mūlakalpa* which runs thus:

"Samudrākhya nṛipaśchaiva
Vikramaśchaiva kīrttitaḥ
Mahendranṛipavaro mukhyaḥ
Skārādyamataḥ param
Devarājākhya nāmāsau yugādhame"

"It is impossible not to recognise in the Kings (nṛipa) Samudra, Vikrama, Mahendra, and "Sakārādya" mentioned in the verse, the great Gupta emperors Samudra Gupta, Chandra Gupta-II Vikramāditya, Kumāra Gupta-I Mahendrāditya and Skandagupta".[23]

Almost similar views have been expressed by Basak, [24] Mookerji [25] and Roy.[26]

It seems Majumdar also later moderated his opinion stating that indications are not wanting that there were some troubles over the succession, and the brother and other sons of Kumāragupta properly set up rival claims for the throne. But the evidence is so vague and uncertain that no definite conclusions are possible. [27]

"But although Skandagupta was apparently successful for the time-being, it is clear that the members of the family dissension were not finally extinguished". [28]

Chattopadhyaya points out that the statement of the Junagadh record that he (Skandagupta) was selected by the goddess of fortune from among other royal princes cannot at once point to a war of succession. But when we find that in the Bhitari record he is credited with having shown mercy to the conquered people and that the Bhitari and Nālandā seals state that Purugupta was the immediate successor of Kumāragupta, we may possibly conclude that there was some trouble over the succession of Skandagupta. "Those who support the fratricidal struggle theory further point out that as the name of Skandagupta's mother is not mentioned in the epigraphs, she was evidently not the chief queen of Kumāragupta and hence Skandagupta was not the rightful claimant to the throne. Though other interpretations are also possible, the argument, it must be admitted, has got some force". [29]

According to S.R. Goyal perhaps the Guptas had not evolved a specific system for the right of succession. [30] It appears that during the last years of Kumāragupta's reign, pulls from various directions affected the question of succession: "the emperor and the army favoured Skandagupta, the queen Anantadevī and a powerful ministerial party supported the cause of Purugupta and in some provinces local officers whitted the ambition of princes such as Ghaṭotkachagupta. In such a condition, dominated by factional power-politics, a close contest for the throne was inevitable. Fortunately for the empire, Skandagupta, the unique hero of the Gupta dynasty, who had the blessings of his father and the support of imperial army on his side emerged victorious in it".[31]

P.L. Gupta has doubted the information supplied by Smith with regard to Kumāragupta-I's coins with the dates 136 on one of his coins, 135 on two of his coins and the date 134 on another coin belonging to him.[32]

In this regard Dashratha Sharma's statement may be quoted that:

> "We should not disbelieve V.A. Smith's testimony about Kumāragupta-I's coin with the 'well-preserved' date 136, unless there is something to discredit it. Secondly when we find Skandagupta on the throne in the very year his father died, we must have very cogent reasons to assert that there was an interregnum. We have found no such reason forthcoming so far. All that can be asserted perhaps is that some others laid a claim to the throne".[33]

P.L. Gupta also states that besides his two sons Purugupta and Skandagupta, Kumāragupta-I had in all probability a third son also and most likely the eldest, called Ghaṭotkachagupta, who appears to have contested the succession after Kumāragupta-I. [34] He also states that after the death of Kumāragupta-I, Ghaṭotkachagupta occupied the throne for sometime before Skandagupta could come back from his victorious campaign, "This would have happened sometime, after 130 G.E. (449 A.D.) but before 136 G.E. (455 A.D.)". [35] Thus P.L. Gupta gives a wider interregnum of 5-6 years which may not be acceptable to us without concrete evidence.

He also quotes the views of some scholars that Kumāragupta-I abdicated the throne in favour of his son and retired to religious life after crowning his prince on the basis of the evidence of Apratigha type of his coins and the stories of *Kathāsaritsāgara* and *Chandragarbh-Pariprichchhā*.

On the other hand, Sohoni states that probably Kumāragupta was killed in a battle with Hūṇas near Bhitari itself "otherwise, there is just no reason why this particular site was selected to commemorate Kumāragupta the verse referring to this point follows immediately".[36]

Another point which merits attention is that whether Purugupta succeeded Kumāragupta-I or Skandagupta i.e. Whether he ruled for about 2 years in the interregnum after Kumāragupta around A.D. 453-455 or after Skandagupta in A.D. 467-469. The Bhitari seal inscription of Kumāragupta (III) states that "Mahārājādhirāja Śrī Puru Gupta was the son (putra) of Mahārājādhirāja Śrī Kumāra Gupta by his wife and queen Mahādevī; and 'one was meditating on the feet of his father in devoted loyalty to him (tatpādānudhyāta). [37] Thus this inscription mentions Purugupta immediately after Kumāragupta and does not mention Skandagupta which has been taken by some scholars to prove that Purugupta was the immediate successor of his father. Mookerjee is of the view that on this point epigraphic practice is not conclusive. He has quoted several inscriptions where the omission of an immediate successor of his father is made, and assigns the dates A.D. 467-469, after Skandagupta's reign to Purugupta's rule.[38]

Basham endorses this view when he states that Skandagupta issued coins on both the lighter and *suvarṇa* standards and Purugupta's coins are on suvarṇa standard only and also the coins of all successors of Skandagupta, Purugupta must have followed Skandagupta and not preceded him. Skandagupta could be the only Gupta king who could introduce the *Suvarṇa* standard in Gupta coinage which was followed by all future Gupta kings.[39]

Sinha, however, refutes this view by stating that "Purugupta may have been the first Gupta emperor to issue coins on *Suvarṇa* standard. Skandagupta first issued coins on lighter standard following his father but seeing that the reversion to the national tradition of *Suvarṇa* standard by Purugupta had become popular, he also issued coins on *Suvarṇa* standard, which was followed by all later kings". [40] He even assigns these coins to Budhagupta and states that there is no coin which can be attributed to Purugupta with certainty and we cannot with absolute confidence say that Purugupta issued coins on *Suvarṇa* standard.[41] At the same time he attributes Prakāśāditya's coins on *Suvarṇa* standard to Purugupta which fall nearest to Kumāragupta-I.[42] This makes the confusion worse confounded.

In consideration of the discussion made above we are of the opinion that eventhough the rival princes revolted against Skandagupta, they did not achieve conclusive victory over the claims of Skandagupta. An interregnum of two to three years may be postulated in this regard. Further we may state that Purugupta may have ruled twice for brief periods, once during the interregnum after the death of Kumāragupta and secondly after the death of Skandagupta. We may not like to change the reign period of Skandagupta (A.D. 455-467) till further evidence comes to the fore.

The Issue of Legitimacy of Skandagupta

We know of Skandagupta's title *Kramāditya* which suggests his hereditary right to the Gupta throne which is criticised by P.L. Gupta on the ground that no rightful successor to throne feels the necessity of proclaiming that he had come to the throne by his right of inheritance to which Dashratha Sharma answers by quoting the practice of mentioning the expression 'Kramaprāpta-nṛipaśriyaḥ" from the contemporary Vākāṭaka seals which resembles the title *Kramāditya.* [43]

Some scholars have taken the view that Skandagupta could have legitimately ascended the throne if his father had selected him but at the same time they have questioned the evidence of such a selection stating that there is nothing even to prove that he was his father's favourite. The expression "Pitṛi-parigata-Pādapadmavartī".[44] "who subsisted (like a bee) on the wide-spreading waterlilies which were the feet of (his) father", does not by itself indicate that Skandagupta was his father's favourite, "it rather reflects his own anxiety to show that he was attached to his father". But we also know from the Bhitari inscription that Skandagupta had been given the best of education as shown in the lines 8-10 of the Bhitari inscription, "By whom, having, with daily intense application, step by step attained his object by means of good behaviour and strength and politic conduct, instruction in the act of disposition (of resources) was acquired, (and) was employed as the means of (subduing his) enemies who had put themselves forward in the desire for conquest that was so highly welcome (to them)".[45] When the empire was shaken to its foundations by the Puṣyamitras, Skandagupta was deputed to subdue the enemy. [46] Dasharatha Sharma quotes a parallel example from Mughal history where one finds Shah Jahan giving priority to Dara Shikoh in all matters, internal as well as external, which affected the interests of the empire. He further quotes from verse 5 of the Bhitari inscription which states though punningly, that like all others of the empire, the emperor Kumāra (Kumāragupta-I) was also highly impressed by the greatness of Skandagupta's military and diplomatic achievements.[47]

The Verse 5 is as follows:

> "Prasabham-anupam (ai) r = vvidhvasta -śastra-pratapair Vina (yasa) mu (Chittas'cha) kshānti-śauryar-nnirūḍham Charitam = amala - Kīrtter = ggīyate yasya śubhram diśi diśi paritushṭair = ā - kumāram manushyaiḥ"[48]

"The resplendent behaviour of whom, possessed of spotless fame - inherent, (but increased) by......... and patience and heroism which are emphatically unequalled, (and) which destroy the efficacy of the weapons (of his enemies), is sung in every region by happy men, even down to the children".[49] The word 'ā-kumaram" has been translated by Fleet as "down to the children" which has been interrupted by Dasharatha Sharma "even by Kumāra (gupta) himself" as cited above.

But it makes no sense, if the subjects sing his praise it is all right but it looks ridiculous if king also sings his praise. Moreover, the question of King Kumāragupta singing his praise does not arise as the King (Kumāra) was dead when the inscription was written is clear from ' Pitari divamupete' when his father had attained the skies from the succeeding verse which starts in the same line i.e. line 12 of the inscription.[50]

If we do anvaya (arrangement) of the sentence it comes thus: "Yasya amalakīrtteh śubhram charitam diśi diśi paritushaṭai manushyaih, ākumaram giyate". "whose, the resplendent behaviour of whom, possessed of spotless fame, is sung region to region by happy men from his adolescent period" So "ākumaram" should only mean "from his adolescent period".

The main argument raised against Skandagupta's legitimacy to the accession of throne is based on verse 6 of the Bhitari Inscription which is as follows:

> "Pitari divam = upe (te) viplutām vanśa-lakshmīm
> bhuja-bala-vijit-ārir=yyaḥ pratishṭhapya bhuyah
> jitam=iti paritoshān-mātaram sāsra-nettrām-
> hata-ripur-iva kṛishṇo Devakim=abhyupe (ta)ḥ ".[51]

"Who, when (his) father had attained the skies, conquered (his) enemies by the strength of (his) arm, and established again the ruined fortunes of (his) lineage; and then, crying 'the victory has been achieved', betook himself to (his) mother, whose eyes were full of tears from joy, just as Kṛishṇa, when he had slain (his) enemies, betook himself to (his mother) Devakī". Obviously his mother is not named which is verily against the usual practice in the Gupta inscriptions where without exception king's mother's name is mentioned along with that of the father.[52]

We do not agree with the view of the scholars who take his mother's name to be Devakī on the basis of the analogy described in the inscription as Skandagupta's name is different from Kṛishṇa. This analogy may explain the similarity of circumstances rather than of names. Some scholars even questioned the status of Skandagupta's mother - Whether she was a properly wedded queen of Kumāragupta or not.[53] According to one extreme view she was Kumāragupta's concubine. A.L. Basham even stated that she was a Śūdra woman and for this he drew our attention to line 15 of the Bhitari inscription "...... Gītaiś 'cha stutibhiś' cha vandaka - jano yam prāpayatyāryyatām" "whom the bards raise to distinction with (their) songs and praises".[54] Sohoni has pointed out that there is absolutely no basis for this conjecture. It is a fact that even in Skandagupta's Supia inscription, his mother has not been named - an omission which was being constantly found in all places when it should not have taken place. This cannot be without significance. But it does not mean that Skandagupta was shy of his mother's name being mentioned. His returning to her after his successful crucial campaign was specifically mentioned in the verse 6 of the Bhitari inscription, and this is sufficient to indicate that he attached great importance to that event. "There is, undoubtedly some element of obscruity about Skandagupta's mother but it would be incorrect to assume that she had no royal or official status. One must await further factual evidence".[55] We agree with the view of Dasharatha Sharma who has contested the view of A.L. Basham that Skandagupta was the son of a Śūdra concubine and that he had been raised to the status of an Ārya by the panegyrics of the bards by stating that if it were a fact "Would Skandagupta be tomtomming it through one of his own inscriptions"? Secondly he has cited examples from literature and inscriptions to show the word 'Ārya' has been used in the cultural sense and not in the ethnic sense of the term.[56]

There can be a possibility of a disputed succession but need not necessarily mean the occupation of the throne by a disputer even though he might have issued his own coins; Ghaṭotkachagupta can be one of the likely claimants to the throne. He is known to us from the Tumain inscription as his father's Viceroy in G.E. 116 (A.D. 435) and also from his Leningrad Museum Coin. His defeat by Skandagupta did not probably take much time; nor does it mean that the disputer's rights were more legitimate than the rival rights to set aside". To realise the truth of this statement one has just to think of the Mughal war of succession in which Shuja, Murad as well as Aurangazeb issued their own coins even when Shah Jahan was alive, though none of them was his father s choice".[57]

R.C. Majumdar was perhaps the first historian who made a strong case against Skandagupta's legitimacy but he regarded it as no more than a hypothesis.[58]

However, P.L. Gupta is of the opinion that Skandagupta had no natural claim to the Gupta throne; neither Kumāragupta-I was able to favour him as his successor nor Skandagupta himself was in a position to claim himself 'the choice of his father'.[59] Skandagupta never uses the term 'tatpādānudhyāta' in relation to his father which is found in all inscriptions of the Gupta dynasty, used by the rulers for themselves in relation to their father.[60]

Dasharatha Sharma takes this statement of P.L. Gupta as dogmatic and has suggested to wait for some new discoveries, "for the discovery of one of Skandagupta's seal perhaps". But till then "it is best to regard this "amalātmā", 'pitṛiparigata-pādapadmavarti' 'Guptavaṁśaikavīraḥ' and 'Guptavaṁśaja' ruler Kramāditya as the legitimate heir to his father's throne"[61] though the same epithets have been taken by P.L. Gupta to overemphasise his claim to the Gupta lineage and to throne for which normally there should be no need until and unless his claims were too feeble or impaired.[62]

There seems to be some ambiguity about the question of legitimacy of Skandagupta to the Gupta throne; to clear this we have to wait for new evidence. But we have no doubts that he was the most capable of all the brothers and his father had utmost trust in his valour and heroism which was shown by him in various conquests over his enemies and the goddess of fortune made him her choice obviating all other princes of royal blood. Hence his claims as quoted above are quite genuine.

His achievements

Whatever may be the case whether disputed or undisputed succession of Skandagupta to the Gupta throne, he excelled all other princes in the matter of qualities and was the most deserving person to succeed his father. This has been allegorically stated in his Junagarh Rock Inscription in the fifth verse which is as follows:

> "Krameṇa buddhyā nipuṇam pradhārya
> dhyātvācha kṛitsnān = guṇa - dosha - hetūn
> vyapetya sarvān = manujendra - putrān
> Lakshmī svayaṁ yaṁvarayāṁ = Chakāra.[63]

"Whom the goddess of fortune and splendour of her own accord selected as her husband, having in succession (and) with judgement skilfully taken into consideration and thought over all the causes of virtues and faults, (and) having discarded all the other sons of Kings (as not coming up to her standard)".[64]

The Gupta empire faced serious troubles towards the end of the reign of Kumaragupta-I and Skandagupta who succeeded him had to struggle hard, is clearly established by the Bhitari Pillar Inscription of Skandagupta in which reference has been thrice made to this position.

(i) Vichalita-kula-lakshmī-stambhanāyodhyatena.[65]
(ii) Pitari divamupete viplutām Vaṁśalakshmīm
bhujabalavijitāriryaḥ pratiṣṭhāpya bhūyaḥ.[66]
(iii) Prachalitam vaṁśam pratiṣṭhāpya".[67]

Fleet has taken the enemies to be the *Puṣyamitras* while Divekar suggested the reading 'Yudhyamitrāṁś' - Cha' which means 'enemies in warfare'.[68]

So far the accepted reading is 'Puāyamitrāṁś' - cha' in the verse 4 of the Bhitari inscription which is as follows:

"Vichalita-kula-lakshmī-stambhanāyodyatena,
kshiti-tala-śayanīye yena nītā triyāmā !
Samudita-bala koṣān Puṣyamitrāṁś'-cha jitvā
Kshitipa-charaṇa-pīṭhe sthāpito vāmapādaḥ"[69]

"By whom, when he prepared himself to restore the fallen fortune of (his) family, a (whole) night was spent on a couch that was the bare earth; and then having conquered the *Puṣyamitras* who had developed great power and wealth, he placed (his) left foot on a foot - stool which was the King (of that tribe himself).[70]

The last part of the verse has been translated by R.S. Tripathi as "placed his left foot on the royal foot-stool" i.e. ascended the throne.[71]

We do not agree with the reading of Divekar and are satisfied with the Fleet's reading about the Puṣyamitras. The enemies cannot be without name as they are shown to have developed great power and resources. Sohoni has stated "If Fleet's reading were correct, it might be surmised that the difficulties faced by Skandagupta were due to this tribe. If Dr. Divekar's reading was correct, no specific tribe or community was named in this stanza which only contained a broad statement about enemies".[72] It is clear from the context that the enemies referred to are with developed power and wealth "Samuditabala-Koshān" which refer to some particular tribe and cannot be enemies in general. The *Puṣyamitras* have been identified with a tribe living along the banks of Narmadā and it has been pointed out on the basis of numismatic evidence that there was a serious disturbance of peace in the vicinity of Makalā, in the middle of the fifth century A.D. and we may connect it with the rising of the Puṣyamitras in that region.[73] Goyal has also connected them with the Makalā region, a country on the Narmadā and has remarked that the Puṣyamitra invasion assumes a new significance in the power-politics of Deccan in the middle of the fifth century A.D., between the two power blocs, the Guptas and the Nalas on the one hand, the Vākāṭakas and the Kadambas on the other hand.[74]

He seems to have suppressed the revolt of enemies from within the family who may have been other royal princes aided by his maternal uncle after conquering whom he had reported to his mother who was full of tears in her eyes (out of joy) as Kṛishṇa had approached his mother Devakī (after killing his maternal uncle Kaṁsa who was an enemy of his family), the allusion to which is provided in the Bhitari inscription.[75]

Another reference to his enemies is allegorically found in the Junagarh Rock Inscription:

> "Narapati-bhujagānām mānadarpotphaṇānām
> pratikṛiti-Garuḍājñām nirvvishi (m) ch= āvakarttā"[76]

"Who plucked (and utilised) the authority of (his local) representatives, who were so many Garuḍas (and used it as) an antidote against the (hostile) kings who were so many serpents, lifting up their hoods in pride and arrogance".[77] This alludes to the rebellious Varmans, aggressive Vākāṭakas and their supporters as stated by Goyal.[78] Skandagupta rose equal to the occasion and with the help of his local representatives, such as Prabhākara, 'who were so many Garuḍas', he once again established his authority in the restive provinces.[79]

We know about the dreadful Hūṇa invasion from the Bhitari inscription probably held near Bhitari (near Ghazipur, Uttar Pradesh), itself. According to some scholars, however, this campaign might have been fought in regions much further west. But "the commemoration of heroism in this battle through erecting a monument at Bhitari would show that Hūṇas had encroached upon Gupta empire, right near its home territories".[80]

The encounter with the Hūṇas is described thus in line 15 of the Bhitari inscription: "Huṇair -yyasya samāgataṣya Samare dorbhyām dharā Kampitā bhīmāvartta - karasya".[81] "By whose two arms the earth was shaken, when he, the creator (of a disturbance like that) of a terrible whirlpool, joined in close conflict with the Hūṇas.[82] We do not agree with Sohoni's interpretation who states that "what was meant was the ground or earth shook when the Hūṇas fell on it, in the course of battle".[83] If we arrange the context it would come to "Yasya Hūṇair (saha) samare samāgatasya dorbhyām dharā Kampitā" "By whom when he came into conflict with the Hūṇas in warfare, the earth was shaken by (the valour of) his arms" which shows his ferociousness in action when he faced the Hūṇas in the battle- field. The defeat inflicted upon the Hūṇas proved so decisive that for nearly half a century the Gupta empire was immune from their depredations.[84]

In the Junagarh Rock Inscription we come across another group of enemies known as the *Mlechchhas* according to the reading taken by Fleet in the inscription.[85] We have pointed out in our earlier work that this has been questioned by D.C. Sircar as some letters are illegible. So how could Fleet take this reading without putting even a question-mark?[86] Sohoni states that there is no basis for Fleet's conjectural reading of "Mlechchha-deśeshu" and suggests that it should be better read as "Sarva-deśeshu" "in all countries". As the scholars have equated the Mlechchhas with the Hūṇas, Sohoni further states that "there is little to support the conjecture that this military campaign against Hūṇas lay in an extreme western province of the Gupta Empire".[87] Whether the Mlechchhas are the same as the Hūṇas or were a different tribe, both the records (Junagarh and Bhitari) claim that Skandagupta completely defeated these enemies. The fact that in both the inscriptions, the reference to the fight with the Hūṇas and Mlechchhas is preceded by a reference to the falling fortunes of the family, supports the identification.[88]

P.L. Gupta has, however, identified the *Mlechchhas* of the Junagarh Inscription with the Kidarite Kushāṇas.[89] We may mention that the word *Mlechchha* has been indiscriminately used at different times for any foreigner.[90] If in the Epilogue (*Bharatavākya*) of the play *Mudrā-Rākshasa* by Viśākhadatta, it has been used for the Śakas or Kushāṇas, in the present context it can be used for the Hūnas who were a major source of disturbance during Skandagupta's reign.

We also know from the Kahaum Inscription that he had driven off the Hūṇa menace completely and complete peace prevailed in his empire by A.D. 460, the period when the above inscription is dated i.e. 141 G.E. (A.D. 460). The inscription refers to his "Rājye Śakropamasya Kshitipa - śata - pateḥ Skandaguptasya śānte" "In the tranquil reign of

Skandagupta, who resembles (the god) Śakra, (and) who is the lord of a hundred kings".[91]

Still some scholars have averred the opinion that there was another Hūṇa invasion in his life-time and that he lost his life to stem this tide of the Hūṇa invasion. [92] That his heavy-weight coins were debased due to bad economy affected by the repeated invasions. This argument of economic strain on the basis of numismatic evidence holds no good at present as it has been shown that "the heavier coins of Skandagupta do not contain any greater percentage of alloy than that of his lighter variety".[93]

B.P. Sinha states that the neglect or probable damage of the temple at Daśapura was caused during the Hūṇa invasion of 454-456 A.D.[94] But he does not agree to the suggestion made by Altekar that during the last years Skandagupta lost Malwa to the Vākāṭakas as the Vākāṭakas being pronounced Brāhmaṇas would be the last to damage an important Hindu temple. [95] Further he states that " the non-mention of Skandagupta in the Mandasor inscription of Prabhākara dated Malwa year 524 (=A.D. 467-468) does not necessarily mean that Skandagupta was not his overlord".[96]

Another event worth mentioning during the reign of Skandagupta is the repair of the burst dam of Sudarshana lake by Chakrapālita who was Governor of Saurashtra, presumably under the direction of the emperor Skandagupta at enormous expenditure of wealth, as stated in the Junagarh Rock Inscription.[97]

Skandagupta was a great warrior, an able administrator and promoter of the welfare of the people. Had he not crushed the Hūṇa menace, they would have rampaged the entire empire. The Junagarh Inscription in its second part states "of him (Skandagupta) who destroyed the pride of (his) haughty enemies; who is of great glory; who is the banner of his lineage; who is the lord of the whole earth; whose pious deeds are even more wonderful than his supreme sovereignty over kings......."[98]

He showed mercy to his defeated enemies and was never arrogant though his glory increased day in and day out, has been mentioned in his Bhitari record. [99] He appointed his governors with a careful thought for days and nights as in case of Parṇadatta who was selected by him to govern the land of the Saurashtras. [100] in the lines 5-6 of the Junagarh Inscription it is stated that "while, he, the King, is reigning, verily no man among his subjects falls away from religion; (and) there is no one who is distressed, (or) in poverty, (or) in misery, (or) avaricious, or who, worthy of punishment, is over - much put to torture" " Thus having conquered the whole earth, (and) having destroyed the height of the pride of (his) enemies, (and) having appointed protectors in all the countries", [101] he is stated to have taken special care for the appointment of a governor in Saurashtra who is not only loyal but is also endowed with all qualities of head and heart which have been elaborately discussed in the inscription.[102]

It may be inferred from the above-mentioned description that Skandagupta felt greatly disturbed even when he got himself saddled on the Imperial throne and had to reorganise the entire central and local administration by appointing his choicest men at all key positions. It is likely that rival brothers who might have struggled to capture the throne when he was engaged in the battle-field against the enemy to repel whom he had been deputed by his father who had died in the meanwhile, whether at the capital or in the battle-field, we are not certain.[103]

Skandagupta was known for his policy of religious toleration which gave free scope to private philanthropy creating endowments in favour of different religions then prevailing in the country.[104] He was himself a Vaishṇava by faith. He installed an image of Śaṁgin (Viṣṇu) with the endowment of a village for the increase of the religious merit of his father.[105] He is mentioned as *paramabhāgavata* on some of his silver coins.[106]

The erection of the temple of Viṣṇu is referred to in the second part of the Junagarh Inscription.[107] Similarly we have the evidence of the worship of the Sun, the Jain Tīrthaṅkaras and the Buddhist wheel of law (Dharma-Chakra) from the Indore Copper-Plate Inscription, the Kahaum Stone Pillar Inscription, and the Mathurā Stone Inscription of the year 135 respectively. [108] Ancestor worship is indicated on the basis of the Supia Stone Pillar Inscription of Skandagupta.[109] But Vasudeva Upadhyaya states that this Pillar was established for the worship of "Ṣaṣṭī Mātā" for the protection of children. This worship is done among the Hindus on the sixth day of the newly-born child to wish for the protection of his life.[110]

Skandagupta had a prosperous reign though in the earlier years of his rule or even for accession to the throne he faced difficulties both from inside and outside the family. A fallacy had been created by scholars on the basis of his heavy-weight coins as pointed out earlier that there had been strain on treasury, without verifying the real fact. P.L. Gupta remarks: "Had there been any strain on the treasury, for what-so-ever reasons, the coins undoubtedly, would have been debased, but without any increase in their weight".[111] The analysis of coins obtained by B.P. Sinha from the British Museum turns the table other way to prove that he had more prosperous times in his later days, the fact which is repeated in the inscriptions of his time. The heavy weight coins contain more gold than the light weight coins. The former contain 76 to 79 per cent of gold while the latter only 67 to 74 per cent of gold. According to P.L. Gupta, it seems that "gold had gone cheap in the later days and to maintain the ratio between the gold and silver coins, Skandagupta had found the necessity of increasing the weight as well as the percentage of gold in his coins".[112]

Whether Skandagupta had a son or not to continue his line, is not settled beyond doubt.[113] The last known date of Skandagupta on his silver coin is 148 (467-468 A.D.) [114] and according to P.L. Gupta he died probably in about the same year.[115] We fail to understand how Sinha can state on the basis of the Gadhwa Stone Inscription dated in the same year (148) that "Skandagupta might have continued to rule till 470 A.D.".[116]

Notes And References

1. P.L. Gupta, *IG*., pp.321, 327.
2. Fleet, *CII*. III. No.14, p.56.
3. B.P.Sinha, *DHM*., pp.1, 13.
4. Fleet, *CII*. III. pp.47ff.
5. B.P. Sinha, *DHM*., pp.4,5, 13.
6. *Ibid*., p.20 f.n.21.
7. U.N. Roy, *GRY*., pp.312-313.
8. R.K. Mookerji, *GE*., p.99.
9. A.S. Altekar, *GGCBH*., p.XXXIV.
10. R.K. Mookerji, *GE*., p.99.
11. P.L. Gupta, *IG*., p.122.
12. V. Upadhaya, *GA*., p.160 line 3 of the inscription.
13. P.L. Gupta, *IG*., p.338. Also see pp.343-344 f.n.47.
14. Ibid., B.P. Sinha, *DHM*., p.40.

15. S. Chattopadhyaya, *EHNI.*, p.217.
16. B.P. Sinha, *DHM.*, p.34.
17. *Ibid.*
18. *Ibid.*, pp.1, 34; A.L. Basham, *BSOAS*, XVII, p.367; P.L. Gupta, *JIH*. XL, Pt.II, pp.250-51.
19. B.P. Sinha, *DHM.*, p.34.
20. *Ibid.*
21. H.C. Raychaudhuri, *PHAI.*, 504-505 as quoted.
22. *Ibid.*, pp.507-509.
23. *Ibid.*, p.509.
24. R.G. Basak, *HNEI.*, pp.76-80.
25. R.K. Mookerji, *GE.*, p.93.
26. U.N. Roy, *GRY.*, pp.300-307.
27. R.C. Majumdar, *VGA.*, pp.175-176.
28. R.C. Majumdar, "The Successors of Kumāragupta-I" in S.P. Gupta (ed), *RPHI.*, p.99.
29. S. Chattopadhyaya, *EHNI.*, p.219.
30. S.R. Goyal, *HIG.*, p.267.
31. *Ibid.*, p.273.
32. P.L. Gupta, *IG.*, pp.186-187.
33. Dasharatha Sharma, "The Legitimacy of Skandagupta's Succession" in *JIH*. Vol.XLIII. Pt.1, April, 1965.
34. P.L. Gupta, *IG.*, p.320.
35. *Ibid.*, p.327.
36. S.V. Sohoni, "Notes and Queries on Bhitari Pillar Inscription of Skandagupta", *JBRS*. Vol.LIII, January-December 1967, p.105.
37. R.K. Mookerji, *GE.*, p.101.
38. Ibid.
39. B.P. Sinha, "The Immediate Successor of Kumāragupta-I" in *RHC.*, pp.110-111; A.L. Basham, *Studies in Indian History And Culture*, pp.143- 144.
40. B.P. Sinha, *Op.cit.*, p.111.
41. *Ibid.*
42. *Ibid.*
43. Dasharatha Sharma, "The Legitimacy of Skandagupta and Connected Problems", *JIH*.XXXVII. Pt.II, August, 1959.
44. Fleet, *CII*. III, Bhitari Stone Pillar Inscription of Skandagupta, L.7, pp.53, 55.
45. Fleet, *CII*. III. p.55.

46. *Ibid.* Lines 10-11 of the inscription.
47. Dasharatha Sharma, "The Legitimacy of Skandagupta's Succession", in *JIH*.XLIII. Pt.I, April, 1965.
48. Fleet, *CII*. III. p.54. LL.11-12.
49. *Ibid.*, p.55.
50. *Ibid.*
51. Fleet, *CII*. III, p.54, LL.12-14.
52. P.L. Gupta, *IG*., pp.329-330.
53. H.C. Raychaudhuri, *PHAI*., p.506, f.n.2.
54. Fleet. *CII*. III. p.56.
    Scholars have given different readings in this context which are still disputed. For details see P.L. Gupta, *IG*., p.339, f.n. 10.
55. S.V. Sohoni, "Notes and Queries on Bhitari Pillar Inscription of Skandagupta", *JBRS*.LIII, January-December, 1969, p.100.
56. Dasharatha Sharma, "The Legitimacy of Skandagupta's Succession", in *JIH*. XLIII. Pt.1. April 1965.
57. *Ibid.*
58. *The Classical Age*, p.26, f.n. 3.
59. P.L. Gupta, *IG*., p.331.
60. *Ibid.*
61. Dasharatha Sharma, *Op.cit.*, *JIH*. April 1965.
62. P.L. Gupta, *IG*., p.331.
63. Fleet, *CII*. III. p.59, L.5.
64. *Ibid.*, p.62.
65. *Ibid.*, p.53, L.10.
66. *Ibid.*, p.54, LL.12-13.
67. *Ibid.*, L.14.
68. S.V. Sohoni, *Op.cit* ., *JBRS* .LIII. p.101.
69. Fleet, *CII*. III. pp.53-54, LL.10-11.
70. *Ibid.*, p.55.
71. R.S. Tripathi, *HAI*., p.261.
72. S.V. Sohoni, *Op.cit.*, *JBRS*..LIII. p.101.
73. T.R. Sharma, *PGNGI*, p.157.
74. S.R. Goyal, *HIG*., pp.273-275.
75. Fleet, *CII*. III. p.54 verse 6.
76. *Ibid.*, p.59 LL 2-3.
77. *Ibid.*, p.62.
78. S.R. Goyal, *HIG*., p.290.
79. Ibid.

80. S.V. Sohoni, *Op.cit.*, *JBRS*. LIII. p.105.
81. Fleet, *CII*. III. p.54.
82. *Ibid.*, p.56.
83. S.V. Sohoni, *Op.cit.*, *JBRS*. LIII. pp.104-105.
84. T.R. Sharma, *PGNGI.*, p.131.
85. Fleet, *CII*. III. p.59, Line 4 of the inscription.
86. T.R. Sharma, *Op.cit.*, p.149.
87. S.V. Sohoni, *Op.cit.*, p.105.
88. T.R. Sharma, *Op.cit.*, p.150.
89. P.L. Gupta, *IG.*, p.333.
90. T.R. Sharma, *Op.cit.*, p.150.
91. Fleet, *CII*. III. p.67.
92. S.R. Goyal, *HIG.*, p.286. Goyal quotes R.D. Banerji and Smith in this regard.
93. *Ibid.*
94. B.P. Sinha, *DHM.*, p.28.
95. *Ibid.*
96. *Ibid.*, p.29.
97. Fleet, *CII*.III. p.64.
98. *Ibid.*, pp.64-65.
99. *Ibid.*, p.55, L.14.
100. *Ibid.*, p. 63 L.9.
101. *Ibid.*, p.62.
102. *Ibid.*, LL.7-9.
103. S.V. Sohoni, *Op.cit.*, *JBRS*. LIII. p.105. Sohoni states that a case can be made that Kumāragupta was killed in the battle with the Hūṇas after making a desperate personal flight. Hence the need for a commemorative monument near the battle-field which was fulfilled by his son Skandagupta by construction of a temple of Lord Vishṇu and donation of a village for the increase of the religious merit of his father.
104. R.K. Mookerji, *GE.*, p.96.
105. Fleet, *CII*. III. Bhitari Stone Pillar Inscription of Skandagupta, p.56 LL.17-19.
106. *CCGD*. CXXXI-XXII.
107. Fleet, *Op.cit.*, p.65, LL-25-27.
108. B.P. Sinha, *DHM.*, p.30.
109. *Ibid.*
110. V. Upadhyaya, *GA.*, p.300.
111. P.L. Gupta, *IG.*, p. 336.

112. *Ibid.*
113. *Ibid.*, p.338; B.P. Sinha, *DHM.*, pp.38, 40.
114. B.P. Sinha, *Op.cit.*, p.34.
115. P.L. Gupta, *Op.cit.*, p.338.
116. B.P. Sinha, *DHM.*, p.35.

Concluding Remarks

It will not be inappropriate, at this stage, to underline a few methodological and analytical innovations in this exercise. We have sifted the material both from the original and the secondary sources which have been well-defined in the work. We have squeezed the entire material to the benefit of the readers, reviewed the evidence and presented it in a precise, concise and coherent form. At many places we have made the corrections at the basic data level itself. As a result new information has come forth and provided scope for a fresh interpretation. We have identified the powers that be on a political horizon before the rise of the Guptas. We have shown on the basis of Allahabad Pillar Inscription how Samudragupta had performed many Vedic sacrifices earlier and that his Northern and Southern campaigns were a prelude to his performance of the Aśvamedha sacrifice. We have looked out for discrepancies and tried to sort them out with due reference to the context. We have attempted to resolve them in a cogent manner. We have settled the controversial issue of the Gupta Era and have set out the Gupta Chronology tangibly for the first time up to the reign of Skandagupta.

The history of the Imperial Guptas becomes obscure after the death of Skandagupta. The scholars have no unanimity of opinion about the successors of Skandagupta, different theories about partition of empire, the succession by Purugupta or Kumāragupta-II, as well identification and status of some rulers pose new problems. So we close down our shutters here and may open up again after thorough study of the data and material available on the post-Skandagupta period of the Imperial Guptas which is marked with internal troubles, foreign invasions and disintegration of empire, finally leading to its decline.

Select Bibliography

Agrawala, V.S., *Studies in Indian Art*, Varanasi, 1965.

Allan, J., *Catalogue of the Coins of Ancient India*, London, 1936, Reprinted, Delhi, 1975.

Allan, J., *Catalogue of the Coins of the Gupta Dynasties*, And of Śaśāṅka, King of Gauḍa, London, 1914, Reprinted, Delhi, 1975.

Altekar, A.S., *The Coinage of the Gupta Empire*, Benaras, 1957.

Altekar, A.S., *The Gupta Gold Coins in the Bayana Hoard*, Bombay, 1954.

Altekar, A.S., *State and Government in Ancient India*, Bombay, 1972.

Altekar, A.S., *Gupta Kālīna Mudrain* (in Hindi), Patna, 1954.

Bagchi, P.C., *India and China*, New York, 1951.

Bajpai, K.D., *Indian Numismatic Studies*, Delhi, 1976.

Banerjee, Manabendu, *A Study of Important Gupta Inscriptions*, Calcutta, 1976.

Banerji, R.D., *The Age of the Imperial Guptas*, Benaras, 1933.

Basak, R.G., *The History of North-Eastern India*, Calcutta, 1967.

Basham, A.L., *The Wonder That was India*, Bombay, 1963.

Basu, Jogiraj, *India of the Age of the Brāhmaṇas*, Calcutta, 1969.

Beal, S., *Si Yu Ki*, Buddhist Records of the Western World, 1883, Reprinted, Delhi, 1969.

Beal, S., *Life of Hiuen Tsiang by the Shaman Hwui Li*, London, 1911.

Biswas, Atreyi, *The Political History of Hūṇas In India*, Delhi, 1973.

Brown, C.J., *The Coins of India, Varanasi*, 1973.

Chakraborti, H., *India As Reflected in the Inscriptions of Gupta Period*, Delhi, 1978.

Chattopadhyaya, Brajadulal, *Coins and Currency Systems In South India , C.A.D. 225-1300*, Delhi, 1977.

Chattopadhyaya, S., *Early History of North India*, Calcutta, 1968.

Chaudhary, R.K.(Ed.), *Inscriptions of Ancient India* (For Gupta Inscriptions, pp.112-138) Meerut, 1983.

Dandekar, R.N. and Others (Eds.), *Sanskrit And Indological Studies* - Dr. V. Raghavan Felicitation Volume, Delhi, 1983.

Dandekar, R.N., *The Age of the Guptas and Other Essays*, Delhi, 1982.

Diskalkar, D.B., *Selections From Sanskrit Inscriptions*, New Delhi, 1977.

Elliot, H.M. and John Dowson, *The History of India* - as told by its *own Historians*, Vol.I, Allahabad. (Reprint)

Elliot and Dowson, *History of India*, Vol.I, Hindi translation by M.L. Sharma, Agra, 1973.

Fleet, John Faithful, *Corpus Inscriptionum Indicarum*, Vol.III, Inscriptions of the Early Gupta Kings, And their successors, Varanasi, 1963, (Reprint).

Ganguly, D.K., *Aspects of Ancient Indian Administration*, Delhi, 1979.

Giles, H.A., *The Travels of Fa-hien*, Cambridge, 1923.

Gokhale, B.G., *Samudra Gupta*, Bombay, 1962.

Goyal, S.R., *Prāchīna Bhārata Kā Rajanītika Itihāsa evam Samakālīna Rājavaṁśa* (in Hindi), Allahabad, 1969.

Goyal, S.R., *A History of the Imperial Guptas*, Allahabad, 1967.

Gupta, P.L., *The Imperial Guptas*, Vol.I, Varanasi, 1974.

Gupta, P.L., *Gupta Sāmarājya* (in Hindi), Varanasi, 1970.

Gupta, P.L., *Coins*, Delhi, 1969.

Gupta, S.P. (Ed.), *Readings in Political History of India*, Delhi, 1976.

Jha, D.N., *Revenue System - In Post-Maurya And Gupta Times*, Calcutta, 1976.

Kane, P.V., *History of Dharmaśāstra*, Vol.I, Part I, Poona, 1968.

Khosla, Sarla, *Gupta Civilization*, Delhi, 1982.

Kosambi, D.D., *Indian Numismatics*, Delhi, 1981.

Majumdar, R.C. And Altekar, A.S. (Eds), *The Vākāṭaka - Gupta Age*, Delhi, 1967.

Maity, S.K., *Early Indian Coins and Currency System*, Delhi, 1970.

Maity, S.K., *Economic Life in Northern India*, Delhi, 1970.

Maity, S.K., *The Imperial Guptas and Their Names*, Delhi, 1975.

Majumdar, R.C., *History of Ancient Bengal*, Vol.I, Calcutta, 1971.

Mohan Chand (Ed.), *Ancient Indian Culture & Literature*, Delhi, 1980.

Mookerji, R.K., *Ancient India*, Allahabad, 1956.

Mookerji, R.K., *The Gupta Empire*, Delhi, 1973.

Mukerji, B.N. and Others (Eds), *Sri Dineśa Chandrikā*, Studies in Indology, Sri. D.C. Sircar Festschrift, Delhi, 1983.

Narain, A.K. (Ed.), *Seminar Papers on the Local Coins of Northern India*, (c.300 B.C. to 300 A.D.), Benaras, 1968.
Pathak, H., *Cultural History of the Gupta Period*, Delhi, 1978.
Panthari, B.P., *Bhārata Kā Swarṇa Yuga* (in Hindi), Lucknow, 1974
Pargiter, F.E., *Ancient Indian Traditions*, Reprinted, Delhi, 1979.
Pathak, V.S., *Ancient Historians of India*, Bombay, 1966.
Philips, C.H. (Ed), *Historians of India, Pakistan and Ceylon*, Oxford University Press, 1967.
Puri, B.N., *History of Indian Administration*, Vol.I. Bombay, 1968.
Rai, G.S., *The Kāvya-Mīmāṁsā of Rājaśekhara*, Varanasi, 1964.
Rapson, E.J., *Catalogue of the Coins of*: The Andhra Dynasty, The Western Dynasty, The Traikūṭaka Dynasty And the "Bodhi" Dynasty, London, 1908, Reprinted, Delhi, 1975.
Rapson, E.J., *Indian Coins*, Varanasi, 1969.
Raychaudhuri, H.C., *Political History of Ancient India*, 7th Edition, Calcutta, 1972.
Rodgers, C.J., *Catalogue of the Coins*, Calcutta, 1895, Reprinted, Delhi, 1985.
Roy, S.N., *Historical and Cultural Studies in the Purāṇas*, Allahabad, 1978.
Roy, U.N., *Gupta - Rajavaṁśa Tathā Usakā Yuga* (in Hindi), Allahabad, 1977.
Sadhu Ram, *Some Important Inscriptions of Aśoka, The Guptas, The Maukharis and Others*, Delhi.
Sarma, I.K., *Coinage of the Sātavāhana Empire*, Delhi, 1980.
Sastri, K.A. Nilakanta and G. Srinivasachari, *Advanced History of India*, Bombay, 1973.
Sastri, Ramateja, *Kauṭilīya, Arthaśāstram* (with Hindi Commemoration), Kashi, 1964.
Sen, B.C., *Some Historical Aspects of the Inscriptions of Bengal*, Calcutta, 1942.
Sharan, M.K., *Tribal Coins - A Study*, Delhi, 1972.
Sharma, R.P., *Rājaśekhara Aura Usakā Yuga* (in Hindi), Patna, 1977.
Sharma, T.R., *Personal And Geographical Names in the Gupta Inscriptions*, Delhi, 1978.
Sinha, B.P., *Dynastic History of Magadha*, Delhi, 1977.
Sinha, B.P., *Readings in History and Culture*, Delhi, 1978.
Sircar, D.C., *Indian Epigraphical Glossary*, Delhi, 1966.
Sircar, D.C., *Indian Epigraphy*, Delhi, 1965.

Sircar, D.C., *Select Inscriptions,* bearing on Indian History Civilization, Vol.I, 2nd edition, Calcutta, 1965.

Sircar, D.C., *Select Inscriptions* bearing on Indian History an Civilization from the Sixth to the Eighteenth century A.D., Vol.2, Delhi, 1983.

Sircar, D.C., *Studies in Indian Coins*, Delhi, 1968.

Smith, L. Bardwell (Ed)., *Essays on Gupta Culture*, Delhi, 1983.

Smith, V.A., *The Oxford Students History of India*, Oxford University Press, 1933.

Thakur, Upendra, *The Hūṇas in India*, Varanasi, 1967.

Thapar, Romila, *A History of India*, Vol.I, Pelican, 1972.

Thapliyal, U.P., *Foreign Elements in Ancient Indian Society*, Delhi, 1979.

Tripathi, R.S., *History of Ancient India*, Delhi, 1967.

Upadhyaya, B., *Sanskrit Sāhitya Ka Itihāsa*, Kashi, 1968.

Upadhyaya, V., *A Study of Ancient Indian Inscriptions* (in Hindi), Patna, 1970.

Upadhyaya, V., *Gupta Abhilekha*, Patna, 1974.

Upadhyaya, V., *Bhāratīya Sikke* (Hindi), Prayāga, Sam.2005.

Visveshwara, Acharya Commemoration, *Kāvya Prakāsha of Mammaṭa*, Varanasi, Sam. 2024.

Watters, Thomas, *On Yuan Chwang's Travels in India*, 2 Vols., Delhi, 1973.

Williams, Monier, *A Sanskrit-English Dictionary*, Delhi, 1963.

Index